Critical Neuroscience and Philosophy

"A 'scientific re-examination of the mind-body problem' is certainly a 'difficult task' and Tomasi seems to navigate the rough water with a safe methodological approach. The book provides the reader with a comprehensive overview, which exhibits a remarkable balance in the presentation of disputed topics. In addition, the author provides the necessary tools to have both people with science or philosophy backgrounds acquainted to the topic.

Neuro-lovers will appreciate and learn from the presentation of the numerous neuroscience 'sub-branches,' together with details on the methodological approaches used in the neuroscience research. Philosophers will enjoy the freedom and degree of theoretical abstraction, unusual in neurobiology books. Tomasi does in fact analyse the 'mind-body problem' with a critical appraisal that combines the rigidness of the scientific method with the speculative insight and thoroughness of the philosophy. The combination of the two sources of knowledge makes this book a fundamental tool for those who share the need to bridge the (apparent) gap between science and philosophy.

Another key adjective for describing the book is multidisciplinary. The author spans from logic to quantum mechanics, from medicine to informatics, from religion to ethics, from theory to practice. In all the cases the rigor in defining critical words makes even a lay reader feel like taken by the hand during the journey."

—Francesco Orzi, *Professor of Neurology, Sapienza University of Rome (retired), and member of the Accademia dei Fisiocritici, Siena, Italy*

"*Critical Neuroscience and Philosophy* is impressive in many ways—from the scope and variety of information analyzed to the inspiration that scientists, philosophers, and the wider public will find in it. Bringing together and setting apart mind and body, self and world provokes thought and evokes emotion. I read it without stopping for breath. Razor sharp till the last detail in such a difficult field as neuroscience in its relationship to existential problems."

— Aneta Karageorgieva, Professor, *former Chair of the Department of Philosophy, Sofia University "St. Kliment Ohridski" Bulgaria, and member of the Bulgarian Society for Analytic Philosophy*

"*Critical Neuroscience and Philosophy* really resonated with me because every day at OVR I work at the intersection of the mind and body. I find myself in a feedback loop while pursuing a deeper understanding of their connection as well as how to manipulate both of them simultaneously to produce positive human outcomes. This involves looking for links between science and philosophy, emotion and reason, data and intuition. Tomasi's book detailing how all of these elements are inextricably linked to one another, from a neurophysical standpoint is poignant. It is especially relevant today, not only to my work, but to how we collectively think about human factors across all fields, markets, and disciplines."

—Aaron Wisniewski, CEO, *OVR Technology LLC, USA*

"David Tomasi has written simply the book we were waiting for, the book that the current fields in medicine and psychology badly need. Human beings are an amazing paradox between mind and body. The author has written really a remarkable book about the correlation between mind and body, from a neuroscientific standpoint. Soul and matter seen by a comprehensive and critical perspective of results in neuroscientific research and epistemological issues. Particularly notable is the choice to critically discuss also the applicability of these results and concepts in clinical medical and philosophical-existential situations.

This outstanding volume offers a complete overview of theoretical information about brain, psyche, culture, bringing an ideal combination of scientific and clinical perspectives to the topic. The author had the courage to say that philosophy is a basic approach toward neuroscience and more and more to speak about subjectivity in science, in terms of the unique truth we can reach and of the unique solution we have actually available to the mind-body problem.

This book offers a paradigm shifting approach, scientifically based, empirically researched, for all professionals in search of more refined ways to conceptualize connection between mind and body, up to reach the Triple-S Model: Self, Soul, Spirit as a solving review of the fascinating mystery of our double nature, spiritual and embodied at the same time.

A book that will change the way you think, a must read that I can't recommend more highly."

—Erica F. Poli, Psychiatrist-Psychotherapist, *IEDTA, ISTDP Institute, EFP Group, Milan, Italy*

"In *Critical Neuroscience and Philosophy - A Scientific Re-Examination of the Mind-Body Problem,* David Tomasi takes on a substantial task in this work on 'Critical Neuroscience.' He presents, in an entertaining, taxonomic style, the most important areas of scientific inquiry to 'determine the theories, methods, techniques, and technologies used in recent times to investigate the connection between neural activation and multiple aspects of human nature' (p.16). His work covers a vast and complex range of topics, starting from the purely biological physicalist viewpoint, discussing the most common subfields of neuroscience that deal with the neural underpinnings of psychological processes, with special attention to the central nervous system, particularly the brain. In the same way that Michel Foucault produced an archaeology of the human sciences in his book *The Order of Things*, Tomasi aims to achieve something similar in this book of sifting about neuroscience.

His work is critical in the epistemological sense of 'providing a critique and criticism of the assumptions, methods, methodologies, theories, techniques, and technologies, and observation/interpretation of results in neuroscientific research,' but is also critical in the sense of 'applied and applicable to critical clinical-medical and philosophical-existential situations' (p.13). But a central and consistent plank of his position to keeping in mind the principle that it is 'impossible to completely avoid the subjective element in any scientific field dealing with human nature' (p11), and challenges throughout the prevailing 'Cartesian "dualistic" model of the primacy ("supremacy") of human thought' (p.41). And a significant conclusion on his part is his view that there is adequate 'evidence of the existence of states, of multiple modalities of perceiving, understanding, and being, and of a non-physically-based-matter-based form of mind, soul, spirit' (292).

What I appreciated about the work is that he achieves his aim of promoting the idea of Philosophy as a basic approach toward Neuroscience and consistently attempts to be flexible in 'moving the focus back and forth from the smallest detail to the bigger picture' (p.261). He also retains a certain pragmatism in the work and never claims to have found a final solution (to the mind-body problem) because 'we do not think that there could be such thing as something "final" given' (p.291).

I like the way that he consistently maintains, as a clinician, a view of another essential aspect of this study, i.e. the promotion of appropriate ideas for the amelioration of care for our patients' (269). And, again in the same clinical vein, he looks for a balanced outcome in the subjectivity/scientific objectivity debate saying that 'we should not try to eliminate (our) subjectivity in reaching a more objective understanding, and in the the same way we should not try to eliminate the place effect in reaching a more universal (statistically intended) healing method' (292).

This work is very helpful to me as a practicing psychotherapist. Whilst I have a basic understanding of neuroscience and the mind-body debate, this work has substantially widened and deepened my understanding of the critical issues and debates in this area. A knowledge of neuroscience is essential in my clinical practice and David Tomasi's taxonomic approach provides a valuable and concise insight into a vast area of clinical knowledge usually only accessible by experts in the neuroscience field."

—Tom Conlon, Clinical Director, *Tri-Factor Health, Ireland*

David Låg Tomasi

Critical Neuroscience and Philosophy

A Scientific Re-Examination of the Mind-Body Problem

palgrave
macmillan

David Låg Tomasi
University of Vermont Integrative Health
University of Vermont Medical Center,
Department of Inpatient Psychiatry
Burlington, VT, USA

ISBN 978-3-030-35353-7 ISBN 978-3-030-35354-4 (eBook)
https://doi.org/10.1007/978-3-030-35354-4

This Palgrave Macmillan imprint is published by the registered company Springer Nature Switzerland AG.
The registered company address is: Gewerbestrasse 11, 6330 Cham, Switzerland

Acknowledgments

Being critical is an essential part of every investigation. Writing a book on Critical Neuroscience and Philosophy certainly falls into this prescription, and the process behind it is the direct result of the many contributions and insights from all the wonderful people I have met through the years. A very special thank you goes to Rachel Daniel, Senior Editor, Madison Allums, Editorial Assistant and all the staff at Palgrave Macmillan for their continuous support, professionality and enthusiasm in this project.

I would like to acknowledge the fundamental contribution and support of Tom Conlon, Alexander Gungov, Allison Kaigle Holm and Sten Holm, Aneta Karageorgieva, Paola Lusini, Giuseppe Manganelli, Francesco Orzi, Erica Poli, Stephen Sammut, Joan Walker, Aaron and Sam Wisniewski. I extend my gratitude also to all clinical and academic staff, students, faculty at the University of Vermont and the University of Vermont Medical Center, the Vermont Academy of Sciences, the Accademia dei Fisiocritici and the International Academy of Sciences.

While opinions in regard to the mind–body problem vary almost as much as the shapes of snowflakes, in this analysis I also discovered that this diversity is not only incredibly complex and vast, but it represents the very nature of the discourse, while determining the best parameters of clinical intervention. In this regard, I cannot thank enough all the patients I had the privilege and honor to work with. They have shared their beautiful stories, their successes and their struggles, their minds and their hearts, in ways that have enriched my experience as a psychotherapist as well as, most importantly, a human being.

Last but absolutely not least, I would like to acknowledge the understanding, the support and the encouragement of my family both in Europe and in the United States, especially my parents Lino and Rosa, my wife Livija and my sons Lucas and Adrian.

This study, as every other study, is dedicated to you.

David Låg Tomasi

Contents

List of Figures

List of Tables

CHAPTER 1

Introduction

From Gr. kritikḗ (tékhnē)

"A Scientific Re-Examination of the Mind-Body Problem" is the subtitle of this study on critical neuroscience. It presents a very difficult task, a struggle many scientists and philosophers have been dealing with since time immemorial. Is there anything that can be said about such an incredibly complex topic, anything at all which can be added to the conversation? Our main goal is to present a brief summary of the most important theories and experimental studies conducted so far on the correlation between mind and body. We will focus in particular on information and analysis from neuroscience and philosophy, but also provide new ways—*new* in terms of recent scientific discoveries and millennia-old philosophical traditions, and all the combinations thereof—of interpreting both. Given these premises, we will provide a general outlook on the most important sub-branches of neuroscience, including their connections to other scientific and therapeutic fields such as psychiatry, psychology and engineering, and at each intersection analyze their epistemological component, validity and applicability to our 'human nature'. Debating the existence and possible correlation, even in a casual sense, between mind and brain or more broadly intended between a (concept of) self and a physical (body) entity is therefore at the center of this discussion. The very definition of 'Critical Neuroscience' is in this sense a product of progressive differentiation and

D. L. Tomasi, *Critical Neuroscience and Philosophy*,
https://doi.org/10.1007/978-3-030-35354-4_1

specialization of areas within neuroscience and philosophy. Of course, we will admit right at the beginning of this discussion that we have a specific point of view toward certain aspects of the debate. We not only present such 'bias' openly and directly but also embrace it in attempting to discuss with the highest possible degree of objectivity and evidence-based scientific analysis and methodology all those elements which can be and should be put under very rigorous scrutiny using the most reliable and verifiable cutting-edge scientific techniques and technologies. We hereby claim, however, that there are areas—or at least certain elements in each area—of the debate which cannot (they do not necessarily *fail to*) be analyzed with the exact same methods used in our general scientific inquiry. This is not because we should rely on 'less scientific' parameters for this type of analysis, but exactly through scientific analysis we have come to the realization—and we are certainly not the first ones to reach such conclusion—that there are much deeper and greater parameters to follow when addressing issues as big as the mind–body problem. Certainly, we completely reject any attempt to discard the scientific method used in any research study and experiment to examine quantifiable data and information pertaining to observable matter. This is fundamental as well as ethically mandatory in our day and age, where we are often faced with pseudoscientific claims on, for instance, the nature of man and the universe or on certain therapeutic techniques in the medical field, which are not only based on unscientific claims but can often lead to unpleasant and risky clinical outcomes.[1] Given these premises, why would there be a need of embracing a 'bias' if we are indeed to follow the most objective, evidence-based, experimental scientific methods offered to us? There are several reasons, which we will thoroughly examine in the following chapters. To present a general summary of these reasons however, we can start from the very concept of subjectivity in interpreting certain results. As with phenomena such as the placebo/nocebo effects (in medicine), the observer effect (in physics) or double hermeneutic (in social sciences), we need to keep in mind that not only it is impossible to completely avoid the subjective element in any scientific field dealing with—at least—human

[1] In regard to the latter, we have briefly examined, in the following chapters, certain theoretical frameworks to determine the validity of the methods utilized in this context, but for a more comprehensive analysis of this vast topic, we refer to the bibliography at the end of this chapter, as well as to our previous publications, especially "Medical Philosophy" (Tomasi 2016).

nature (although it is certainly possible, at least to a certain extent, to monitor and quantify the possible effects in terms of confounding variables, skewed results or comorbidity factors), but we could use such element to our advantage, in order to better understand certain phenomena. This is important, especially when some results appear to be unclear outliers or simply do not conform to the current scientific paradigm. The purpose is also to reach a higher level of understanding of the 'big picture' (as we do not claim here to be able to see the 'whole' picture, at least *not yet*), as opposed to a confined, limited understanding of only this or that factor or variable. Of course, an analysis of a method is very hard to conduct *from within*. In other words, investigating the scientific method used for all aspects of human nature using the scientific method (as commonly defined by the parameters used until very recent times[2]) could represent a paradox, an oxymoron or even a double-bind, a 'Catch 22' or a 'regression problem' in our effort. Of course, not every investigative strategy which starts and develops 'from within' is to reject a priori. Some deductive and inductive aspects of concepts such as 'introspection', especially in the psychological sense, appear to be quite useful and effective in this context, as long as we understand the difference in magisteria between the areas where this strategy is justified and the areas in which the 'standard' scientific method ought to be applied. Certainly, this would account for moving the issue 'one step up', and we are still faced with the main concern of determining the parameters for this differentiation. The strategy we hereby suggest to solve the problem comes from the application of philosophy as a method for science, particularly medical science, and especially in the context of psychosomatic disorders. Of course, this is the core of epistemology, a branch or subfield of philosophy concerned with the analysis of science and the scientific method, and more broadly the parameters, the prerequisites and the boundaries of knowledge, including the acquisition, the attainability and the validity of such knowledge, which is also understanding, comprehension and *realization*, as evidenced by the German equivalent *Erkenntnistheorie*, and *Gnoseologie* as synonyms for *Epistemologie*. We will therefore apply many of the concepts used in classical epistemology as well as the ones presented by criticism to the

[2] We will further discuss these aspects as well. For now, let us limit ourselves to define science as an evidence-based, experimental method structured on the principle of falsifiability, originating from post (not necessarily against!) Galilean, Newtonian, Darwinian worldviews and also product of enlightenment, industrial revolution, positivism, reductionism and materialism.

core of epistemology, whether from philosophical skepticism or from general postmodernism. But we will extend the discourse to one of the most important areas of investigation of the human nature, neuroscience.

Given the exciting results of the 'decade of the brain' and the following 'decade of the Mind', and the philosophical schools of thoughts originating from this type of research, for instance in neurophilosophy, we wanted to present the most recent viewpoint on the validity and application of this research in areas traditionally associated with philosophical thought, such as conscience vs. consciousness, (self)perception and awareness, identity, and personality, meaning, sense, purpose and so on. In this sense, the title "Critical Neuroscience" has a double valence. It is critical in the epistemological sense of "providing a *critique-criticism* of the assumptions, methods, methodologies, theories, techniques, and technologies, and observation-interpretation of results in neuroscientific research", but is also critical in the sense of "applied and applicable to *critical* clinical-medical and philosophical-existential situations". In other words, we are talking about a critical situation or emergency that is both *situated* (in a social/anthropological sense, but also historical/philosophical sense[3] and in regard to embodied cognition) and truly *emergent* (Lat. ex -mergere). As we stated, some crucial elements in this analysis have already been presented in other works, especially in *Medical Philosophy* (Tomasi 2016), although in this context they have been further elaborated under the lens of neuroscience and, in some areas, have been used as critical variables to be examined again in the exploration of a philosophical defense for certain positions, especially in regard to the mind–body problem. It follows that some viewpoints that were simply presented in *Medical Philosophy* to provide a general survey of the possible philosophical positions on, for instance, the nature and essence of being human are hereby discussed as integral parts of a "guided suggestion-suggested guide" for a better understanding of our self, especially in the context of treatment modalities. Thus, beyond the defense of the aforementioned scientific methodological basis necessary for experimental science, we also question here some assumptions Western science has had over the centuries. Assumptions which were relevant during the age of enlightenment, and even more so in modern times, where science has often turned into, in the words of Iranian philosopher Seyyed Hossein Nasr, "a replacement for religion" (Hossein Nasr 1968). This is a position similar to Giorgio De Santillana, especially in

[3] With special reference to Hegel and Heidegger, particularly in the concept of 'Being', presented in *Sein und Zeit*.

regard to the valence of pre-Socratic and ancient philosophical-religious traditions and ideas on the origin and development of our nature, but very interesting parallels, especially in regard to brain–computer interface, biological vs. virtual perception, and artificial intelligence, with the work by Marco Somalvico at the *Politecnico di Milano.*

The concept of 'primacy of perception' is part of the scientific-philosophical debate on consciousness under the framework of existence over essence—thus presenting themes akin to the ones found in existentialism—or 'matter over mind', albeit with possible subjectivity over objectivity. This dialectical and comparative conceptualization originating in post-Cartesian speculations is the basis for perspectives on 'bodily intentionality' and 'motivation'. According to this viewpoint, we '*see* the world through *sight*', that is, through the e*vid*ence via sensory modalities. Our 'bodiliness' is thus responsible for the way we interpret the world, although interpretation and perception are not separate(d) at all in that they define two different aspects, or more precisely two different moments—not necessarily in a logico-temporal (chronological) way, but in a perspective-like *orientation*—of a cognitive or mental process. In other words, our thoughts, emotions, feelings, perceptions and behaviors cannot be 'reduced to matter' because they already are or they 'emerge from' matter. This view is a central axiom in some areas of contemporary philosophy of mind, whether of embodied cognition, as in Foglia and Wilson (2013), or in a more emergent materialist-based typology, as in Bunge (2013). In our analysis we will take into account these perspectives and investigate the validity of a process-based epistemology of method, that is, if we can make (internal) sense of the development of the scientific view of the human mind, and also examine the weight of history in such development, including the traditional and religious components (Fasol 2011). In this context though we also want to stress the idea of possible comparation-comparison between "old men → being ≠ modern men → becoming". We will further discuss this suggestion, but for now it will suffice to ask ourselves whether the theoretical assumption of the existence of a truly scientific development, intended as evolution of not mere methods but real understanding, is possible, and if it is possible, if it has to be limited to modernity, conceptualized as (again) post-enlightenment, post-industrial revolution and possibly post-Nietzschean (but not necessarily post-postmodernist) outlook on life. This study, therefore, not only is analytical in the sense of providing an excursus of the debate around neuroscience but also aims at providing new ways to look at ourselves, our nature, the world we live in and we are part of. To better investigate the problems at hand we must therefore pay special attention to the following sub-branches of Neuroscience or related fields:

- Affective Neuroscience
- Artificial Intelligence
- Behavioral Neuroscience
- Cellular and Molecular Neuroscience
- Clinical and Medical Neuroscience
- Critical Neuroscience
- Cognitive Neuroscience
- Computational Neuroscience
- Cultural and Transcultural Neuroscience
- Cultural, Cross-cultural and Trans-cultural Psychiatry
- Developmental Neuroscience
- Ethnopsychology and Psychological Anthropology
- Evolutionary Neuroscience
- History of Neuroimaging
- Medical Anthropology and Ethnomedicine
- Neuroanatomy
- Neuroanthropology
- Neurocritical care or Neurointensive care
- Neuroeconomics
- Neuroengineering
- Neuroethology
- Neuroheuristics or Neuristics
- Neuroimaging and Neurophysics
- Neuroinformatics
- Neurolinguistics
- Neurophysiology
- Neuropsychology and Neuropsychotherapy
- Neurosurgery and Neurology, including Behavioral Neurology
- Neurotheology and Psychology of Religion
- Paleoneurology
- Psychoneuroimmunology, Psychopharmacology and Psychobiology
- Sensory Neuroscience and (Theory of) Perception
- Social Neuroscience
- Sociobiology
- Systems Neuroscience

The list above, in alphabetical order and by no means complete and omni-comprehensive, presents the most important areas of scientific inquiry we will use to determine the theories, methods, techniques and technologies used in recent times to investigate the connection between neural activation and multiple aspects of human nature. In some instances we divided the same field into two separate discussions (for instance, 'History of Neuroimaging' and 'Neuroimaging and Neurophysics') to better address specific elements of the discipline. In other cases we combined multiple fields into one group (e.g. 'Cultural, Cross-cultural, and Transcultural Psychiatry', 'Medical Anthropology and Ethnomedicine' or 'Neuropsychology and Neuropsychotherapy') to evidence similarities of scope and practice, and compare methods and theories. In other parts of the volume we indicated when synonyms are used for the same field or different nomenclature has been used to indicate different approaches in the same field (e.g. 'Neuroheuristics or Neuristics' or 'Ethnopsychology and Psychological Anthropology'). Furthermore, we have subdivided each discipline based on its connection with the main topic of the chapter, and we have also drawn parallels between the thereby presented fields to promote a multidisciplinary approach.

We also want to stress that we want to provide a *re-examination of* the mind–body problem and not a *solution to* the philosophical mind–body problem. The reason behind this choice is choice itself. In other words we want to point out that each individual, subjective, situated, contextual view does actually constitute a solution to the mind–body problem. Of course, this perspective has to do with verifying whether subjectivity, in the phenomenological and possibly even in the postmodernist sense, provided warrant for the lack of absolute or universal truth(s). To be sure, we are not espousing this perspective, not on ontological grounds, but on logical basis. In fact, if we are to take, as an example, two opposites in this context, we could think of a materialist-reductionist form of neuroscience on one side, and a contextual-critical examination of the assumptions posited by the former on what defines (constitutes, in terms of 'meaning/function- building blocks', our self). The very decision of separating these two approaches and literally putting them on opposite sides is in itself a form of dualism. The same can be said about having an either monist-physicalist or monist-idealist perspective on one side and a holistic-omnicomprehensive view on the other. They do in *appearance* contradict each other in the sense that, in purely practical terms, in a monist-physicalist viewpoint the core focus is on the *physis*, and in the monist-idealist on the complex series of conceptual terms including mind-spirit-self, and so on. In any case, there cannot be balance between the two poles, as it is matter (M) > mind (m) or mind (m) > matter (M); thus, on the opposite side we find what could be considered an equivalency, albeit not an identity in necessary terms, between the two poles. In even simpler terms, this would account for:

$$M > m;$$
$$M < m;$$
$$(M \equiv m) \neq (M = m)$$

The following options though are logically ***not*** possible, due to the defining premises of the aforementioned positions:

$$\left[(M > m);(M < m)\right] = \left[(M \equiv m);(M = m)\right];$$
$$\left[(M > m);(M < m)\right] \equiv \left[(M \equiv m);(M = m)\right]$$

In this sense we can appropriately say that there is at least an oppositional distinction between monistic and dualistic perspectives and this distinction represents in itself a form of dualism on principles of non-identity and non-equivalency. Whether ***not choosing*** between different views is equal to choosing between ***non different*** views is what we will explore in the central part of this work. Of course by 'exploring' we certainly do not mean 'avoiding the problem'. On the contrary we want to raise awareness on the importance of choice vs. non-choice in contextual terms; this exploration is not a form of 'cop out' in the face of the difficulties of the mind–body problem but is actually an 'opt in' because of these difficulties. In any case the very concept of choice presents a vast complexity which we will analyze starting from the purely biological-physicalist viewpoint, thus discussing the most common subfields of neuroscience that deal with the neural underpinnings of psychological processes with special attention to the central nervous system particularly the brain (Fig. 1.1). We will continue by discussing general and particular philosophical understandings of self and consciousness with a particular focus on free will and phenomenological explanation

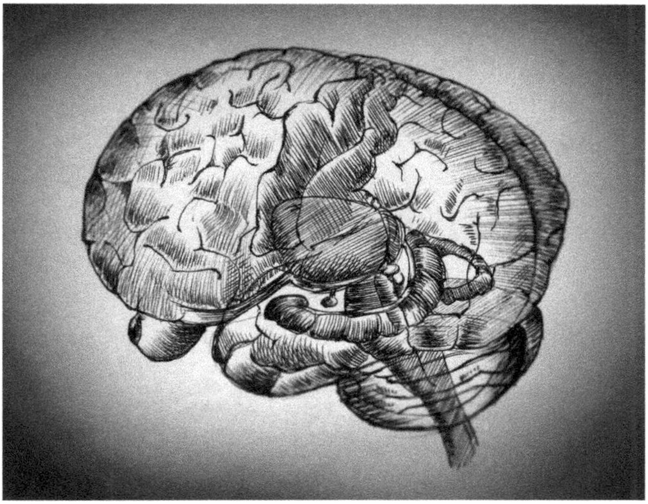

Fig. 1.1 A left dorsolateral view of the brain, with focus on the cerebral cortex with the four lobes, the cerebellum and the limbic system. Highlighted are the central proprioceptive neural areas with the central sulcus dividing the primary motor cortex and the somatosensory cortex

Furthermore, given what we said (and we will further explore) about the context (social, cultural, etc.) in which neuroscience operates, itself influenced and influencing brain activity *and* philosophical speculation, we will raise some questions connected to fields apparently very distant from either philosophy and neuroscience, such as public and healthcare policy (especially in regard to medicine and nursing), politics and even legislation processes. Certainly, we will also challenge some of the assumptions in the modern neuroscientific viewpoint, especially in regard to the hierarchical elements of description, namely, that only based on neuroscientific techniques such as neuroimaging we can (read: we are logically allowed to) infer that there is a 'democratic' process happening on a neural—not neurological, as it is part of the same hermeneutics we are discussing—level that makes all the processes therein collaborate 'at the same level of value/valence efficacy of action' and there is absolutely no evidence of an external/internal, certainly ulterior or even transcendental factor or 'entity' that influences (or even designs) those processes, functions and structures 'from above'. Thus, we will argue that there is a rhythmic, circular movement between monism, dualism (even trialism, etc., possibly pluralism or universalism) views and the hermeneutic interpretation of the applicability of neuroscience to philosophy and vice versa. We could view this viewpoint as 'tending to balance', or 'homeostatic' between what we could call 'Mother Nature and Father Nurture'. With a certain risk of falling into psychologistic explanations or psychoanalytic overinterpretation, we will also examine the debate on the mind–body problem under the same investigative frameworks, that is, we will try to understand how much of 'fear components' are present in choosing one philosophical-scientific viewpoint over another one, and defending it with extreme passion and, unfortunately, a lot of narrow-mindedness, its premises. This can be said about pretty much everything from reductionist-materialistic views of the self to the so-called quantum mind, from social cognition and embodied cognition to esoteric understandings of mystical experiences, etc.

Given these premises, the decision to list all these philosophical-scientific models and neuro-related disciplines in a text on critical neuroscience in the context of medical treatments comes from our intention to guide the reader through the specific viewpoint of each field, by itself first, and in combination with others after. Of course, the main goal is to present new ideas and a possible new paradigm, according to the Aristotelian saying: "The whole is greater than the sum of its parts." We also want to argue that a new point of view is only possible when the most important elements of every previous perspective (thus, every field of scientific inquiry) has been understood, mastered and applied to verify its strengths and weaknesses and the most effective areas of applicability. That is precisely why 'Critical Neuroscience' as a distinct field is presented only after examining other disciplines, in the last chapter of this work, *Philosophy as Basic Approach Toward Neuroscience*. And philosophy is the *true glue* which keeps all these discussions together and views them from above and from below. Of course, we attempted to present multiple philosophical approaches and schools of thought, but—as we previously mentioned—we embraced our bias fully to subjectively indicate[4] where our mind, or our brain, stands.

[4] Whether ours was a *free choice* or not, we will leave to the discussion on free will in the dedicated chapter.

References and Further Readings[5]

Bibliography

Bunge, M. 2013. *Medical philosophy. Conceptual issues in medicine.* Singapore: World Scientific Publishing.

Tomasi, D.L. 2016. *Medical philosophy. Philosophical analysis of patient self-perception in diagnostics and therapy.* New York: Ibidem Verlag/Columbia University Press.

References

Foglia, L., and R.A. Wilson. 2013. Embodied cognition. *Focus Article* 4 (3): 319–325.

Hossein Nasr. 1968. *Science and civilization in Islam.* Cambridge: Harvard University Press.

Further Reading

Fasol, U. 2011. *Dio? No grazie, siamo scienziati!* Verona: Sentinelle del mattino. Available at: https://vimeo.com/23885035

CHAPTER 2

The Exact Science of the Hard Matter

2.1 General Aspects

How reliable is philosophy in comparison to science? This question has been posed under the assumption according to which those two fields are (a) very separate, both in terms of history and epistemological essence, especially in regard to methodologies and goals, and (b) philosophy, especially in the context—that is, theoretical framework and/or mindset—of analytical philosophy, which sees itself as a 'support' of sciences, particularly 'hard' sciences, to shed light or clarify assumptions, ideas and thoughts, often relying on formal logic and cognitive structures used in natural sciences. Whether we are to support such assumption or reject it under, for instance, ideas from general continental philosophy, postmodernism, deconstruction-deconstructivism, traditionalism, perennialism, and so on, we could ask ourselves whether such assumption could at least be useful to investigate the core concepts we are discussing here, namely, the connection between mind and brain, and the possible neural underpinnings of cognition, memory, perception, consciousness and behavior. Consciousness in particular appears to be the connecting link between multiple disciplines in the attempt of just explaining the most basic ontological questions related to this complex concept. For instance, the American philosopher Donald Phillip Verene thoroughly analyzed the work by Hegel, especially in regard to the possibility of understanding consciousness. As we have seen, according to his interpretation of the

D. L. Tomasi, *Critical Neuroscience and Philosophy*,
https://doi.org/10.1007/978-3-030-35354-4_2

German philosopher, the two moments of *Ansich* are brought together by the work of consciousness. However, this is a constant battle because of the very nature of *Ansich*, which contains the very same essentials of the *[εν]νοούμενον* in contrast to the *φαινόμενον*. Given that we are talking about a 'Phenomenology of (the) spirit', this *Ansich* is therefore a posited object or event *in existence with the absence of* sense or perception, coded in sense, perception and understanding, representing the three failed attempts to bridge the gap between these opposites, which cannot be apprehended as a function of the object, but they lead to self-certain truths. Of course, this search for the truth is dialectical even beyond the Hegelian conception, and somewhat akin to the usage of the term in 'Dialectic Behavioral Therapy' or in socio-political discourse aimed at mitigating or at least 'making sense' of two opposites. In the context of 'The exact science of the hard matter', we could think of the mind-brain problem at the center of critical neuroscience. More specifically, we are thinking of epistemological efforts to justify biologically, for instance, the natural *emergence* in evolutionary-economic terms, of human morality, or (on the opposite view) to claim that human morality is in itself enough proof of the existence of a divine maker. We can think of swinging on an exercise ball/platform to increase the core muscles of the abdomen as a metaphor for trying to balance the absurd, or conflicting views (ex: science vs. religion) or the absence of any values or universal principles. Following this metaphor, the opposite is someone who 'melts' on a rigid, but stable platform (i.e. people who 'blindly' follow dogmas, to make themselves feel at ease). In any case, we would need to define the parameters of an illusion or (self) delusion of existence. This act of (re)defining or (re)learning the deep meaning of (in the example above) morality, we would be faced with a 'toward self/center movement', 'emergent intuition', 'self-consciousness-of self', which, in the Hegelian analysis, represent the primary path to a truth which is (a) absolute and (b) beyond (interpreted under the framework of *Jenseits*) illusion. This path leads to the true *Meinen* in a Hegelian sense, which allows this emergent truth to emerge from (we could argue) both 'consciousness asleep' and 'sleep of consciousness' through the *Unglücklichkeit* of skeptical-stoic perspectives. This is the linguistic and meaning-making/originating phase before the (re)absorption of this 'self-affirming consciousness-based reality' into the *verkehrte Welt*.

2.2 Examining, Controlling and Verifying

2.2.1 Neuroanatomy

To start our conversation of the multiple subfields of medicine, neuroscience, psychology and related areas, we will first focus on the best example of the matter-at-hand, that is, anatomical observation. In Greek, *ανατομή* indicates 'dissection', a compound formed by *ανά* or 'through' and *τέμνω*, to cut. Our focus is on the analysis of the brain, more specifically of neural structures underlying our cerebral mechanism. Thus, neuroanatomy is the discipline which allows us to conceptually organize and functionally comprehend the nervous system. Certainly, human neuroanatomy differs from other animals' neuroanatomy in many aspects (particularly the contrast between radial symmetry and bilateral symmetry), and we should be aware of both the strong similarities and resemblances between multiple species, as well as the evident differences.

Throughout history, the brain has been understood either in dualist connection/correlation or monist integrity, as well as—sometimes paradoxically—in an integration/interaction or mutual combination. In particular, we need to note that in vertebrates, the peripheric nervous system and the internal cerebral structures are extremely complex, and neuroanatomy is therefore a specific scientific discipline, albeit originating as an offspring of/from neuroscience. Aside from the specific components of the Central and Peripheral Nervous Systems however, the development of the very idea of this type of system in natural philosophy, and more generally in medical science, is of great importance. In regard to the very function of the nervous system, we must note that in antiquity, the core concept of *πνεῦμα* was representative of both the physical and the spiritual/etheric components of what we could nowadays translate as 'spirit', 'breath', 'mind' or 'ghost'. In this sense, the German term *Geist*, the Italian *soffio vitale*, and possibly the Russian *дух* are among the best translations. The pre-Socratic tradition linked the term to concepts such as archetype, soul, or origin; an invisible, immaterial and yet 'instilling life in/to the matter' akin to the *ἀρχή*. If Stoics more closely defined this word as 'spirit' and linked it to 'pre-destined', 'written', 'mandated', 'human destiny under divine influence',[1] this view was certainly shared by both

[1] In this sense we should also note the Hebrew רוח הקודש, *ruah hakodesh* for 'Holy/Sacred Spirit'.

Greco-Archaic as well as Classical-Roman perspectives, in particular given the double meaning of 'blowing/breathing/inhaling and exhaling/dying' of the Indo-European root **(s)peis-*; in Italian/Latin *spirare*.[2]

We will later return to this fundamental aspect of the discussion, but in the context of neuroanatomy it is important to note that in antiquity the nervous system was viewed mainly as the (mere) system of channels/canals through which the pneuma would flow, after being created by the brain from the blood. Of course, a system of calcium ion channels, potassium ion channels, sodium ion channels, voltage-gated ion channels, is still at the center of contemporary neuroscientific research. However, in ancient times, the brain was both the site of mind/soul (at this level the distinction is not yet well defined) and reason/understanding/cognition. A similar view was pretty much shared until the third century BC and the Alexandrine school, in particular Herophilos and Erasistratos. Another Greek, Alcmaeon, defined the brain as the primary controller of the body, thus removing the primate of the heart over human senses. However, the very first anatomical study specifically analyzing the human brain was found in Egypt, the 'Edwin Smith Papyrus'. Following the work by Galen and Rhazes (Abū Bakr Muhammad ibn Zakariyyā al-Rāzī) many Greek philosophers (whom many today would possibly refer to as 'neuroscientists'), there was a general halt to further development in the field of neuroanatomy, especially in medieval and early Renaissance Christian Europe, due to the impact of severe restrictions to human dissection.[3]

Further developments to this science—as well as to human anatomy in general—were at the center of the work by Taddeo Alderotti (who also developed a very interesting philosophical analysis of Hippocrates) and his disciple Mondino de' Liuzzi (who developed the very first complete modern anatomy textbook/atlas), Guy de Chauliac, Andrea del Verrocchio and Leonardo da Vinci, Berengario da Carpi (Jacopo Barigazzi), Andreas van Wesel (Vesalius), Bartolomeo Eustachi (Eustachius), Niels Stensen, Marcello Malpighi, François Pourfour du Petit, Thomas Willis (who also coined the term neurology), Domenico Mistichelli, Antonio Maria Valsalva, Giovanni Battista Morgagni (founder of modern pathology),

[2] With the obvious connection to terms such as *sperare*, and *speranza* (hope).

[3] Of note, this situation lasted until Pope Sixtus IV (1414–1484) decided to 'change policy' and lift the ban, thereby promoting new and impressive development in the field throughout the Renaissance. Furthermore, in 1832 the British Parliament finally passed the Anatomy Act to allow the legal dissection of executed murderers.

Filippo Pacini and others. This scientific background is the basis of modern (neuro)anatomical research, including the works by Samuel Thomas von Sömmerring, Marie François Xavier Bichat, Alfonso Corti, Pierre Paul Broca, Camillo Golgi, Henry Gray, Carl Wernicke and Korbinian Brodmann. Certainly, these developments would not have been possible without a thorough examination of the structural, functional and theoretic links and connections between human and other animals' brains, such as the cerebral analysis in mice, fruit flies, zebrafish, all the way to simpler—in a purely neuroanatomical sense—systems such as the ones found in the *Caenorhabditis elegans.* This transparent nematode was used multiple times, especially since the first outcomes in the research studies by Sydney Brenner,[4] to investigate neuronal development and use this information as a model system for humans. More in detail, since the nervous system in this roundworm is relatively stable in terms of structural composition from one subject to the next, neuroscientists and neurobiologists were able to precisely map the neuronal pathways in the *Caenorhabditis elegans.* Of note, this species is found to have about 300 neurons active in its system. In this regard, we must also note that there have always been competing models when it comes to the very determination of neuronal functions, and this aspect is important not only within the structural analysis of neurotransmission, but has impacted the discourse on a multiple of theoretical approaches in neuroanatomy, including the comparison between neural activity in the cerebrum as opposed to neural activity in the brain stem and the cerebellum (Fig. 2.1).

In terms of structural composition, we should note that at the center of neuroanatomy the division between *tissue level* and *organ level* is of fundamental importance. The first identifies the nervous system as being mainly composed of many types of glial cells (responsible for maintaining homeostasis, providing support, nutrition and protection for the neurons and the brain in general, including via the production of myelin) and neurons, as well as of the extracellular matrix. The organ level of neuroanatomical analysis instead defines the nervous system as composed of brain regions or brain (cerebral) areas. Within these cerebral areas we can also identify the four lobes, the limbic system, the gyri and sulci, and so on. As neurons

[4] Of note, Émile Maupas first described this species of roundworm in 1900. In particular regard to the research by Brenner, it is important to remember the so-called European Plan and the American Plan, with the genetic lineage determining cellular function in the first, and cell migration as primary chronological-developmental factor in the latter.

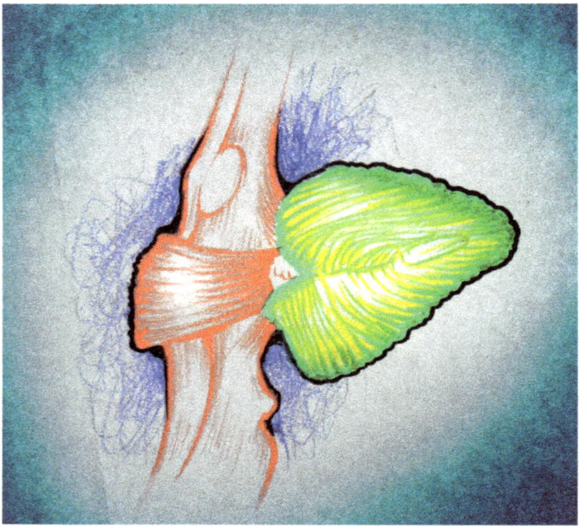

Fig. 2.1 The brain stem in the central nervous system (in red, including pons, medulla oblongata and spinal cord), seen here with the cerebellum (in green)

are the information-processing cells of the nervous system—build upon nerves, themselves created of axons—and neurotransmitters are the chemical/electric transferring/signaling components of such system, many studies in neuroscience focus on the role and activities of these elements, but we must also note that these building blocks, albeit fundamental, are only parts of the whole. The nervous system is traditionally divided into central nervous system or CNS and peripheral nervous system or PNS, the former consisting of the cerebrum/cerebellum, the retina, the spinal cord and the connection pons/medulla oblongata, and the latter—further subdivided into the somatic and autonomic nervous systems—consisting of the connecting nerves between CNS and the rest of the body. Furthermore, the neurons of the somatic nervous system are divided into afferent or carrying sensory signals from the sense organs to the CNS, and efferent or carrying motor signals out to the muscles. Finally, the sympathetic and the parasympathetic nervous systems are the two remaining subdivisions of the autonomic nervous system, itself made up of nerves containing afferent and efferent fibers. To clarify the common terms used to spatially define specific locative aspects in neuroanatomy, we will focus on the most important descriptors, staring from dorsal (from *dorsum*) and ventral (from *ventrum*), with the first term denoting the upper/top side and the latter the lower/bottom side. The term rostral derives from the Lat. *rostrum* and is therefore indicating the front of the object—in our case, the body—and the caudal (from *cauda* or tail) the back. Other terms used in this context are superior and inferior—which need to be determined by taking into account the cephalic flexure in this context—and medial and lateral with respect to the midline. In spatial terms, the three spatial directions can be horizontal, sagittal, and traverse, while the coronal sections can be horizontal, oblique or transverse.

Aside from purely physical observation, neuroanatomical advances have been made possible through the development of technologies such as cell staining, genetically encoded markers, histochemistry and brain imaging methods such as computed axial tomography, diffuse optical imaging, event-related optical signal, magnetic and functional magnetic resonance imaging, magnetoencephalography, positron emission tomography, single-photon emission computed tomography and cranial ultrasound. More specifically, dyes are used in cell staining—in the rough endoplasmic reticulum, or acidic polyribosomes, but also cell bodies, glia, dendrites and axons—in order to evidence the aspects of neuronal structure or cytoarchitecture in the nervous system. In the case of genetically encoded markers, variable amounts of red, green, and blue fluorescent proteins in the brain are used to differentiate between neighboring neuronal areas to map neuronal connections. A somewhat similar operation to study axonal connectivity, albeit created through a light beam, is used in optogenetics, especially in mice. In histochemistry instead, the biochemical reaction properties of enzymes are used to analyze the function, structure and architecture in the brain. Given the extreme importance that interdisciplinary approaches such as psychoneuroimmunology have provided to the study and practice of medicine—especially in mind-body and behavioral medicine—a particular subgroup of histochemistry is worthy of being discussed, albeit briefly, in this context. Immunocytochemistry or ICC is used to anatomically visualize the localization of a specific protein or antigen in cells,[5] axonal fascicles, neuropiles, glial processes, blood vessels (as well as specific intracytoplasmic or intranuclear proteins), by using selective

[5] Of course, given that Immunocytochemistry is performed on samples of intact cells with most of their surrounding extracellular matrix removed, this laboratory technique differs from Immunohistochemistry or IHC.

antibodies against a variety of chemical epitopes of the nervous system. In the case of neuroimaging, this type of technologies allows researchers to investigate not only the structure of the brain, but also its function and related pharmacology for therapeutic purposes. We will discuss the two main areas of neuroimaging, structural and functional imaging, in the next chapter.

Before we continue our discussion, we should also mention that in the context of neuroanatomical research, there are other methods besides both brain imaging and dye-based methods to investigate the cerebrum, the cerebellum and associated areas. For instance, we find viral-based methods to analyze the activity of viruses in crossing synapses and replicating in cerebral cells. Connectomics instead studies the (human, but not only) nervous system via an ever-increasingly complex mapping of nervous connections. These neural connections are organized in wiring structures or diagrams and called *connectomes*. Another type of technology bases its research strengths on neuroimaging, but it focuses on controlling-computational systems within the nervous systems, thus incorporating multiple fields such as communication theory, information technology—some argue even some basic aspects of integrated information theory or IIT—and computational (heuristic) intelligence, but also game theory, applied informatics, operational research and signal processing. The technology used in this area is therefore referred to be the constituent part of computational neuroscience or computational cybernetics and is therefore closely linked to disciplines such as artificial intelligence.

2.2.2 *Cellular and Molecular Neuroscience*

As neurons are fundamentally cells sharing information through electrical and chemical signals, the study of neuronal activity, including morphology and physiological properties at a cellular level is at the center of Cellular Neuroscience. The first analysis in this discipline is the distinction between typologies of neuronal cells and their classification in structure and function, a very important task to further understand fundamental phenomena such as neurogenesis, synaptogenesis, neuro-synaptic plasticity and specific features of signal transduction processes, for instance in seven-transmembrane receptors as in the case of G protein–coupled receptors (Fig. 2.2). From the perspective of morphology, neuronal cells can be unipolar/pseudounipolar (including Unipolar brush cells), bipolar, and multipolar depending on the structure and mutual relation between axons

Fig. 2.2 A rendering of the partial process part of the activation/deactivation cycle involving G protein–coupled receptors (GPCRs) and related G protein–mediated signaling. Of note, the cyclic adenosine monophosphate signal pathway and the phosphatidylinositol signal pathway are two principal GPCR signal transduction pathways

and dendrites. Furthermore, neurons with long-projecting axonal processes beginning in the gray matter of the central nervous system are referred to as Golgi I (for instance, Purkinje and Pyramidal cells), and, together with the short-axoned—sometimes with no axons at all—Golgi II represent multipolar neurons such as Granule cells. If the axon cannot be distinguished from dendrites, these neurons are generally called anaxonic. Aside from afferent and efferent neurons, a third directional subclassification defines local and relay intraneurons communicating between sensory or motor neurons and the central nervous system. Among the wide range of neuronal types, we also find many other neurons such as Anterior horn cells, Basket cells, Betz cells, Lugaro cells, Medium spiny neurons, Renshaw cells and Spindle cells. In terms of cytochemistry and neurotransmitter production, other nomenclature-based subdivisions are Anandaminergic, Cholinergic, Dopaminergic, GABAergic, Glutamatergic, Monoaminergic, Nitorxidergic, Peptidergic and Serotonergic neurons. Moreover, glial cells—including Schwann cells in the peripheral nervous system—represent a very specific and peculiar class of cells, as they play a fundamental role in the nourishment of neurons and in the modulation of synapses. Beside specific elements from molecular biology and pharmacology, in cellular neuroscience the multiple techniques utilized to investigate neurons include Ca2+ imaging, confocal imaging, intracellular recording, electrophysiological techniques such as patch-clamp and voltage-clamp, and two-photon laser scanning microscopy. These techniques allow for a very detailed analysis of neuronal function, excitability, synaptic transmission, myelinating processes, action and graded potentials, and cellular response (involving Na+/K+-ATPase and related 'pump') to polarization and depolarization through threshold, rising phase, falling phase, undershoot phase and recovery. Furthermore, cellular neuroscience also involves the examination of the combination between the chemical and electric components or, more appropriately, the ontological basis, of the synapses. On a more practical level, of course, it has been observed that chemical synapses are slower than electrical synapses. This analysis also covers the structure of the receptors, transmembrane protein molecules, which can be ionotropic and metabotropic. To be more specific, ionotropic receptors such as GABAA and GABAC form an ion channel pore, and are a combination of a receptor and an ion channel, while metabotropic receptors like GABAB are indirectly connected to ion channels on the cellular plasma membrane through signal transduction mechanisms, usually G proteins. In this context it is important to at least mention processes such as the excitatory postsynaptic potential or EPSP and the inhibitory postsynaptic potential or IPSP. In the first case, we observe an excitatory postsynaptic depolarization. In the latter instead, the postsynaptic receptors allow Cl– ions to enter the cell or K+ ions to leave the cell. The synaptic activity is an integral part of molecular neuroscience as well, with a special attention to structures, processes and mechanisms within the nervous system as well as (epi)genetics and molecular biology. In fact, as variations in gene expression might affect a wide array of brain functions, molecular neuroscience studies epigenetic mechanisms of regulation, DNA methylation and histone modification, also in order to identify/isolate specific markers. This approach has an immediate, practical application in the study and possible prevention/treatment of late-onset neurodegenerative diseases like

Alzheimer's or neurodevelopment diseases such as autism. For instance, the progressive loss of both cognitive functions and memory affecting patients with Alzheimer's disease (among the primary causes and/or underlying conditions for dementia) has a complex etiology linked to several neurological and/or genetic hypotheses, including the cholinergic, the amyloid and the tau hypothesis. A GABAergic mechanism as well as a dopaminergic component is found in Parkinson's disease, for instance, in the specific loss of dopaminergic neurons in the substantia nigra. Another example is the absence of normal inhibitory inputs from medium spiny neurons of the basal ganglia as the cause of Huntington's disease. In the case of excitotoxicity instead, the primary damage to neuronal cells comes from an excessive stimulation of NMDA and AMPA (in the same family of ionotropic glutamate receptors such as kainate) by glutaminergic storm.

As we have seen, cellular and molecular Neuroscience are fundamental fields not only from a research background, but because of the immediate benefits for a deeper understanding of mind-body connection and general well-being, starting from the very structures or 'building blocks' observed in neuroanatomy (Fig. 2.3). As cells are indeed biological 'rooms' (from

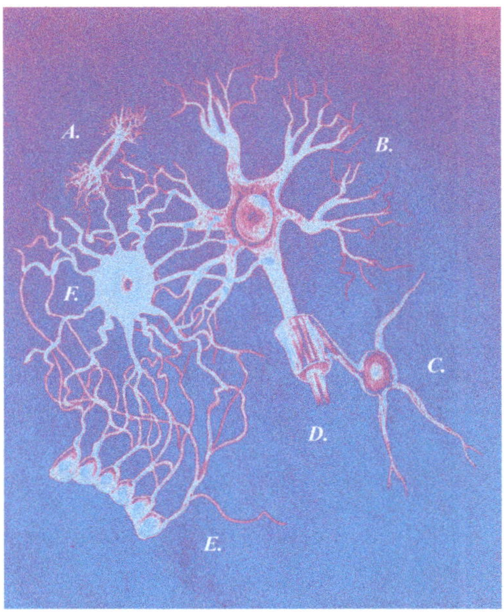

Fig. 2.3 A general view of the most important neural elements, in particular microglial cell (**a**), multipolar neuron (**b**), oligodendrocyte (**c**), myelinated axon with Schwann cell and node of Ranvier (**d**), ependymal cell (**e**), and astrocyte (**f**)

Lat. *cellula*, a microstructure derived from *cella*) and are therefore the basic morphofunctional unit of living organisms subdivided into plasm (or *cyto*plasm, from the equivalent Greek term), membrane (plasma or cytoplasmic membrane) and organelles (again from the minor adj. derivative of Lat. *Organum*), injuries to cell structure and extracellular matrix are the primary cause of structural patterns of disease, as they affect tissues and organs (Alberts et al. 2002). In our case, injuries of that kind at the level of the central nervous system represent a challenge for the development and maintenance of normal neural functioning. More specifically, as cells membranes are constantly morphing to adjust to the homeostatic processes within organisms, connectivity and permeability are among the fundamental factors and functions in the preservation of health at the cellular level. When homeostasis is deficient, or—in other terms—these health levels in the body are not observed, patterns of cellular injury arise. For instance, the deprivation of oxygen supply or hypoxiation causes damage to the sodium-potassium pump (known throughout the cellular processes, but especially relevant in the delivery of action potential through axial neurotransmission). This might cause sub-lethal—thus, reversible—damage including swelling, a common epiphany of the vast majority of cellular injuries. However, when the cause of such hypoxic process is not eliminated, the damage becomes irreversible (McCance and Huether 2014). More in detail, when a reduced blood supply (as in ischemia) leads to arteriosclerosis, cells shift from an aerobic to an anaerobic state (therefore, metabolism). This in turn affects tissues and increases metabolic acidosis, with a lethal outcome for cells. At this stage, we talk about a permanent pathologic change, or irreversible cell injury. Of course, any analysis in neuroscience would not be complete without a thorough examination of the influence of nature over nurture. A good example in this context is represented by genetic and epigenetic factors, such as derangements related to congenital malformations, metabolic errors. These factors are responsible for the morphing patterns in the cytoplasmic membrane's shape and structure, as well as in its receptors or delivery-transport mechanisms. Beside the damages on the cell membrane (as in apoptosis), injuries can happen at the level of DNA patterns. The alteration in these patterns is at the basis of misregulation of gene expression found in cancer (You and Jones 2012). Therefore, the progression from reversible to irreversible cellular injury arises through the alteration of the genes controlling the epigenome, thereby disabling the repair functions of the DNA. Of course, all these observations need to be understood *in context*, that is, without missing the outer effectors in the dialectic between nature and nurture. As this chapter is about 'the exact science of the hard matter', we need to focus on our definition of this hard matter and make sure to use the appropriate tools to investigate it. Moreover, with the same level of attention with which we apply those tools in our scientific research, we should also be constantly reminded of the bias that we, at least to some extent, always carry, as children and heirs of certain positivistic attitudes—and we use this word with a degree of purpose—we inherited from previous scientific paradigms, especially post-enlightenment-based worldviews. Gadamer in particular was very critical of certain aspects of the Enlightenment, thereby viewed as "prejudice against prejudice" or "the refusal to recognize the significance of our own insertion in a tradition that, at some level, we already understand" (Kearney and Rainwater 1996). In this context, reality is the framework where the individual lives, feels, thinks and acts through *Wirkungsgeschichte* (effective-activating history) that underlies any potential "fusion of horizons" we hope to achieve (Tomasi 2016). Thus the particular, contextual, personal and individual element is to be understood under the investigation of bias and prejudice, which is closely connected to the concept of authority according to the German philosopher.

2.2.3 Developmental Neuroscience

In the field of developmental neuroscience, pre-natal development is at the center of the investigation of the processes beyond the creation, maintenance, function-forming and structure-generating aspects of the nervous system. As an intersection between classical neuroscience, psychology and biology, this field also incorporates developmental cognitive neuroscience as a discipline covering the neural development from the ectoderm giving birth to the (mammalian) central nervous system all the way to the specification/migration and self-organization of neurons allowing for axons and dendrites to further extend and develop into more and more complex synaptic-based communicating systems. To be sure, in a dualistic approach the mind would have to play a fundamental role—either passive or active—in this very development. This perspective, as we will see in later chapters, is also providing a philosophical basis to the theoretical justification of phenomena such as the *placebo* and *nocebo* effects, but also shaping the core of neurogenesis and mirror neurons. In fact, the very creation of separate forebrain (prosencephalon), midbrain (mesencephalon) and hindbrain (rhombencephalon), to the further subdivision of telencephalon, diencephalon, mesencephalon, metencephalon and myelencephalon (and their future developments, respectively, cerebral cortex and basal ganglia, thalamus and hypothalamus, pons and cerebellum, and medulla oblongata) stems from the processes, activity-independent and activity-dependent mechanisms, some of which are truly cognitive, in the sense of sensory motor underpinnings to behavior, analyzed in developmental (cognitive) neuroscience. As with other fields in neuroscience, beside animal models, brain imaging is used to investigate steps and biological landmarks of neural development. Magnetic resonance imaging or MRI is used to analyze both gray and white matter, for instance, to better monitor and understand the stages of evolution of the cortical areas as well as the myelination process which allows synaptic transmission to be functional. In this context, Multicomponent relaxometry or MCR is the technology used to more precisely measure the levels of myelin in the brain. This translates into an improved ability to account for possible medical problems, such as epilepsy (including pediatric epilepsy), which is suspected to involve altered myelination in the efferent or afferent white matter pathways adjoining epileptic foci (Spader et al. 2013). The same applies to disorders such as multiple sclerosis, also associated with gross white matter damage and demyelination. Without examining in depth specific stages of development, for which we refer the reader to the bibliographical information at the end of this chapter, we need to mention the importance of the neuronal plate, more specifically the creation of the neuro-ectoderm from an undifferentiated ectoderm through the activity of the mesoderm. Furthermore, developmental aspects such as neural induction, neurulation and further differentiation are processes central to developmental neuroscience. Regionalization, neuronal (radial, tangential, axophilic, multipolar) migration (including corticogenesis), somal translocation and patterning due to differences in concentration of signaling molecules. The discussion would not be complete without neurogenesis, a process which serves as one of the fundamental effectors in our analysis and one that carries enormous philosophical weight. We will discuss these aspects more in depth in the next chapters; for the present moment let us remember that this process occurs during embryogenesis, most specifically through the creation of excitatory and inhibitory neurons from neural stem cells and progenitor cells in all animals. It is also important to remember that this process happens throughout the lifespan of an individual, and is therefore not limited to specific stages of neurodevelopment (Fig. 2.4), including

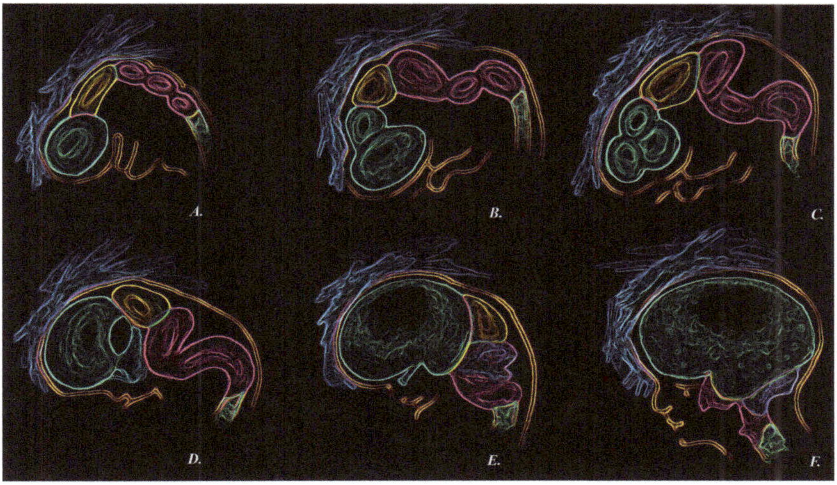

Fig. 2.4 Analysis of embryonic neural development and brain growth after 25 days of gestation (**a**), 35 days (**b**), 40 days (**c**), 50 days (**d**), 100 days (**e**), and after 9 months (**f**). Highlighted, prosencephalon with optic vesicle (in green), mesencephalon (yellow), rhombencephalon (fuchsia), and spinal cord (light green)

the final stages of the creation and structuration of the prefrontal cortex, but continues long into adulthood, mostly in the dentate gyrus of the hippocampus and the subventricular zone. These considerations open up the discussion on strategies to be used in the case of 'amelioration and improvement' of one's ability, capacity and skill level.

As previously mentioned, we will further discuss what this means in philosophical-theoretical terms of the mind-body (brain) problem and, on an ethical-moral level, whether these goals are justified and justifiable through specific parameters. In any case, a number of studies have suggested[6] that both psychotherapy and pharmacological intervention—especially in the case of depressive states—target neurogenesis and are therefore fundamental in our understanding of human nature. In this context, neurogenesis appears to be part of a fundamentally defining phenomenon occurring on a neurological level, starting from the brain but extending to the nervous system as a whole, and possibly to the combination of multiple

[6] In this regard, Please see references at the end of this chapter, especially the meta-analysis offered by Beauregard and O'leary (2008), Heintzelman and King (2014) and James (2016).

systems in the (human) body, a multisystem like the ones presented by interdisciplinary fields such as preventive and behavioral medicine, as well as psychoneuroimmunology with special focus on the nervous system, endocrine system, cardiovascular system and stress response, in particular the HPI (hypothalamic–pituitary–interrenal) axis and the HPA (hypothalamic–pituitary–adrenal) axis. Of course, the neuronal activity and influence on and through these systems are closely related to homoeostatic and self-preserving processes. Neuronal self-preservation is in fact obtained (among other processes) by the regulatory activity of the trophic factors, as hypothesized by Rita Levi-Montalcini and Victor Hamburger. As it is well known, Levi-Montalcini was instrumental, together with Stanley Cohen, in purifying the first trophic factor, the Nerve Growth Factor or NGF, which, together with Ciliary neurotrophic factor or CNTF and Glial-derived neurotrophic factor or GDNF are among the fundamental elements of the processes of preservation and growth as well as gene expression and more generally development. In a discussion on developmental neuroscience, we should not forget the development of the myoneural or neuromuscular junction, represented by a chemical synapse with acetylcholine transmitter and acetylcholine receptor (AchR) allowing the connection between a motor neuron and a muscle fiber. More specifically, *prepatterning* represents the process whereby AchR is created by the muscle cells and further expressed in the central regions. Other studies on the synaptogenic processes have also involved other elements and parts of the central nervous system, from the synaptogenic signals related factors including, but not limited to, Agrin, astrocytic factors, SynCAM and neuroligins.

2.2.4 Evolutionary Neuroscience

From the concept of development to the one of evolution, neuroscience often works on multiple levels, and interdisciplinary fields. This is again the case of evolutionary neuroscience, which combines theories and experimental, laboratory, and field observations and experiments in order to better investigate the (biological) evolution of the nervous system. Therefore, evolutionary neuroscience has drawn a lot of information from biology, anatomy, and chemistry, and in its comparative aspects, has used phylogenetic comparative methods together with theoretical elements from evolutionary developmental biology (informally referred to as 'evo-devo') as well as from other subfields in anthropology, biology, psychology and sociology. Evolutionary Neuroscience thus studies the development, on evolutionary terms, of the processes and structures of the nervous system, analyzing functional, modular, relational, cognitive and behavioral aspects under the specific scientific framework—in itself also a theoretical assumption—that the nervous system, more in detail the (especially mammal) brain, can regulate behavior due to the design natural evolution has put into action in order to solve adaptive problems faced by organisms. In this sense, concept and (diagnostic) terms such as 'maladaptive behaviors' used in psychology, psychotherapy and psychiatry have their 'evidence-based' foundations in these biological underpinnings. In fact, we could argue that the study of genetic transmission and variation have had a direct effect (we could also infer an 'affect') on the way modern psychiatry views mental health disorders, that is, 'neurobiological disorders'. We will certainly return to this topic in the next chapters, especially given the ethical consideration related to the conceptualization and ontological-ethical debate linked to psychiatric diagnostics. Certainly this field is at the center of an ongoing discussion on natural selection and evolution in Darwinian terms, which contributes to the specifics of the theoretical assumptions linking biological underpinnings to the neurological base of multiple layers of (especially animal) interaction. In this sense, evolutionary neuroscience presents its closeness to developmental neuroscience, especially in the analysis of the connection between processes and phenomena such as mirror neurons; psycho-linguistic evolution and language acquisition; modular processing and lateralization of functions; social and cognitive skills, including aspects like emotion regulation, relation, empathy, social bonding and cohesion; feedback control and feedforward control all the way to self-awareness, self-image, self-adaptation—again as possible product of biological adaptation, in this framework—and possibly self-conceptualization, including self-esteem. Of course, all these elements are deeply connected to the development of areas such as the cerebral cortex under the lens of neuroplasticity. Furthermore, the wide range of questions from the areas of philosophy of mind, most specifically theory of mind, and embodied cognition find in evolutionary neuroscience an appropriate target audience. From a phenomenological viewpoint, the genesis-generation of consciousness moves from the embodied body-subject to the world-object. These concepts, heritage of Merleau-Ponty's philosophy, are the attempt to overcome/surpass the Cartesian dualistic model of the primacy (supremacy) of thought. In his view, the φαινόμενον is actually a correlate of the body and our sensory perception, which translates, in our analysis, into the neurological underpinnings of 'perceptual awareness'. The latter, in the classic Hegelian interpretation was actually an overlapping-rhythmic connection between spirit and body. More in detail, when consciousness is not (yet) present, and the subject is therefore unconscious, the Spirit did not 'receive/acquire' a conscious mind. Thus, spirit is not yet ready to move on to the Hegelian second stage of self-consciousness, which is defined by the presence of the mind and the absence of comprehension of unity and connectedness of reality. In this sense, consciousness is not full. In modern terms we would say that 'full' consciousness happens only through 'shared knowledge' in a very similar sense to perspectives found in artificial intelligence and computer science perspectives. Besides the etymology of the word itself, it is also true that only in the third stage of consciousness described by the German philosopher does the Spirit progressively reach the ultimate meaning of reality. This reality is *beyond* separation and division. This is why thinking of Hegelian philosopher in terms of "thesis, antithesis, and finally synthesis" is a misnomer, albeit helpful in the very beginnings of the study of Hegel. Hegelian reality is in this sense presenting itself both in subjective and universal elements. From the perspective of neuroscience applied in therapy, these concepts are themselves relying on specific posits involving the sequential–causal functions of biology, not in oppositional terms, but in binary cooperation. For example, many studies have focused on the differences in the development of parietal cortex organization in monkeys, thus presenting—via deductive approach, albeit with some very strong inductive elements—a link between the ultimate goal/purpose/function and its biological rendering/representation. To be sure, among the causal assumptions promoted (a term used here without any polemic intent) in this context we can certainly understand that the nervous system was created/born—we will leave some open and uncertain definition for now—and further developed/evolved with the purpose of controlling behavior in an adaptive sense. In humans, differences in these developmental underpinnings of behavior have been monitored via electroencephalogram (EEG). This technique allowed researchers to verify the connection between the developmental processes in the brain and the systematic increase in wave frequency and structural complexity (especially during wakefulness), as well as the increasing connection between cortical areas.

2.2.5 History of Neuroimaging

Under the term 'neuroimaging', a wider range of technologies has been used in order to investigate anatomical and pharmacological structure, as well as the function of multiple brain areas. This discipline will be at the center of further discussion on the deeper understanding of (human) mind. More specifically, we will compare correlation and causation to verify claims of neurological/neural 'underpinnings' or 'substrata' of aspects such as thought process and cognition, emotions and feelings, conceptualization and self, and many others, especially from a human perspective. In this chapter, we will briefly examine the development and typologies of neuroimaging techniques offered to the scientific community. From a historical viewpoint, the 'founding father' of neuroimaging and a fundamental personality in the very history of modern medicine, the first person to ever develop a neuroimaging technique was the Italian physician, physiologist, archaeologist, professor and Senator Angelo Mosso, who also invented the sphygmomanometer, the ergograph, the hydrosphigmograph and the plethysmograph. From the perspective of neuroimaging, the biggest contribution came from the 'human circulation balance' invented in the nineteenth century by Mosso, a technology which allowed for a non-invasive measurement of the redistribution of blood during emotional and cognitive activity. More in detail, Mosso was able to identify and analyze some of the most important variables still used in modern brain imaging techniques. For instance, he promoted the experimental and simultaneous recording of multiple physiological parameters focusing on the 'signal-to-noise ratio'. The main hypothesis underlying the experiment conducted by this great scientist is still one of the most important tenets of modern neuroscience, that is, the connection, in terms of direct correlation, between processes and physiological activity, more specifically between intellectual and/or emotional activity and increase in blood flow in specific (functional) areas. From this perspective, functional changes determined alterations in blood flow and organ volume change on a brain level. Mosso measured these variations by converting brain pulsation into plethysmographic waves (Sandrone et al. 2014). The studies by Mosso strongly influenced research in neuroscience and psychology, especially in clinical and diagnostic settings. However, until recently some of the most important aspects of these studies were almost unknown—a possible exception is the brief mention of Mosso's work by William James—due to the absence of appropriate translation of the original manuscripts, written in Italian. Between 1875 and 1895, a series of discoveries by scientists like Johann Hittorf, Julius Plücker, Eugen Goldstein, William Crookes and Wilhelm Röntgen led to the use of X-ray technologies in the medical imaging fields. Following these discoveries, the new technology of ventriculography led to the development of pneumoencephalography, introduced in 1919 by Walter Dandy. A few years later, angiography or arteriography was introduced by Egas Moniz, also using fluoroscopy as an X-ray-based investigative technique, whose medical applications and related uses were further developed by Eduardo Pereira and Sven Ivar Seldinger, author of the Seldinger wire technique. In the 1960s and 1970s, the development of computerized axial tomography (usually referred to as CAT or CT scans) allowed for combinations of multiple X-ray images taken from different angles. CT scans thus create tomographic images non-invasively and more precisely than standard X-ray technologies. The first tomographic techniques, albeit single-image, were first utilized by Alessandro Vallebona in the early 1900s and were further ameliorated by Godfrey Hounsfield and Alan Cormack. Of note, both positron emission tomography or PET and single-photon emission computed tomography or SPECT are types of CAT. These techniques consist of radioactive neuroimaging technologies developed, in the case of PET, by Niels A. Lassen, David H. Ingvar and Erik Skinhøj following the studies by David E. Kuhl, Luke Chapman and Roy Edwards. In the case of SPECT, we need to mention that there were several stages of development, as this technology was at first replaced with higher-resolution computerized axial tomography or magnetic resonance imaging techniques. Of note, these two types of brain imaging technologies produce a static image of structure, while SPECT allows for an analysis of dynamic brain activity. A specific type magnetic resonance imaging technology, the functional magnetic resonance imaging or fMRI, produces an even better quality of images, with the plus of being non-invasive and safer for the subject/patient, as it does not use radioactive isotopes. Of course, in neuroimaging, multiple combinations of techniques and technologies are used. A good example is the PET-MRI and the PET-CT scanner. Other common techniques are multichannel electroencephalography or EEG, magnetoencephalography or MEG, as well as hybrid technologies used not only to analyze structure and function, but also to elicit a response, as in transcranial magnetic stimulation or TMS. This method derives from the studies by pioneers like Luigi Galvani, Michael Faraday, Lucio Bini and Ugo Cerletti. The comparison between the use and effects vs. side effects of TMS and electro-convulsive therapy or ECT has been at the center of both neuroscience and psychiatry. Beside psychological applications such as personality traits, characters, brain-typing, nature vs. nurture, lie detecting techniques and others, one of the most important aspects of neuroimaging is the debate around the monitorability of wakefulness, awareness and consciousness. From a clinical standpoint, this has immediate outcomes in the medical decision making processes related to comatose states, vegetative states vs. minimally conscious states, locked-in syndrome, and others to the very definition and medical, ethical and legal implications of diagnoses such as 'brain death'. We will analyze this debate more in depth in the next chapters, but we want to at least provide some useful definitions of the terms utilized in this context. To begin with states of consciousness, in clinical settings it is important to differentiate between (Table 2.1):

As we will point out several times in this work, there are a series of *observations on observations* to make. Neuroimaging is truly 'neural imaging' and thus contains images *of* neural activities, not necessarily those activities themselves. As part of scientific evidence, the 'emergent view' is pure observation, without (rational) explanation. This view, embraced by Bluhm (2014), among others, can be complete only with an appropriate philosophical interpretation. To clarify, there are two different targets: target (a) is rationalism, with the emphasis on etiology, diagnostics and symptomatology, thus focusing on the causal mechanisms inclusive of anatomical observations, as in neuroanatomical analysis (Fig. 2.5, as an example) for instance of what is observed; and target (b) embodied by empiricism to connect with possible prognostic and clinical (treatment) strategies in the context of neuroscience applied to medicine/psychiatry and psychology. This perspective stretches into the realm of medicine due to its criticism of a 'pyramid of evidence' whose hierarchy is based on a descriptive atheoretical empiricism, which is, especially within psychiatry fully statistical (with reference to the DSM) and thus on the "Average of the Averages" (Higgins et al. 2014). Certainly, there are many theoretical frameworks to define—in quantitative, but at times also in qualitative terms—the 'levels' or 'stages' of consciousness. An interesting analysis in this sense is represented by Integrated information theory (IIT) developed by Tononi; the Orchestrated objective reduction (Orch-OR) by Penrose and Hameroff; the Holonomic brain theory by Pribram and Bohm; and many others, including the contemporary neuroscientific-philosophical perspectives by Chalmers, Metzinger, Damasio, Edelman, Kaku, Koch, Pinker, Popper and Roth; we will further expand this

Table 2.1 States of full/partial/limited consciousness vs. vegetative states according to the General Medical Model

Locked-in syndrome or LIS:
A term first used by Fred Plum and Jerome Posner to indicate a subject being isolated due to quadriplegia and pseudobulbar palsy, albeit having awareness, sleep-wake cycles and clinically meaningful behavior. In a specific case, the syndrome is classified as Total locked-in syndrome (TLIS) or completely locked-in state (CLIS). Of note, in LIS the eyes of the patient are paralyzed as well.

Minimally conscious state:
In this state, the patient has intermittent periods of awareness and wakefulness and displays some meaningful behavior, that is, demonstrating purposeful or voluntary behavioral responses to auditory, tactile, visual or noxious stimuli.

Persistent or continued vegetative state (PVS or PCS):
A clinical diagnosis in which a wakeful unconscious state lasts longer than a few weeks, with no evidence of sustained, reproducible, meaningful behavior as well as no evidence of language comprehension and/or expression; intermittent sleep-wake cycles; sufficient preservation of autonomic functions to permit survival with adequate medical care; bowel and bladder incontinence; variable preservation of cranial nerve and spinal reflexes.[a] Furthermore, subjects in this state are not aware according to three factors: "a) the presence of stereotypical ocular, facial, and motor movements seen only with the destruction of higher cortical structures, accompanied by the absence of any voluntary movements; b) PET studies that show decreased glucose metabolism in these patients; c) post-mortem autopsy studies showing extensive structural damage that is probably incompatible with consciousness." (Giacino et al. 2002)

Chronic coma:
Often simply referred to as coma, this state indicates lack of consciousness/awareness and sleep-wake cycles, failure to initiate voluntary actions—displaying only reflexive behavior—and to respond normally to stimuli such as light, pain and/or sound

Brain death:
This is a clinical/medical, ethical and legal definition upon which multiple debates are still trying to provide clarifying attributes. Generally speaking, this condition indicates a full and complete loss of brain function including involuntary processes required/necessary to sustain life, and requires a certification process including: "1. Identification of history or physical examination findings that provide a clear etiology of brain dysfunction. 2. Exclusion of any condition that might confound the subsequent examination of cortical or brain stem function. 3. Performance of a complete neurological examination including the standard apnea test and 10 minute apnea test. 4. Assessment of brainstem reflexes. 5. Clinical observations compatible with the diagnosis of brain death. 6. Responsibilities of physicians. 7. Notify next of kin. 8. Interval observation period. 9. Repeat clinical assessment of brain stem reflexes. 10. Confirmatory testing as indicated. 11. Certification and brain death documentation." (Goila and Pawar 2009)

[a]According to the definition provided in: The Multi-Society Task Force on PVS (1994). Medical Aspects of the Persistent Vegetative State—Second of Two Parts. *The New England Journal of Medicine (NEJM)* 330: 1499–1508

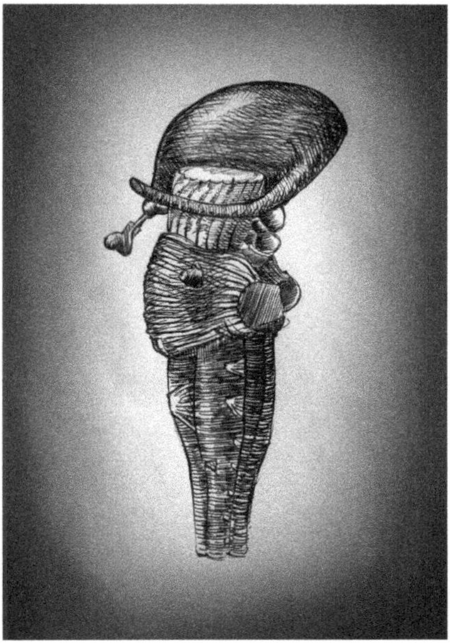

Fig. 2.5 Lateral view of the brain stem with mesencephalon, pons and medulla oblongata. Aside from sensorimotor functions and regulatory activities of the brain stem toward the respiratory and cardiac activity, this area of the central nervous system is relevant for the regulation of sleep patterns and wakefulness/awareness/consciousness and thus plays a primary role in the brain stem stroke syndrome at the root cause of locked-in syndrome

discussion with a special focus on the mind-body problem in Chap. 5 *Perception and Cognition*.

Our analysis of the technological features of neuroimaging and the related ethical questions regarding states of consciousness bring us to an incredibly wide range of possible discussions. If we already ascertained that the very definitions and clinical assessments upon which the scientific-medical community builds their diagnosis of, for instance, persistent or continued vegetative state are very complicated and often controversial, we also need to pay attention to the complexity of clinical assessments in general. A discussion on the ontological elements of clinical diagnosis has been at the center of previous works (Tomasi 2016) and in part re-analyzed in Chap. 4 *Medicine On, Of and Off the Brain*. In this context, we want to focus on the importance of a deeper understanding of medical errors. Given the fundamental essence of a clinical understanding of the parameters that define life and death in a medical sense, we do believe that every neuroscientific effort in this sense should pay enormous attention to this topic.

In fact, addressing the problem of a possible link, from a correlational to a causational standpoint, between deaths and (reported) medical errors is indeed a very hard task. A good analysis in regard to these aspects can be found in the 1999 report by the US Institute of Medicine, named 'To Err Is Human: Building a Safer Health System'. This report suggested that 2–4% of all deaths in the United States were *caused* by medical errors. Aside from all the metaphysical, spiritual or religious consideration related to the very assumption of the chronologically-embedded precision and necessity of death ('we all die when our time has come, not a second before, not a second after'), we should certainly make sure that we understand the core problem of such claim, especially given the risk of falling for conspiracy theories spanning from mind-control to 'big pharma' vs. 'big medicine' and similar. To better frame the problem, we would like to refer to the analysis of Mario Bunge (2013) in regard to the Bayesian theorem as applied to medical errors. In particular, the author notes that "Frequencies, along with averages and variances, are *collective properties*. [...] probabilities are properties of *individual facts*. [...] the probability concept is theoretical, not empirical like that of frequency, i.e. probabilities are assumed or calculated, never directly measured" (Bunge 2013, p. 99). In 1999, the 'To Err Is Human' report set a minimum goal of 50% reduction in errors over the next five years. Although an immediate reply to the promises of the reported has been given, in a negative sense, multiple times until now, we should certainly ask ourselves whether talking about a 'death (incidence) rate' in this context makes any sense at all. Generally speaking, death rate is often defined as mortality rate, thus as a measure of the number of deaths in a specific population, cluster, cohort or subgroup, scaled to the size of that population, per unit of time, expressed in units of deaths, per 1000 individuals, per year. In other words, the consideration of the fact that the value for death (should) follow a binary code (e.g. 1, alive 0, dead) is in itself pretty explanatory of the problematic analysis of values comprised between these two. In fact, at any point, we could never find a subject—in this case, a patient—who is at a 0.5 level, "halfway between being alive and being dead" (Tomasi 2016). These aspects are of fundamental importance when analyzing the problematic aspects of medical errors in vulnerable populations such as cognitively impaired, traumatized or comatose persons, as well as elderly or terminally ill patients. Considerations such as the ones presented above are very helpful in determining whether the US healthcare system has met, or will meet in the near future the needs of a safer health system. Although the report's promises were clearly not kept, as the medical and epidemiological literature has clearly shown, the US healthcare system has certainly benefited from it in most of the areas addressed by the report (Mujkanović 2016). Certainly, we could argue that from this point of view the healthcare system in the United States is far from reaching the levels of other geo-cultural areas—especially Europe, and especially in the context of fair, equitable/free and universal healthcare—but the efforts set by the continuing work of our social policy-making framework appeared to be improving, at least until the end of the year 2016 and the new political scenario following federal elections. In fact, in 2001 the 'Crossing the Quality Chasm' report was a very appropriate follow-up to 'To Err Is Human: Building a Safer Health System' in the sense that it addressed the six aims necessary to improve quality of care delivered to our patients. These aims were defined as safe, effective, patient-centered, timely, efficient and equitable. Among the most important contributions of these two reports to the general amelioration and improvement of the US healthcare system we can therefore mention the sensitization of the public opinion, including the personalities at the top of the political debate and decision/policy-making strategies, toward these issues, and the increasing initiatives and campaigns focused on achieving these goals (Mujkanović 2016). Among these initiatives, we can certainly mention the 'Institute for Healthcare Improvement's 100,000 Lives Campaign' in June 2006, which, albeit still making claims such as "124,000 deaths were prevented in over 3,000 hospitals through patient-safety initiatives in a period of 18 months" (Institute for Healthcare Improvement 2006), is a proof of the constant scientific efforts to collect, monitor and utilize in a statistical, epidemiological (which translates into direct patient care) way to increase the well-being of every patient.

2.2.6 *Neuroengineering*

Neuroengineering, also called Neural engineering, is a combination of multiple theories, techniques and technologies used in the setting of applications for the analysis, monitoring, repairing/replacing, enhancing/

expanding/exploiting and amelioration/augmentation/improvement of neural systems. Furthermore, neuroengineering comprises the research on the connection, combination and interface between organic and synthetic-artificial-non-living systems, communication pathways and constructs. In this sense, this discipline overlaps with mechanical neuroprosthetics research and comprises brain–computer interface (BCI), direct neural interface (DNI), mind–machine interface (MMI) and brain–machine interface (BMI). Therefore, much of neuroengineering focuses on approaches drawn from general neuroscience as well as artificial intelligence, clinical neurology, computer engineering, cybernetics, electrical engineering, nanotechnology, neurohydrodynamics, materials science, neural tissue engineering, robotics and signal processing. Neuroengineers combine qualitative and quantitative research methods to investigate the connection and composition of physical and mechanical structures and forces to quantify strength, operation, organization, capacity, ability, load, vulnerability, stress and possible overload from the laboratory to the artificial-virtual realm. These translational aspects of the research are especially relevant in neuromechanics, but offer great insights also into chemical/molecular neuroscience. In fact, studying the electro-chemical properties of neurotransmission, as well as the multitude of stimuli acting on the (de/re)polarization processes linked to the action potential—not only chemical and electrical, but also magnetic, mechanic, optical and specific receptor-related—produces new knowledge for the benefit of a deeper understanding of the nervous system and the application of such understanding in clinical settings, especially in the field of neurocritical or neurointensive care, which we will discuss in Chap. 3 *Between Psyche and Mind.*

In the context of neuroengineering, we have the duty of mentioning neurorobotics, an interesting field at the interface between neural and artificial, where 'neurorobots' (also called neurodroids) *serve*[7] humans by 'helping out' in the study of motion-locomotion-deambulation and motor (skills) control, learning and memory processes, and visuospatial perception-apperception-proprioception also (or especially, given the clinical application) in the context of possible damage as in the case of 'pseudospectral protocol' used for the production of artificial seizures to study epilepsy. Of course, artificial/man-made creations including robots and androids elevate the status of possible requests, whether cognitive,

[7] Following the etymology of the Czech word *robota*, as well as the inference of origin/declination from a (human) being in -ειδής at the base of android.

computational or operational, researchers first and the general public after can ask to these machines. With this increase in effective abilities, new and more complex ethical questions arise, together with the true possibility of enhancement and augmentation of (human) neural systems. Whether we talk about basic deep brain stimulation or artificial intelligence, we are faced with the opportunity to change some of the aspects which throughout the history of medicine, science and philosophy defined our sense of self, our personality, our identity. The aforementioned techniques will allow for an increase in memory and recollection capacity (with DBS) and remorph/rearrange specific personality elements (although not necessarily categorizable as 'traits' in the strict psychological sense), reduce inhibitory processes, and increase motivation, self-reliance and focus (of course, in categories previously decided, whether fully based on evolutionary biology-based 'push' or not). These procedures, when deemed appropriate, for instance, in the case of psychiatric and/or neurological disorder that had previously impaired one or more elements in the aforementioned areas, can dramatically increase the level of well-being and general psychophysical health in an individual. Repairs of spinal cord injuries for instance focus on natural processes to foster nerve regeneration via the stimulation of the PNS environment starting from the level of the damaged nerve. Similar, yet more complex regeneration processes are at the center of biomolecular therapies which implement the research on neurotropic factors, thus following the studies by Levi-Montalcini and others. More in detail, neurotrophins such as the famous nerve growth factor (NGF), as well as brain-derived neurotrophic factor (BDNF), neurotrophin-3 (NT-3) and neurotrophin-4/5 (NT-4/5) can positively impact the neural development and promote its branching and (re)growth. Similar processes are also made possible by eliciting neural responses via acidic and basic fibroblast growth factor (aFGF, bFGF), ciliary neurotrophic factor (CNTF), glial cell line-derived growth factor (GDNF) and the glycoprotein of the extracellular matrix fibronectin, which binds components in the extracellular matrix such as collagen, fibrin, and heparan sulfate proteoglycans. Further developments in this sense are represented by the research-based use of neuronal cytoskeletal components, antiapoptosis factors, regeneration-associated genes (RAGs, including adhesion molecules, Cap-23 and GAP-43), as well as by therapeutic blockade of inhibitory biomolecules in the CNS following glial scarring with hondroitinase ABC and blocking NgR, ADP-ribose. Fundamental research studies in this sense, and including more advanced methodologies (injection molding, ink-jet polymer printing magnetic polymer fiber alignment, phase separation, solid free-form fabrication, and others) combining nerve guidance channels (constructed conduits for reducing growth inhibition from scar tissue and thereby sprouting axons directing growth) and multiple stimuli, are found in the publications by Schmidt and Leach (2003) and Tate et al. (2003). The main target of advanced techniques is obviously to artificially replicate the neural structure, more specifically the bioarchitecture of aligned fibers or channels, starting from their internal matrix. Grafting is used in neuroengineering especially in the case of repairs by biocompatible-autologous tissue/nerve grafts harvested from another anatomical area, for instance, to improve the clinical presentation following injury or inflammation (bone vs. muscle areas) through nerve rehabilitation. Scaffolding is also possible with acellular grafts, in combination with nonautologous tissue, other components found in the matrices and the fundamental immunosuppressants. The latter are not necessary when using artificial-synthetic structures such as methylcellulose-based materials, although biocompatibility might be an issue in surgery. The same applies to delivery devices such as microspheres, osmotic pumps, polymer matrices and silicone reservoirs.

Finally, another fundamental technique in neuroengineering is the area of gene therapy to investigate and develop new strategies for the implementation of growth factors and regeneration via cell adhesion molecules, stem cells and olfactory ensheathing cells as viral or non-viral vectors. Of course, this type of research presents a series of ethical challenges. In the case of stem cell research, we need to stress that in this context we are not dealing with random molecules, but cells; not random cells, but human cells, which might or might not contain embryonic, human, stem cells. First of all we need to remember the scientists whose hard work and intelligence contributed to this fascinating area of medical science, in particular the German Ernst Haeckel, the Italian Mario Capecchi and the American James Thomson. Haeckel was the first to use the term 'stem cell' to identify the fertilized egg gestating into an organism. Capecchi was the co-winner of the 2007 Nobel Prize in Physiology or Medicine for the 'knockout mice' discovery and method (i.e. creating mice in which a specific gene is 'turned off'). Thomson was the first scientist to create the first human embryonic stem cell line. There are certainly many other extremely important names in these discoveries and applications, as we all 'stand on the shoulders of giants', but these three names help us frame the issue from three perspectives:

- Scientific definition
- Study on a molecular, laboratory and animal level
- Possible application on human subjects

Given these premises, we need to narrow our scope of investigation if we want to provide an accurate and unbiased analysis, if not a possible clear answer to the questions many people ask. To start, we need to remember that the main controversy on stem cell is related to research in human subjects, its possible therapeutic application and the moral prerequisites and outcomes in the social arena. Leaving aside the very complex debate on the very essence and characteristics that make us human—let us not forget, that, in defining where life starts and how biological life might be created from a/the soul, the term 'animal' comes from the very Latin word for soul—most of the debate revolves around human stem cells. In adults, there are four main areas where stem cells can be obtained more in particular:

- Adipose tissue, requiring liposuction
- Bone marrow, requiring harvesting or extracting
- Blood in general, requiring apheresis
- Umbilical cord blood, requiring extraction right after birth

Although the research, use and application of stem cells extracted in these areas are not immune to complex scientific, ethical and social debates, the biggest controversy actually deals with one specific form of stem cells, the embryonic cells, which are specially important as they are pluripotent and therefore potentially able to differentiate into any of the three germ layers developing in embryogenesis: endoderm, mesoderm and ectoderm. As is often the case, the controversy is made much bigger than necessary—that is, necessary for the understanding and/or to understand—by opinions and commentaries which do not share much with either scientific advancement or moral guidance. Certainly, we could argue that we do not have yet enough data about the exact effectiveness and efficacy of any stem cell therapy, but we could say with fair judgment that there is an enormous potential for the treatment of a multitude of medical problems such as autoimmune, bone, gastrointestinal, heart, and neurodegenerative diseases as well as cancer. If adult stem cells and embryonic stem cells were exactly the same in terms of structure and ability to morph into other tissues, then the controversy would be far less preeminent in the contemporary debate. However, there are very big and important differences, especially since in the case of embryonic stem cells we are actually dealing with an embryo. Thus, we cannot avoid questions about structure, development, beginning and end, and the meaning of life, human life in particular.

Unfortunately, extreme views on both sides of the political spectrum contribute to the issue. Therefore, we find often narrow definitions such as 'Pro-life' spanning from ethical, social and religious stance against abortion together with opposition to embryonic stem cells, in-vitro fertilization, adoption and gender identity. It is certainly impossible to clearly address the multitude of issues arising in this context, although there might be philosophical difficulties, for instance, in trying to understand how someone who claims to be 'Pro-life' could also support private gun ownership for 'self-defense' and increased spending for the military. Life is certainly important at birth, but continues to be important throughout the lifespan, so any device aimed at killing other human beings appears to be an oxymoron—at least theoretically, as we are certainly aware of the multiple layers of both interpretation and application—in terms of defending life. On the other side of the sociopolitical spectrum however, we often find an absolute discard and contempt for very deep aspects of human nature, elements of meaning and purpose that go far beyond what science (in the sense of evidence-based laboratory study and research) can provide (Mujkanović 2016). The question of 'how something happens' (i.e. it is created/produced, develops and morphs/changes and grows/expands or gets injured/damaged) on material, biological terms is certainly the dominion of science, and we should therefore support scientific research in every possible way. The question of 'why' instead is the realm of philosophy, of spirituality and of religion. Many argue that, from a religious (and most certainly dualistic) perspective what really matters is the spirit, not the body, and we should therefore avoid limiting scientific research in any kind of way, and simply "let science do its job." Of course, in regard to the 'Sanctity' and 'Sacredness' of life, we (the scientific community, but the general population as well) simply do not know (yet) enough about embryos, and we certainly are very far away from understanding human life in its totality and complexity, certainly not in terms of when exactly it begins, which is the core of the debate of conception versus birth. It is therefore very hard to assess the validity and ethical, intrinsic 'goodness' of embryonic stem cell research, as a '*no*-answer' is not equivalent to a '*non*-answer'. Therefore, 'inaction' is not 'absence of action', and our decision *not to* do research could have negative effects on our ability to improve medical science, and thus do greater good for the whole humanity. In any case, it is also true that:

- Embryonic cells originate in embryos which have been donated by individuals or couples after in vitro fertilization.
- No human embryonic stem cell lines can be created without explicit consent from the donor.
- The embryos used to create embryonic stem cell lines are rarely donated to other individuals or couples, and will therefore be destroyed anyway (California Institute for Regenerative Medicine 2015).

Certainly, this does not justify the use of embryonic cells in moral, ethical or religious/spiritual terms, but it constitutes at least a realization of the positive potential of in vitro fertilization for the benefit of scientific research, and therefore for the well-being and healing processes of fellow human beings. Thus, the controversy is moved 'one step up' to the debate on in vitro fertilization, which is beyond the scope of this analysis. This aspect has fundamental repercussion on legislative decisions made by different countries around the world. In Europe, for instance, Austria, Germany, Finland, Ireland, Italy, the Netherlands, and Portugal prohibit or severely restrict the use of embryonic stem cells, while Greece, Sweden and the United Kingdom "have created the legal basis to support this research" (EMBO 2009).

References and Further Readings[8]

Bibliography

Alberts, B., D. Bray, K. Hopkin, A.D. Johnson, M. Raff, K. Roberts, and P. Walter. 2002. *Essential cell biology*. 4th ed. New York: Garland Science.

Beauregard, M., O'leary, D. 2008. *The Spiritual Brain. A Neuroscientist's Case for the Existence of the Soul*. New York, NY: HarperCollins.

Bluhm, R. 2014. *Why evidence-based medicine is bad for biological psychiatry*. Psychiatry Grand Rounds lecture at the University of Vermont College of Medicine, Burlington.

Bunge, M. 2013. *Medical philosophy. Conceptual issues in medicine*. Singapore: World Scientific Publishing.

California Institute for Regenerative Medicine. 2015. *Myths and misconceptions about stem cell research*. Oakland: California Stem cell Agency. Retrieved from: https://www.cirm.ca.gov/

EMBO. 2009. *Stem cell research. Status, prospects, prerequisites.* The European Molecular Biology Organization's official website. Retrieved from: http://www.embo.org/documents/science_policy/stem_cell_research_2006.pdf

Giacino, J.T., S. Ashwal, N. Childs, R. Cranford, B. Jennett, D.I. Katz, J.P. Kelly, J.H. Rosenberg, J. Whyte, R.D. Zafonte, and N.D. Zasler. 2002. The minimally conscious state: Definition and diagnostic criteria. *Neurology* 58 (3): 349–353.

Goila, A.K., and M. Pawar. 2009. The diagnosis of brain death. *Indian Journal of Critical Care Medicine* 13 (1): 7–11.

Heintzelman, S.J., and L.A. King. 2011. The local baby and the global bathwater: Circumscribed goals for the future of the multilevel personality in context model. *Psychological Inquiry* 22: 23–25.

———. 2014. Life is pretty meaningful. In *American psychologist*, vol. 69, 6561–6574. Washington, DC: APA, American Psychological Association.

Higgins, J.P.T., J. Thomas, J. Chandler, M. Cumpston, T. Li, M.J. Page, and V.A. Welch, eds. 2014. *Cochrane handbook for systematic reviews of interventions.* London: Cochrane.

Institute for Healthcare Improvement's 100,000 Lives Campaign. 2006. *Institute for Healthcare Improvement.* Retrieved from: http://www.ihi.org/education/Pages/default.aspx

James, O. 2016. *Not in your genes: The real reasons children are like their parents.* London: Vermilion.

Kearney, R., and M. Rainwater. 1996. *The continental philosophy reader.* London/New York: Routledge.

McCance, K.L., and S.E. Huether. 2014. *Pathophysiology: The biological basis for disease in adults and children.* 7th ed. St. Louis: Mosby.

Mujkanović, L. 2016. *Nursing research studies.* Ethics, community, and healthcare. Hooksett: SNHU.

Sandrone, S., M. Bacigaluppi, M.R. Galloni, S.F. Cappa, A. Moro, M. Catani, M. Filippi, M.M. Monti, D. Perani, and G. Martino. 2014. Weighing brain activity with the balance: Angelo Mosso's original manuscripts come to light. *Brain* 137 (2): 621–633.

Schmidt, C.E., and J.B. Leach. 2003. Neural tissue engineering: Strategies for repair and regeneration. *Annual Review of Biomedical Engineering* 5: 293–347. https://doi.org/10.1146/annurev.bioeng.5.011303.120731.

Spader, H.S., J.A. Grossberg, R.A. Haas, and G.M. Soares. 2013. Fundamentals of the neurologic examination for patients undergoing central nervous system interventional procedures. *Seminars in Interventional Radiology* 30 (3): 240–244. https://doi.org/10.1055/s-0033-1353476.

Tate, D.F., E.H. Jackvony, and R.R. Wing. 2003. Effects of internet behavioral counseling on weight loss in adults at risk for type 2 diabetes: A randomized trial. *Journal of the American Medical Association* 3 (289): 1833–1836.

Tomasi, D.L. 2016. *Medical philosophy. Philosophical analysis of patient self-perception in diagnostics and therapy*. New York: Ibidem Verlag/Columbia University Press.

You, J.S., and P.A. Jones. 2012. Cancer genetics and epigenetics: Two sides of the same coin? *Cancer Cell* 22 (1): 9–20. https://doi.org/10.1016/j.ccr.2012.06.008.

CHAPTER 3

Between Psyche and Mind

3.1 Recollection

The concept of recollection and its Platonic relative *anamnesis* are truly at the intersection of psyche and mind, intended as either two separate entities or two aspects of the same *thing*. Recollecting is obviously connected to reconnecting, remembering and memorizing as we previously evidenced (Tomasi 2016). Furthermore, this activity is at the center of the discussion on "remembered wellness", "faith factor" and "relaxation response" in a spiritual and religious sense (Benson 1997). In regard to the latter, Herbert Benson describes a fundamental nature of human beings as "wired for God" in connection to this process and at the base of the healing process. This position is validated by numerous studies, for instance, in the research by Levin (1994, 1996) in connection to general medical practice and Widerquist (1992) for nursing. The work by Louis Ritz at this intersection between psyche and mind is also the cornerstone of the Center for Spirituality and Health at the University of Florida, which notes that:

> Spirituality deals with what we find eternal, beautiful, meaningful and just, and asks us to contemplate "what should be". Science and technology deal much more with "what is" and how best to predict and manipulate it. The interface between these powerful forces is of immense and immediate importance because their interaction is likely to strongly influence how the next generations shape the future of our world. Careful analyses of our debates

D. L. Tomasi, *Critical Neuroscience and Philosophy*,
https://doi.org/10.1007/978-3-030-35354-4_3

> about healthcare, ecology, economics, politics, nationalism, educational policy and even law often reveal this underlying conflict between "faith" and "fact" based realities.[1]

Of course, any discussion involving psyche and mind cannot avoid some inclusion of the parallel term 'soul'. There are many publications discussing these aspects, for instance, *Whatever Happened of the Soul? Scientific and Theological Portraits of Human Nature* (Brown et al. 1998) or *Did My Neurons Make Me Do It? Philosophical and Neurobiological Perspectives on Moral Responsibility and Free Will* (Murphy and Brown 2007). The underlying hypothesis is that there are multiple connections between an empirical/dialectical *observation* of a 'top down' system which accounts for an increasingly complex structure, cellular and molecular, starting from the DNA, that does not specify itself, but 'recollects' the elements given/created by a 'higher function' which certainly resembles concepts like a 'higher self' or a 'higher power' with obvious implications for the philosophical debates on free will, determinism, materialism/reductionism and the mind-body problem. The focus however moves from experimental neuropsychological research on the implications of agenesis of the corpus callosum, and investigates the related neural functioning as underlying mechanism for higher cognitive processes in humans. The perspective here is truly monistic and materialistic, but certainly not reductionist, as it claims that the immortality of the soul cannot be (empirically) verified, but human beings were created by God with a form/shape in which every part of our body is the origin and creation of our self, and thus we are fully 'embodied', cognition and mind included. Of course, this perspective also does not diminish the belief in a creator God, as the expectation—also in an eschatological sense—is that God will recreate us after our physical death as a complete version of us, starting from the beginning, although concepts like space and time would obviously need to be rethought in connection with a divine being. This 're-embodiment' has obviously a strong metaphysical basis, albeit it can also be viewed from a purely evolutionary biology perspective, or in connection with other forms of religious naturalism and even—as much as this terms sounds very paradoxical to some—computational-digital spiritual atheism (as in Eric Steinhart). Of course, this means that individual soul/mind and individual

[1] University of Florida Center for Spirituality and Health, Louis A. Ritz, Available at: http://www.ufspiritualityandhealth.org/directorsadvisors/ritz.asp

vs. universal spirit are *fully part* of our bodily features, and therefore cannot be separated in essence, as they are not—but they do represent—two different things. According to this view of course, this representation is simply "representing to the perception of the viewer", who thereby examines such evidence. In the platonic interpretation, this has to do with the *anamnesis* at the basis of the theory of recollection. In other words, our emerging evidence is rooted in recollecting ideas our soul always possessed, but though the imprisonment of/by (although the latter would imply a purpose actor in/as) the body. The fundamental point at this level is not so much related to questions on existence of free will, the (immortal) soul or (a) god, but on a concept of 'recollection' based on such philosophical premises, as opposed to 'recollection' in psychological terms. More specifically, following the interpretation of Hegel by Verene, psychology observes reality, and objectifies it in terms of analyzing thoroughly, but without comprehending it in terms of its internal essence-being. The comparison here is between evidence-based observational science and 'self-perceiving' phenomenology. Continuing with this interpretation, this phenomenology expresses this act of recollecting in terms of consciousness being both inside the memory being recalled and outside it as the power of its recall. Therefore, and we have to be clear in this regard, our neural functions cannot be equated with our psychological functions; our brain cannot be the same as our mind. The reason is that it is the mind that actually operates the thoughts behind these lines, and it is actually the mind creating this comparison between itself and the brain, although the brain plays a fundamental role in modulating this activity. Verene even states that in attempting to "extract the mind from such inner process and manifestation and reestablishing it in the brain", psychology becomes very similar to reductionist-materialistic—scientifically very misguided—approaches such as low-level physiognomy and phrenology. In the words of Verene, the concept of a "beautiful soul" is defined by being "its principle understandable to the self of the matter in hand" (Verene 1997).

Psychology puts, rightly so, a lot of emphasis on external behavioral processes, but in drawing parallels between these behaviors and their neural underpinnings, it fails to consider inner existence and essence. In this context, we have to mention the importance of the validation of inner existence and essence from the perspective of possible mechanistic explanations—that is, neural processes—and certain philosophical views. In particular, we are referring to the multitude of offspring studies originated in the *Libet experiment* (Libet et al. 1979) in connection to the existence of (or lack thereof) free will. To quickly summarize the main points evidenced by the results of this experiment, physiologist Benjamin Libet wanted to shed light on the connection, whether causal or not, between the conscious experience of volition and the so-called *Bereitschaftspotential*, usually translated as 'readiness potential' on the base of previous findings by Hans Helmut Kornhuber and Lüder Deecke (1990). In details, these researchers were able to demonstrate that a basic hand movement in the subject studied, between a certain initial nervous activity in the motor cortex *of* the subject and its actual execution *by* the subject, about one second elapses, although our human everyday experience is that the time between action and execution *appears* much shorter, so that we think that *we did* (i.e. decided to activate) the movement. In the Libet experiment the repeated results showed instead that the *Bereitschaftspotential* began about 0.35 seconds *earlier* than the subject's reported conscious awareness that "now he or she *feels the desire* to make a movement" thus demonstrating that, *in spite of appearance* (= perception) humans have no free will in the initiation of (their) movements. This outcome has obviously a very strong impact on the very concept of recollection, since this concept involves time or, better said, an accurate (objective vs. subjective) perception of the passing of time, in which (a) decision of taking action precedes the action itself and (b) recollecting something is meaningful (true) only if this something happened/existed *before* the act of recollection. However, a problem arose from this experiment, that is, the fact that, if free will turned out to be an illusion given these premises, the same couldn't be said for the 'free won't'. With the latter, what is usually intended is the 'ability to do otherwise' or the veto 'power' to prevent the actuation of action—in our case, the ability to prevent movement right before activating such movement, for example, 'at the last moment'. The 'ability to do otherwise' requires a further analysis of the many steps in between and often opposite, and the many conflicting or pseudo-conflicting positions required in soft and hard determinism vs. indeterminism and possibilism vs. impossibilism. A criticism of the assumptions in the conceptualization of free will evidenced in the Libet experiment, more specifically on the controversial nature of the term, 'free will' is a core idea in the work by Peter van Inwagen, with particular regard to the comparison between compatibilism and incompatibilism. William Klemm (2010) instead posits that there are certain flaws, beyond conflicting data and outcome analysis, in the interpretation of similar experiments based on the fact that:

> 1) timing of when a free-will event occurred requires introspection, and other research shows that introspective estimates of event timing are not accurate, 2) simple finger movements may be performed without much conscious thought and certainly not representative of the conscious decisions and choices required in high-speed conversation or situations where the subconscious mind cannot know ahead of time what to do, and 3) the brain activity measures have been primitive and incomplete.[2]

The analysis by Klemm covers a wide range of objections to this type of experiments aimed at establishing the existence or non-existence of free will from the perspective of neuroscience. Among the most important aspects of Klemm's analysis we should evidence that both concepts of decision and conscious realization need to be reassessed in the context of evidence-based experimentation, more specifically understanding that the processes underlying decision-making are multiple and therefore (a) cannot be reduced to a single mental process to be analyzed, and (b) cannot be used as a basis (i.e. as an ontological-experimental justification) for 'all mental life'. Furthermore, Klemm argues that there is a delay between the (conscious) realization/perception/awareness that a certain decision (which, in itself is not instantaneous) has been made and the actual decision made. Moreover, Klemm sees these two elements as representatives of two distinct processes. Klemm also *recollects* some concepts from classical philosophy and psychoanalysis including the *subconscious mind*, which, in his view, is behind certain decision-making processes and "cannot know ahead of time what to do" (Klemm 2010). From the perspective of memory and recollection, this could be certainly applied to procedural memory, responsible for the 'know-how' involved in multiple skills such as motor coordination (certainly in collaboration with other brain areas, including the cerebellum) and repetition or playing an instrument, and is a fundamental part of long-term memory.

As we have seen, different understandings of memory and recollection imply and influence different conceptualizations and categorization of mind and psyche. The direct outcome of these views is a very complex *contemplation* of what it means to 'have' free will, and whether this freedom is something 'we' can 'possess' at all. Using the framework shared at times or at least in part or in some areas by compatibilists and soft

[2] Klemm, W. R. 2010 Free will debates: simple experiments are not so simple. *Advances in Cognitive Psychology, 6*, 47–65.

determinists, we could attempt to define stages, degrees or layers of such freedom and assume that there isn't such a thing as one type of freedom or at least that "*freedom is not free*, at least not philosophically." In other words, we would assume (possibly in conflict with Occam's razor perspectives) that there are other contributors or effectors to the "levels of existence or action" of freedom in the context of 'Will',[3] including time/history, consciousness/awareness and *personal action*. This perspective, shared in part also by Kornhuber and Deecke (2012) claims that there are ways an individual-subject can 'increase' the level of freedom through an improvement of the self (which we argue could bring us back to the *cura animae*) as opposed to self-mismanagement, which would lead to an incremental loss of degrees of such freedom. Obviously, this view is in many ways very similar to some religious perspectives, especially within the monotheistic traditions, on the concepts of 'sin' and 'repentance' involving an external source of judgment and activation, because we, as human beings part of and in this world cannot have 'full' free will, because this would entail a complete freedom *from* nature, which we are part of. Following this interpretation, we hereby mention the very interesting series of studies by Sartori and Defanti (2012) in regard to 'our role' in the definition and activation of the *Bereitschaftspotential* and the very 'activation of existence' of free will. The most important aspect of these studies is that we might actually be 'responsible for our response' in the sense that our belief in the existence of free will has the power to impact the readiness potential positively (with a decrease in time) or negatively (with an increase in time), by virtue of our judgment of this 'positivity' or 'negativity'. In other words, the more we believe that free will exists, the more our readiness potential will increase in speed, in turn making/allowing us to have a more direct control over our decision making processes, and thus our actions.

As previously evidenced, if these studies primarily focused on free will, a similar perspective can certainly be applied to the 'free won't', as in the research by the Berlin group led by John Dylan Haynes (2016) which suggested—those were the exact experimental results in the studies—that in a 'speed-activation' competition between human subjects and a Brain Computer Interface, the 'point of no return' for the ability of humans to veto an action was at 200 milliseconds before the movement. In this

[3] Leaving aside all the ontological consideration around the etymology of 'will' as opposed to the 'libero arbitrio'.

experiment, the computer was programmed to predict the human subjects' movements in real-time, from observations of their *Bereitschaftspotential* evidence in EEG activity. The interesting factor was, however, that even passed this point of no return, thus when the pedal was already set in motion, the subjects in the study were able to *reschedule* their action by vetoing (i.e. not bringing to completion) the already (previously) started behavior/process/activation. These considerations achieve a status of a much higher activating capacity in the context of possible interactions between 'subjective free will' and 'common-communal free will'. With these postulates, we want to stress the importance of a conceptualized awareness of action, even more of being, in relation to action and being at the community level. More specifically, we should better understand how our response to self and by self, in the context of 'activation of existence' and (self) identity comes to influence the level of freedom in choices of belonging to a group, a community or even a 'cause'. In other words, what is the role of our sense of identity and our ability to make free and informed choices in the context of bigger societal decision making patterns? If our sense of self and sense of belonging to a family, a community, a religion, or a nation depends on these parameters, then we will understand that there are multiple layers of activating essence in the action of the single individual within the group, between group members, and at each intersection, especially on ethnic and racial levels. An interesting question in this context would be what are the possible implications of confounding-confusing terms such as *race* and *ethnicity*? Can we operate a 'switch'? Can this 'switch' be turned on or off? From the perspective of memory, recall and recollection, the problem of self-identifying with and within a specific subgroup of people has a lot to do with cultural background, upbringing as well as psychological understanding of our cognitive processes, our emotions and the possible distortions involved. Although we could argue that it is ethically preferable, at least on a social-public level, to be 'politically correct' in preferring the term ethnicity over race in the public speech, as it moves the weight from biology to culture, this process only appears to solve the problem, while in fact it simply moves the issue one step further, and whether this comparative truly means 'further up' or 'further down' remains to be seen. In fact, we could argue that the main issue lies in the perception some might have of being either 'inferior' or 'superior' to other (sub)groups of people. This (mis) perception is from time to time justified using a multitude of scales and references systems—from the purely biological to the socioeconomic and historical—to defend the need for separation in the judgment of specific groups of people (Mujkanović 2016). The multidisciplinary approach of the study of diversity should help us understand how biology, sociology, anthropology, philosophy, psychology and economics are deeply intertwined in our attempt to understand how these premises influence our everyday interaction with each other. Simply changing the scale or references system won't destroy the false assumptions and misconceptions underlying ethno-racial superiorities and inferiorities.

3.2 Talking to the Mind and Mind-Talking

3.2.1 Behavioral Neuroscience

This disciplines combines a variety of approaches provided by multiple areas of investigations, justifying the other names this specialty is known as, including—depending on geographical, political, and cultural areas the definitions might differ—psychobiology or biopsychology, psychophysiology or physiopsychology, comparative psychology and psychopharmacology, all the way to the inclusion of cognitive neuropsychology. As we mentioned elsewhere, the somewhat artificial separation between disciplines that are so deeply connected is in part related to cultural norms as well as artificially produced to satisfy economic needs, for instance, research grant funding or specific scientific/academic needs. Analyzing behavior, especially human behavior, is as fascinating as complex and complicated. Throughout the history of social and 'hard' sciences, there have been an enormous amount of theories, trials, experiments, suggestions, ideas, qualitative and quantitative data and interpretations supported and/or challenged by at least as many philosophical positions. In behavioral neuroscience, the divide between mind and body appears to bear the weight of all these approaches, their difficulties and obstacles, and all the passionate ways in which proponents of monistic and dualistic, perspectives—including the now classical Cartesian duality on one side, and physicalism, idealism and neutral monism on the other—as well as every other position in-between or beyond the 'standard' viewpoints, have been carried on in the pursuit of an ultimate truth of the origin, scope and function of (human) behavior.

In this context, we need to remind the reader once again that both psychology and neuroscience rely on evidence-based approaches, most importantly experimentation on animal models, an activity supported by the somewhat aprioristic assumption—although reductionist materialists might completely disagree with this analysis—that due to the biological and functional (computational, cognitive, etc.) similarities between human brains and the brains of other animals, especially mammals, we can learn how we behave based on the observation of other animals. Throughout history, there have been multiple attempts to identify a specific link between 'the matter of the matter' and the 'other/outer spheres' of behavior. "What is mind? No matter. What is matter? Never mind" is a famous quote by George Berkeley that was often used in the area of a possible connection between philosophy and neuroscience or psychology, especially important from a mind versus brain perspective. The pineal gland (Fig. 3.1) had also a similar connecting purpose according to the Cartesian view, while the very function of the brain has been widely controversial since antiquity, where philosophers-scientists such as Aristotle presented a view according to which the brain allowed specific mechanical-physical features like emotion regulation of the impulses received from the heart.

Among the major schools of psychological thought, we find a purely biological point of view—to some extent originating in Platonic perspectives—as well as the cognitive, social or socio-cognitive, existential and humanist(ic), psychoanalytical and, of course, the purely behavioral and functional(ist-ic). Behaviorism and functionalism have played a very special

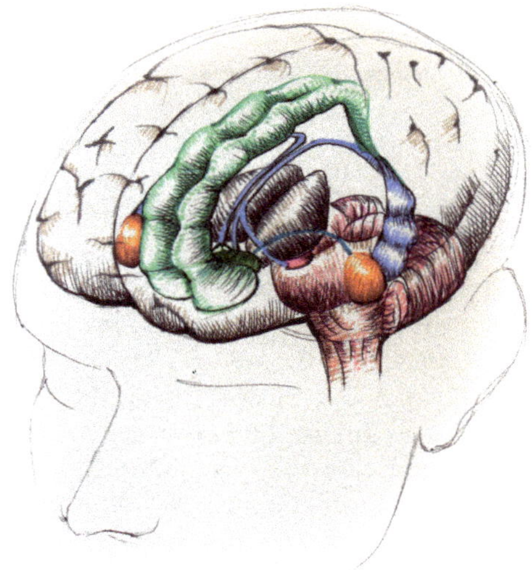

Fig. 3.1 Left superior frontolateral view of the cerebrum with the limbic system with cingulate gyrus (green), hippocampus (blue), amygdaloid bodies (red), mammillary bodies (fuchsia), thalamic areas (brown), pineal gland (light orange), and brain stem (pink)

role in connection to behavioral neuroscience especially in relation—more often juxtaposition or absolute opposition—to type physicalism, itself variously defined as identity theory of mind/type identity theory but also (dualistic) mind–brain identity theory or, alternatively—especially in a philosophical sense—reductive materialism within philosophy of mind. Of course, in the case of behaviorism, "the assumption is that behavior is what is learned, whereas in cognitivism the assumption is that behavior is the outcome of what is learned "(Stevenson 1983). While behavioral neuroscience does not espouse any specific perspective, it is clear that the major theoretical contributions come from functionalistic approaches (especially in the concept of neural underpinnings of behavior) such as analytic, homuncular, machine-state, mechanistic, and psycho functionalism. In this sense, there are types of mental/psychological events in connection/correlation to types of neural/physical events. Of course, the causal element is the main point of discussion, debate, and disagreement in this context. It presents some very complex theoretical issues, in part linked to other diatribes such as the divide between psychology of traits and social cognitivism. Of course, the main behaviorist point of view states that human behavior is a result of environmental or personal stimuli, including classical and operant conditioning and therefore assigning to stimulus-response learning processes on one side, and reinforcement, reward, punishment on the other, a constituting value to our thoughts and actions. In the case of functionalism instead, the claim is that everything—that is, our behavior, but also our (system of) beliefs, our wishes and desires, our intention and will—mainly depends on the related function connecting sensory inputs and behavior. Aside from specific theories, it is important to mention that, from the perspective of behavioral neuroscience and the comparative 'nature vs. nurture' the social-cognitive theory triad 'personal-behavioral-environmental' offered by Bandura offers good material for discussion. In any case, the investigation on the legitimacy and clinical application of these models rely on the conceptualization of parallel concepts such as the self, the 'I' and the person(hood). To use a psychoanalytic framework, we would for instance set 'I'-focused psychoanalysis against Object relations theory. In the first case we encounter the theories by Freud, Hartmann, Rapaport, Spitz, Mahler, and Erikson; in the second case the most important representatives are Fairbarn, Klein, Bion, Winnicott, Guntrip, Stuart, Bowlby and, of course, Ferenczi. Jung and especially Rank have been in some instances presenting a very useful link between these two approaches.

Of course, theories of behavior need to find their evidence-based counterpart in direct and/or laboratory experimental observation. Scientific improvements happen in this context thanks to the continuing efforts of researchers, as well as fortunate events (as in 'lucky', at least for the scientists involved, not necessarily for the subjects-patients) representing a scientific 'jump' in advancement toward understanding, as in the case of Phineas Gage. Other examples are offered by the research conducted by Gazzaniga on corpus callosotomy or the studies by Penfield and Rasmussen on epilepsy. In the case of (artificial-surgical) separation of right and left hemispheres, however, the common view generally sees the left side as 'better' (for instance) at mathematical-logical-analytical processes, but also writing and reading. The right side of the brain instead is 'better' at identifying patterns, schemata and 'seeing the bigger picture'. From the perspective of behavioral neuroscience, we can certainly see how this division is symptomatic (in a theoretical sense) of a partial function, even more a partial understanding of reality. Given this premise, we need to reassess terms such as 'magical thinking', which are often viewed, especially within psychiatry, as indicators or even 'traits' of delusional content, as the subject-patient 'connects dots that are either not connected, or connected in a very different way, or even not existing at all'. While this is true in many cases of mania, hypomania, paranoia and psychosis, we believe we should pay more attention to phenomena that we '*perceive* as connected' and turn out to be indicator of a bigger picture. Of course, delusions are still true in the aforementioned clinical cases, but we argue that there are different stages, layers or even levels of perception; on/*in* some levels connection is only apparent (thus absolutely absent) but in others a deeper connection might be present (read: real, true) but not evident using either our common senses (minus, we could argue the so-called sixth/seventh sense or the sensus divinitatis) or the 'classical' scientific method. In fact, and we mentioned this aspect many times in this work, we should not confuse scientific (in the commons sense) evidence with all evidence, or scientific non-evidence with all non-evidence. In other words, we should be even more 'scientific' when assessing the results of, for example, a research study. If the study does not show results, that is (in this case), any connection or correlation, it simply means *this*. It means that 'the study did not find/yield any connection or correlation', it doesn't mean (e.g. it doesn't logically follow that) 'there aren't any connection or correlation *at all*'. In relation to this *observation*, and in connection to other neuroanatomical areas, we could *think* that perhaps what we can experience with (for instance) the prefrontal cortex is only a very limited part of reality, which has only to do with our rationalization of things. Maybe spiritual events can be perceived with the parietal/temporal lobe, etc. As with every study involving behavioral neuroscience, the *conditio* sine qua non for this type of investigation, is that experiments always contain at least one biological variable, either dependent or independent. As these parameters are used to link biological changes on the nervous system, more specifically on the brain, and behavioral changes, many experiments in behavioral neuroscience involve naturally—that is, as a result of natural biological development vs. decay or accidents, illnesses, diseases and disorders[4]—or artificially (man-made/developed) caused lesions. These include chemical, electrolytic, and surgical lesions (in a clinical setting, some examples are neurotoxins, surgery, and electric shock) and can be temporary, time-dependent, or even permanent (in the case of naturally occurring events). Of course, in all these cases, the theoretical framework linking multiple animal models to humans plays a critical and controversial role in the study of mental health disorders, especially in the research behavioral neuroscience shares with abnormal psychology. This means that this discipline can offer great insight to the treatment of disorders such as Alzheimer's Disease, Huntington's Disease and Parkinson's Disease. In the case of more complex (not necessarily in a purely medical sense, but given the complexity of psychoneurological and social components involved) disorders such a Major Anxiety Disorder, Major Depressive Disorder and Autism Spectrum Disorder, the direct causal effect of neurological areas, functions, structures and processes appears to be in need of further exploration and understanding. Again, this can happen via technologies previously discussed, such as brain imaging (MRI, fMRI, EEG, PET, SPECT, CAT/CT, MEG, etc.), manipulation (including inhibition[5] and psychopharmacological interventions[6]), and stimulation (as in TMS), as well as via measuring methods like electrode-based monitoring (single-unit or multi-electrode recordings), sensitive dye-based optical techniques (examples include voltage-sensitive dyes and voltage-sensitive fluorescent proteins that have been genetically encoded), as well as functional neuroanatomy and genetic

[4] At this level, we use these terms nonspecifically.

[5] For instance, optogenetic inhibition, where researchers observe the expression in the cellular structure of a light-activated inhibitory protein.

[6] As it is the case with agonists, antagonists, activators, and inhibitors interfering with neurotransmission and inducing (new) neural activity.

(sequential) mapping or engineering. To be sure, at the center of the aforementioned mental health disorders there are areas of 'perceptual disconnect' with reality, either in a form of cognitive distortions such as rumination, jumping to conclusions, all-or-nothing thinking and negative self-talk, or due to neurologically-based disruptive feature to communicative and spatial-relational abilities. These parameters are even more defined by the very way we observe the normal functioning of psychological qualities in such context and the causal or mechanistic process inference we use to justify our psychological observation. From a philosophical perspective, the stage of 'Observational Reason' in the analysis of Verene of Hegel's *Phenomenology*, as well as Lavater's physiognomy and Gall's phrenology do not merely represent a mistake in scientific analysis, but constitute a preliminary stage which will eventually lead to observational psychology and the analysis of mind processes. This process is however made increasingly complex by the environmental (in the philosopher's words a true milieu) component, both inner and outer, experienced by the individual-subject/subjected individual. To be more specific there is a reflection of the inner into the other/outer. In fact, Verene goes as far as demanding—and this request is to be found in times that predate the very beginnings of both psychology and neuroscience, as in Hegel's phenomenology—a so-called technology of behavior beyond both behavioral psychology and neuroscience, in which behavior is not merely causally connected to states of mind, feelings, traits of character, human nature or "faculties, inclinations, and passions" (Verene 1997). That is exactly the problem with some behaviorist perspectives in psychology and neuroscience, that is, the forced "flattening of the inner and the outer", which were already denounced by Hegel with regard to physiognomy (Tomasi 2016). At the core of this problem is the false assumption, which is truly a 'presumption without presentiment' that our current scientific method is able, when aptly and thoroughly applied, to investigate the ultimate reality of things, whether conscious or unconscious, and whether conscious or unconscious *actually* exist.

3.2.2 *Affective Neuroscience*

Similarly to what we have said in regard to behavioral neuroscience, with affective neuroscience we generally indicate the analysis of neural mechanisms as they underlie processes of emotion. More specifically, feelings, moods, emotions and aspects of personality are examined both using a neuroanatomical approach as well as by investigating psychological theories of personality. The contemporary debate about the function of neuroscience to study emotions has moved from 'Descartes' Error' as evidenced by Antonio Damasio to a more omnicomprehensive theory of emotions which links their production, activity and effect on the neurological aspects of human nature, thus not separated from high(er) cognitive functions. Before Damasio, important contributions to this research have come from Paul Broca, James Papez, Paul D. MacLean and were subsequently developed by the studies on empathy and the discovery of mirror neurons by Giacomo Rizzolatti, Giuseppe Di Pellegrino, Luciano Fadiga, Leonardo Fogassi, and Vittorio Gallese, as well as by Jaak Panksepp, who actually coined the term 'affective neuroscience'.

The main scope of this discipline is related to the investigation of the neurological underpinnings of emotion. However, while behavioral neuroscience focuses more on the 'external, connecting, social' factors of behavior, affective neuroscience investigates the 'internal, relational, personal- and mood-related' aspects of *affection*, in a broader sense. Thus, there are many shared points of investigation between these two disciplines, although the focus on neural mechanisms of emotions, particularly the limbic system, is particularly related to affective neuroscience. Common research areas in this field include psychological and neural development, cognition, emotion, motivation, self-image, self-awareness—especially in relation to others—and self-regulation, decision making, feelings and feedback processes, inter/social/shared-individual variability, mindfulness and consciousness, normative and norm-based behaviors, ethics and moral judgment, social and cognitive representations, (confirmation) bias, stereotypes, belief system(s), and many others. This discipline is supported by a multitude of theoretical approaches, and its incipit came from the now usual mind-body debate. A good example in this area is to be found in the work by Spinoza, who pointed out that 'even ethics' could be based upon biology, and would therefore have a natural origin, now rooted in the modern cognitivist psychological approach, justifying the neural underpinnings of tribal/primal/primitive social communication and sociality, a type of intersubjective communication which is both affective and effective. From a clinical perspective, a similar theoretical framework can be very useful to foster empathy and understanding for patients' difficulties, especially in the areas of emotional self-regulation and social or relational struggles, including communication. However, we need to understand to what extent we can base our clinical interventions on the assumption that (every?) behavior is led, guided or even promoted, produced by neural mechanism. Furthermore, we need to ask ourselves whether these mechanism can be 'clustered' in (previously observed vs. inferred) patterns. This is not just a semantic, semiological or broadly (in this circumstance, 'merely') philosophical need, but it truly helps framing the direction, both strictly behavioral/emotional (as usual, with neural underpinnings evidenced by neuroanatomical and neurofunctional research, as in Fig. 3.2), as well as inductive vs. deductive, but also containing a structural and possibly utilitaristic-utilitarian component of the 'movement toward and for something or someone' of this pro-motion or pro-duction of behavior, especially in connection to emotions, or "feelings moved *from* and *to*".

This analysis would serve a better definition of a 'mental state', not only from the psychiatric point of view, but also according to constructionist approaches within psychology. In this sense, an entire system of multiple elements—which might be in turn considered as systems themselves—work in cooperation/combination (thus eliciting patterns of neural activation) to produce specific emotions. Of course, there is an immediate need of defining what the '3 B' or 'basic building blocks' would be on a neuroanatomical level (i.e. the brain areas and structures involved) and on a phenomenological/ontological level of definition. An example of the latter might be the seven universal facial representations/expressions of

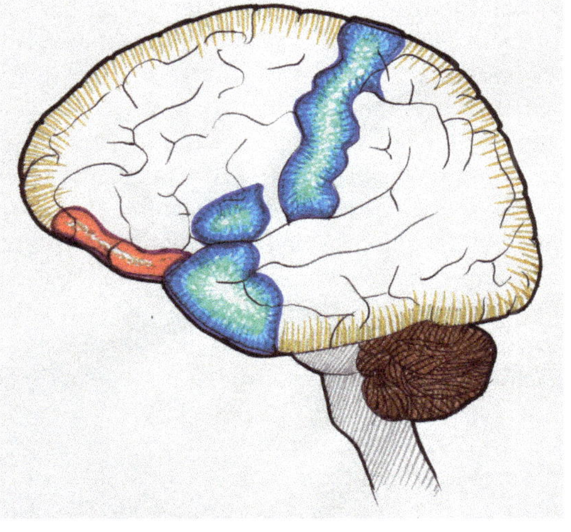

Fig. 3.2 Lateral view of the brain with cerebrum (beige/white) and cerebellum (maroon), with highlighted neural areas involved in emotional processing according to Pessoa (2008): somatosensory cortex, anterior insular cortex and anterior temporal lobe (blue), and orbitofrontal cortex (red, used most frequently)

(human) emotions of anger, fear, sadness, disgust, surprise, contempt and happiness developed by Paul Ekman starting from a Darwinian research viewpoint, and further expanded into the emotions non-necessarily encoded in (universal) facial muscles: amusement, contempt, contentment, embarrassment, excitement, guilt, pride in achievement, relief, satisfaction, sensory pleasure and shame. Of course, the investigation in this sense focuses on the construction of mental states on different levels. First, on the level of bio-physical/mechanical functions and processes (facial expressions in particular, but also body-related metalanguage such as posture, micro/macro movements, all the way to possible psychomotor activation and/or agitation). Second, on the level of neural networks and patterns, to account for specific psychological elements such as attention, cognition, computation, memory and perception. Third, on the level of the internal-external connection between the above and the production, process and communication/delivery of emotions. Starting from the hypothesis that more basic structures and functions account for (which is again, not the same as 'create' or 'produce') multiple, interconnected and distributed patterns of neurological-psychological (also only a correlation, at this level) activation, we thereby infer that an 'affect' originates from structurally 'inferior' components, with the production of an emotion as the last part of the chain connecting the individual to others and to himself/herself. Another point of view challenges this assumption, defining the previously discussed emotions as basic, not only psychologically but also in a biological sense. This approach is referred to as locationist because of the perspective in which a 'single emotional category' contains multiple mental states and relates to either a single brain area or multiple neural networks, but always distinct, although subjectively experienced. Another approach takes another route and views the connection vs. separation within a mutual activation of the right vs. brain hemisphere.

More specifically, the *right hemisphere hypothesis* identifies in the right hemisphere the neocortical structures and subcortical systems of attention and autonomic arousal underlying (human) emotions, more specifically the expression, and perception of synthetic/symbolic, whole-holistic-integrative (the so-called big picture), image-related, nonverbal, pictorial mental frameworks and interpretations of reality. Another point of view, the *valence hypothesis*, distinguishes between the activities of the right hemisphere as processing center for negative emotions and the activities of the left hemisphere as processing center for positive emotions, although there are many 'moderate' positions according to which the two hemispheres are not fully 'compartmentalized', 'specialized' or 'lateralized', but present combined, communicative features in which certain aspects are more ***dominant*** in one hemisphere than in the other. Of course, aside from the evidence-based data collected through the previously discussed scientific methods and technologies, there is a certain level of cultural influence into dividing the brain into a more emotional-creative right hemisphere and a more analytical-logical left hemisphere. While it is true that aspects such as symbolic representation, imagination, intuition, visual-spatial representation, musical perception and/or appreciation, and 'whole picture-insight' have been connected to the right hemisphere, and language understanding/speech production, analytical-logical cognitive processes and reasoning to the left hemisphere with the support of qualitative and quantitative studies reporting on functional activity and underlying processes, a complete lateralization in which only one side does only one thing appears to be somewhat misleading and incomplete. Of course, the same can and should be said about the differentiation in terms of neural structures and functional activity from the perspective of sexual/gender characteristics. From the perspective of affective neuroscience, one of the hypothesis tested in this context, is whether female brains, or brains usually associated with female traits appear to promote a more emotional-empathic behavior as opposed to male brains. Again, sociocultural upbringing, expectations and stereotypes might play a very big role in defining what it means to be a woman or a man in the first place. What the scientific method can provide is the analysis of those brain areas that have been associated (again, underpinning/underlying behavior in connection to neural functions) with certain traits. In this sense, most of the current research has focused on chemistry and structural anatomy, and taken into account specific functions of neural areas. In the first case, among the most relevant examples we find studies on the effects of estrogen and oxytocin on female subjects and testosterone and vasopressin in male subjects. In the second case, the relationship between nervous and endocrine system is especially meaningful to account for differences in functionality and transmission.

From the broader perspective of affective neuroscience however, the vast majority of studies available to the scientific community have so far dealt with specific brain areas such as the amygdalae, the cingulate gyrus, the fornix, the mammillary body, the olfactory bulb, the thalamus and the hypothalamus, the hippocampus (all parts of the limbic system), as well as the basal ganglia, the cerebellum, the insula, the orbitofrontal and prefrontal cortices, and the ventral striatum. Of course, these areas are observed with specific experimental studies (read: hypothesis testing according to the principle of falsifiability) aimed at quantifying affection in connection to aspects such as cognition and emotion, and are therefore overlapping with behavioral and cognitive neuroscience. Combined research and meta-analyses aside, at the very center of this type of studies we find processes like inhibition, modulation, (social-interpersonal) response—especially to 'go and no-go' cues—and psychophysical response to internal-external stimuli of differentiated types. In this regard we should also mention the importance of social constructs in connection with semantic production, especially from the viewpoint of developmental psycholinguistics, as particular responses might be elicited by specific words or combinatory set of words, within specific culture, and familial, ethic, religious, and/or social upbringing and background. From this perspective, finally, of great importance is scientific testing to categorize memory and attentional bias (measured for instance with emotional stroop task) in connection with the (again, relative/interpersonal vs./subjectively perceived) moral-ethical value, weight and valence of negative and positive words on affect. Other very valuable tests used at the intersection of affective and cognitive neuroscience are the Fear-potentiated startle (FPS) and Dot-probe paradigm, with foci on fear conditioning in the first case, and selective attention in the latter.

3.2.3 *Neuroethology*

Given what we have so far discussed in regard to brain anatomy, structure, function and possible neural-neurological correlations and underpinnings of human cognition, computation, affect, behavior and emotion, we have now the hard task to investigate just how much of knowledge gathered from animal models can be applied and is applicable to 'us humans'. This is obviously not to diminish the wonders of all the beautiful animal species that now populate or have populated our planet, especially given that human beings are, indeed, an integral part of the animal kingdom. This is even more important given the possible double hermeneutics in observing and experimenting with either fellow human beings or other animals. We certainly need to be fully aware of all the specific aspects, including differences, of animal behavior to propose, investigate and defend specific neuroscientific claims. Neuroethology in particular studies aspects of animal behavior using approaches, tests, techniques and technologies from a wide range of subjects, with a special focus on biological, comparative and evolutionary perspectives. To better define the main interests, goals and scopes of this fascinating field, we will directly quote Jörg-Peter Ewert, unanimously viewed as one of the most important developers of neuroethology, who identified the following questions (Ewert 1980):

1. How can an (animal) organism detect stimuli?
2. How can the nervous system of an (animal) organism perceive and represent external/environmental stimuli?
3. What are the processes in the nervous system allowing acquisition, storage and retrieval/recall of information about a stimulus?
4. How do neural networks encode behavioral patterns?
5. What are the processes of behavioral coordination and control in the nervous system?
6. What are the premises of the ontogenetic development of behavior in relation to the processes, systems and mechanisms of the nervous system?

As Ewert was 'academically born' as a neurophysiologist, it is evident how many concepts in the above presented questions have a lot in common with the scope of neurophysiology. As we will encounter this very field in another section of this volume, we will here discuss only the most important and field-specific aspect of the contemporary approach in neuroethology. As with classical ethology, we therefore identify the study of animal behavior in natural circumstances, that is, under naturally observable conditions and natural environment/habitat. This study starts from a conceptual and methodological framework shared with both evolutionary biology and psychology, behavioristic perspectives in particular, with new insights provided by comparing behavior and mental processes in human and non-human animals. According to this view, animal behavior is thus a biologically–evolutionarily adaptive trait, and as such presents measurable responses to external inputs observed and tested in laboratory settings. Of course, here the focus is less solely on psychological process—as in comparative psychology—and more on biology and neuroscience. Therefore, neuroethological research aims at investigating the underlying mechanisms of the nervous system via the analysis of species-specific (one side) and comparative (on the other) elements of behavior with the ultimate goal of possibly identifying general and/or universal principles at the basis of such behaviors. Neuroethology thus follows subjects even beyond laboratory/side-specific model investigation, observing animal behavior even beyond single/still components, that is, taking into account the whole life story/history of the organism investigated while keeping track of neuronal activity via natural and non-natural stimuli. This approach is especially useful, albeit not without theoretical challenges,[7] to produce new elements to be used in the development of fields such as artificial intelligence (neural) engineering and robotics.

As Neuroethology shares many elements with classical neuroscience, neurobiology and evolutionary biology, psychology and zoology, it is not surprising that the history of Neuroethology overlaps with the history of these disciplines. The philosophical attitude toward the systemic combination vs. division between human and non-human animals has changed many times in history, but we can say that a truly scientific method in the modern, evidence-based sense of the term has been applied to ethology only in the twentieth century, and as such it has allowed ethology first, and neuroethology after, to be considered separate fields from natural science—at least the classic version of this field, traditionally descriptive *in nature*—and ecology, albeit still part of zoology. The main contributors to this process were Konrad Lorenz, Niko Tinbergen, Karl von Frisch, Erich von Holst, Theodore Bullock, Santiago Ramón y Cajal, Edgar Adrian, Charles Sherrington, as well as Robert Capranica, Jörg-Peter Ewert, Walter Heiligenberg, Alan Hodgkin, Andrew Huxley and Masakazu Konishi. Although the connection between field research and studies in laboratory setting has been investigated throughout the history of Neuroethology, it is evident how some of the techniques used in this field are limited to the latter, also due to the very structure of such experiments and the standards, in terms of observation, control, accuracy and falsifiability, they need to follow. This represents both a challenge (for instance, the validity and relevance of findings in natural habitat) and a strength (experimental design allowing for the development of new techniques and technologies useful to replicate natural-istic environment and mimic behavioral aspects) of modern neuroethology. Given these premises, among the most important foci of investigation in neuroethology, we find a broad investigation on the contributions of neurons, hormones and genes to animal behavior, comparative, cognitive and computational aspects of function, including higher processing centers, visual/spatial

[7] In particular, the difficulty in applying inferences from linear systems of response mechanism to the animal nervous system, which is nonlinear.

memory collection, production, storage and recollection, activity levels in sensory systems, eye and head movement, signaling plasticity and behavioral neuronal complexity, learning processes and skill development (although there is certainly a somewhat anthropocentric bias in this very definition, for instance, in relation to the production and execution of sounds/songs).

3.2.4 *Neuropsychology and Neuropsychotherapy*

Many scholars view psychotherapy as an applied form of psychology, more specifically involving the application of psychological theories and research to provide therapeutic support to mental health patients. Certainly, this is a very limiting and narrowing definition, but it helps us identify some basic concepts upon which to relate the term 'neuro'. In fact, if psychology is said to be the study of the mind—leaving aside the mind/brain problem and the discussion on the existence of a/the soul for now—and thus constituting both a scientific field (whether 'social' and/or 'soft' science as descriptors make sense at all, we will discuss later) as well as an academic discipline, psychotherapy by definition focuses on the θεραπεία and is therefore a form of service, care, treatment and/or healing method and technique. Certainly, as such, the scope of psychotherapy is very close to the one of psychiatry and we should ask whether being, at least theoretically but with clear legal and social outcomes, separated from the field practice by the ἰατρός is actually appropriate in this setting. In this regard, the ongoing debate on the inclusion of mental health in the general medical field under the branch of psychiatry is of fundamental importance from a philosophical point of view. We will further analyze this debate, for now we will just refer to the work by Thomas Szasz in connection to the anti-psychiatry movement, George Graham's irrationality model, the Tidal model, Peter Sedgwick's social construct/value judgment model as well as other philosophical approaches including the perspectives of Paul Feyerabend and Ludwig Wittgenstein.

The major contributions of both neuropsychology and neuropsychotherapy have obviously to do with the relation of the neural activity, most specifically the structure and function of the brain, to psychological processes. Therefore, similarly to what we previously said for affective and behavioral neuroscience, neuropsychology and neuropsychotherapy provide a scientific analysis of the neuro-physiological underpinnings to behavior, cognition, computation, emotion and other aspects related to the (especially human) mind. More in detail, the focus of neuropsychology is on scientific research studies in those areas, using neuroimaging, neurocognitive tests like the Cambridge Neuropsychological Test Automated Battery (CANTAB) or the CNS Vital Signs (CNSVS), psycho-educational tests like the Nelson-Denny test or the Woodcock-Johnson test, as well as neuropsychological tests such as the Benton Visual Retention Test, the Boston Naming Test, the Controlled Oral Word Association, the Wechsler Adult Intelligence Scale (WAIS), the Wechsler Memory Scale (WMS), the Wisconsin Card Sorting Test and others from both purely analytic perspectives as well as for clinical-therapeutic purposes. While the history of neuropsychology overlaps with the history of neurophysiology, neuroscience and psychology (for which we must be thankful to Berti, Bouillaud, Broca, Damásio, Denes, Goldstein, Ladavas, Lurija, Pizzamiglio, Velichkovskij, Wernicke and others), the history of neuropsychotherapy is a relatively recent development of specific neural-based investigations within or in (clinical, evidence-based) support of classical psychotherapy by researchers such as Alessandrini, Bellamoli, Birbaumer, Cozolino, Fahlböck, Gauggel, Grawe, Laaksonen, Ranta, Rohracher, Rossouw, Schiepek and Zoccatelli. The application of brain research in clinical settings has had an enormous impact on the very way we perceive psychological-psychotherapeutic interventions, especially in regard to neuroplasticity. In this context, we could infer that the mind-body problem found a direct, applied, medical application in the context of both better understanding disturbance-specific negative neural changes and ameliorating—with all the philosophical, especially ethical implication that the term entails—one's health conditions from the perspective of therapeutic and lifestyle changes (thus including perspectives from positive psychology, behavioral medicine and motivational interviewing) with long-lasting effects. In this sense, concepts such as conditioning, learning and (positive/negative, perceptual-conceptual, associative, kindness, masked, repetition and response) priming are at the center of the multifold relations between physical-mental but also social-environmental stimuli, responses and their psychological mechanisms of action, including the relation to the activity of specific neurotransmitters (Fig. 3.3).

To clarify, neuropsychotherapy provides support-solution to problems of discordance/disconnect by focusing on the very interface between mental processes and their neurological counterparts from a therapeutic perspective. According to Grawe (2004), therapeutic success happens through the appropriate *Bedürfnisbefriedigung* (roughly translated as

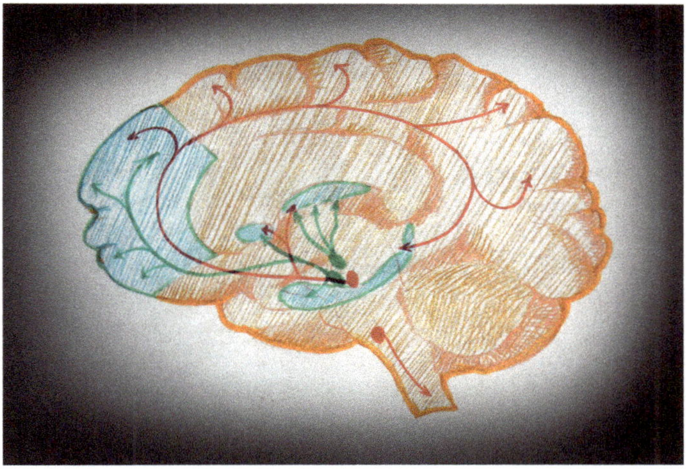

Fig. 3.3 Left lateral midsagittal section of the brain with serotonin pathways (red) and dopamine pathways (green). The two red dots represent the Raphe nuclei, and the two blue dots indicate the ventral tegmental area, while the aquamarine-highlighted areas indicate frontal cortex, nucleus accumbens, striatum and hippocampus

'satisfaction of needs' even beyond Maslowian viewpoint) of orientation/control, pleasure/avoidance, attachment, self-elevation/protection, as a preventive-protective mechanism for the possible development of mental health disorders, which are in turn the product of failed *Konsistenzregulation* or consistency regulation. Following Grawe's consistency theory (1998, 2004), the (human) organism strives from a form of union-unity-integrity defined as *Übereinstimmung* or *Vereinbarkeit*, thus referring to conformity and compatibility of the concurrent neural and psychological processes. The higher the levels of conformity and compatibility, the healthier—in a thereby addressed connection between body and mind—the organism (itself a combination of the two). Thus, neuropsychotherapy fosters a better patient-provider relationship, which in turn focuses on using the explicit function mode of the brain to induce implicit changes by enabling/empowering the patient to raise awareness of her/his perceptions and achieve her/his motivational goals. From a neurological perspective, strong motivational activity via concrete positive life experiences produces healthier structures and processes in the brain, starting from the activation of the dopamine system. In this sense, the clinician becomes (as it was in the traditional-historical sense) a mentor, guide, teacher and role model for the patient. The therapist thus promotes healthy decisions by working on the "[I] want/ [I] will" parameters of the patient, given that (a) people learn things they want to learn more easily and (b) the conscious goal is not top of the neurological hierarchy, but is dominated by higher-settled, unconscious (implicit) goals. Therefore, if people make a conscious (explicit) decision, this has already fallen implicitly (Growe 2004).

In the NPT panorama by Rossouw (2014), neuropsychotherapy starts from the neuroscientific investigation on environment, biology, chemistry and psychology to focus on cognitive and functional re/habilitation, neuropsychoanalysis, neuro-behavioral psychotherapy, therapeutic assessment, and affective reconsolidation approaches. More specifically, neuropsychotherapy combines/incorporates psychodynamic, Gestalt, and cognitive approaches where "Each psychological treatment program has components of empathy, supporting, understanding, observation of difficulties, integrity, learning, and a basic understanding of the effects of placebo, etc." (Roussow 2014). In this combination of multiple theoretical perspectives and diverse psychological schools of thought, we need to be reminded of the very specific orientations of humanistic and positive psychology. In the case of a humanistic approach, philosophical traditions close to phenomenology, hermeneutics, constructivism, postmodernism, transpersonalism, existentialism and phenomenology—Camus, Husserl, Merleau-Ponty, Sartre, in particular—serve as psychological basis (which is assumption) for the debate and follow a more qualitative and individual, singular, subjective-based approach in research (Waterman 2013). In the case of positive psychology instead, the traditions go further back in the philosophical tradition, even tradition with a 'bigger "T"'. Certainly, in humanistic psychology we also find some references to (a)theological debate, and even hermeneutics, but it is in the positive approach that we truly find the Classical tradition of Aristotle and the Hellenic, eudaemonist background, as well as an empirical, pragmatic and quantitative methodology using sample size research samples and focusing on well-being strategies, motivational interviewing, empowerment, exercises and action-taking (Tomasi 2016). Positive psychology also means finding ground for debate in Kierkegaard, Nietzsche and Tillich, the medieval tradition of Christian philosophers, as well as Stuart Mill (more in detail, the inductive approach and the five methods), Russell, Popper and Fromm (Waterman 2013). A possible combination of humanistic and positive approaches is welcomed in the context of 'identity as self-discovery', which is at the center of Waterman's eudaimonic identity theory (2013). In the context of neuroscience and psychology applied to therapy there are constant 'cognitive swings' between the principle of (medical-clinical) contiguity and contingency, which are the product of behavioristic frameworks, including measurable outcomes and continuous progressive testing, as part of predictability and falsifiability. Neuroscience benefits even more from cognitivism especially due to the further developments of situated cognition (Engeström and Middleton 1996). In neuroscience, this results in a much more attentive analysis of processes such as information and memory processing, including Short-, Mid-, and Long-Term Memory (Long Term Potentiation process based on neuronal synchronicity) demyelination-based effectors on procedural developments. Moreover, if this cognitive approach can shed even more light on phenomena such as confabulation and source misattribution, metacognition is in itself a cognitive form of self-perception. This self-perception can indeed be understood under the framework of the Greco-Roman heritage, in particular Socratic humanism. However, we could certainly argue that the primacy of self in the context of self-discovery has morphed multiple times in history, with often contradictory outcomes. The concept of γνῶθι σεαυτόν has moved from an oracular Delphic tradition with promethean components to the empowered centeredness of the human self in the Renaissance and finally to the "cogito ergo sum" of Cartesius and the critical reactions of Gassendi, Kierkegaard, Lichtenberg, Sanders Peirce, Macmurray and Williams, including the "sum, ergo cogito" by Monte, Hendricks, and others, fundamentally based on Nietzschean offsprings. This perspective also opens the door to translated interpretation to philosophically/cognitively interpret reality and our perception of it and our role in it through art, myth, poetry, rhetoric and eloquence following the models of Cicero, Quintilian, Socrates, Pico della Mirandola and Giambattista Vico. Furthermore, these philosophical methods are part of the debate on valuable self-knowledge strategies which include rhetoric and poetry exactly as part of a 'form of memory', also in the work by Verene (1997), as possible guidance for human activities, in a phronetic sense. Gadamer (1993) also relates to this viewpoint, in the sense that these human actions, behaviors, cognitions and decisions are through a comparison between an equilibrium brought to us humans by nature, passive in nature, and controlled, monitored by nature, versus an opposite equilibrium, entirely depending on our actions, and thus linked to the moment(-um) of human choice. This means that when the wholeness of man-nature-reality-existence is broken, diseases, psychological or physical, appear. Neuropsychology and neuropsychotherapy have to embrace also these perspectives to be effective. In other words, when we feel either overwhelmed or underwhelmed, we are stressed because of the (perceived or objectively actual) increasing number of chores, tasks, stimuli and stressors we face or because we lack enough 'things to do or be stimulated by' and we literally 'lose track of perspective, of pattern, of purpose'. This brings us to the exact etymology of the Italian *noia* or the French *ennui*, which represent the negative aspects of the *taedium vitae*, a true 'esse in odio' which brings boredom to a threat. Given these premises, we can certainly understand how psycho-physical health is the opposite of disease versus continuum, and how maintaining balance and homoeostatic tension means comprehending whole/full identity as opposed to split/divided/dissociated identity, particularly in psychiatric-psychological terms, via diagnosis and psychotherapy focused on grief and loss. That is why psycho-physical health and lack thereof are truly philosophical concepts, as in the Latin bonus/bonulus/bellus and the Greek καλὸς, to define health as a (phenomenologically understood) phenomenon, an epiphany, possibly even a manifestation or teophany as we will see in Chaps. 5 and 6, related to *Erfahrung* and *Erlebnis*, in connection to human suffering, meaning, healing process and the existence of absolute parameters.

3.2.5 Ethnopsychology and Psychological Anthropology

The debate 'nurture versus nature' has been at the center of philosophical and psychological research since the very beginnings of these disciplines, and the interwoven, interdisciplinary fields of ethnopsychology and psychological anthropology (the latter term coined by Francis Hsu as a change to 'culture and personality') appear to be well equipped to try to address the questions arising from this debate. In attempting to understand human beings, especially in relation to other animals and the environment, the notion of culture and ethnicity plays a fundamental role. In fact, whenever we try to assign specific values and principles from an investigative or therapeutic standpoint, we are faced with questions of validity, universality and applicability of such concepts. Furthermore, this type of investigation truly challenges broader assumptions such as evolution/devolution/involution, identity, as well as time, history and space. Anthropocentric and ethnocentric views have been challenging foundations of ethnopsychology and psychological anthropology in the early stages of ethno-folk psychology with Wundt's attempt to determine the cognitive abilities in a cross-cultural sense,[8] as well as in (Freudian) psychoanalytic anthropology and have developed thanks to the work of Piero Coppo (ethnopsychiatry), George Devereux (cross-cultural research on mental illness), Erik Erikson (culture, ethnicity, identity), Jakob Friedrich Fries and Gottlob Schulze (psychic/mental anthropology), Erich Fromm and Géza Róheim (philosophical approach and field studies), Abram Kardiner (mental health vs. illness in ethnic minorities), Gananath Obeyesekere (psychological conflicts from the perspective of ethnicity and culture), Melford Spiro (psychoanalytic concepts as well as investigation of socio-cultural symbols and institutions as defense mechanisms), John and Beatrice Whiting (authors of 'the Six Cultures Study'), as well as the series of research studies from psychoanalytic/psychological and psycho-anthropological perspectives by Vincent Crapanzano, Cora DuBois, Gilbert Herdt, Douglas Hollan, Viktor Frankl, Geoffrey Gorer, Carl Gustav Jung, Clyde Kluckhohn, Robert LeVine, Carl Rogers all the way to the anthropological research by Franz Boas and Claude Lévi-Strauss. Of course, the anthropological focus is toward ethnic, historical, social and evolutionary aspects, while the psychological focus covered transcultural (often acultural) and transnatural (but not non-natural, especially from the perspective of neurological underpinnings) elements of behavior, cognition and emotion.

Although often separated in academic setting (we will follow this trend by expanding the discussion in the section *Cultural, Cross-cultural, and Trans-cultural Psychiatry*), the field of ethnopsychiatry is also deeply connected to ethnopsychology and psychological anthropology. In this context we need to mention the work by the Italian neurologist and psychiatrist Franco Basaglia; the French psychiatrist and neurologist Henri Collomb; Roberto Beneduce, an Italian Professor of Cultural Anthropology, and Medical and Psychological Anthropology; György Dobó, a French-Hungarian (with German-Jewish and in part Romanian heritage) ethnologist and psychoanalyst; and Frantz Omar Fanon, a writer, theorist, philosopher, psychiatrist, and political activist with a very interesting heritage (a descendant of enslaved Africans and indentured Indians on his father's side and of black Martinician and white Alsatian descent on his

[8] Although the differentiation between the 'totemic' stage, the 'age of heroes and gods' and the 'enlightened age of humanity' also contains a level of 'utilitarian/modernist/functionalist' bias.

mother's) and a passion for the connection between culture, ethnicity, psychology and politics. Fanon wrote seminal works in this field, including *Black Skin, White Masks* (1952), *A Dying Colonialism* (1959), and *The Wretched of the Earth* (1961), the latter with a foreword by Jean-Paul Sartre. From the perspective of ethnopsychology and psychological anthropology, these researchers played a fundamental role in better understanding and possibly defining personality in terms of the characters vs. characteristics shaping and shaped by one's culture, ethnicity, social/familial environment (including expectations, rules, regulations and norms) and biologically-based elements. In this analysis we find multiple approaches, including the basic-modal personality approach by John and Beatrice Whiting and Cora DuBois, the configurationalist approach by Ruth Benedict, A. Irving Hallowell and Margaret Mead, the National character sociological analysis by Alex Inkeles and Clyde Kluckhohn and the cognitive-anthropological approach by Roy D'Andrade, Ward Goodenough, A. Kimball Romney and Anthony Wallace. The definition of 'culture as personality' belongs to the configurationalist viewpoint, in the sense that each experience is absorbed and combined through the morphing, shaping, translating and transmitting activity of the symbolic structure. However, this very activity can be analyzed and thus interpreted with a cognitive approach as well, especially in relation to the networks of neural connections.

3.2.6 *Neurotheology and Psychology of Religion*

An important distinction needs to be made whenever we talk about the relation between religion, theology and any form of science, whether 'soft' or 'hard', in the modern conception of the term. First of all we have to relate to possible double hermeneutics (Giddens 1987) and the very conceptualization of the field and the practice. Conceptualization is under this analysis also translation and transfer, conversion, application and adaptation, perhaps even condensation via the use of a scientific method and methodology to something which can hardly be defined and/or possibly comprehended and understood through the same mechanism or conceptual-neurological framework of either neurology/neuroscience or psychology/psychological science. In this regard I would certainly refer to Plantinga's 'Warrant' (Plantinga 2011, 2015) to better frame the problem. In the case of psychology of religion, we are faced with the application of these methods and conceptual frames of interpretation to spirituality and religion, their traditions, practices and epiphenomena as well as to the people related to the above, whether because of their practice or lack thereof, and whether the former and the latter can be quantified in terms of awareness of choice, a (free, free-will-based or lacking, ontologically non-existing, opposite to) choice between belief and disbelief; between creed, credence, faith and faithfulness; between atheism, theism, deism, agnosticism, defense, apologetics, theodicy and many more aspects and perspectives, attitudes and actions. Certainly, given the psychological nature of this discipline, the focus here is on psychological processes, as opposite to social practices and organizational forms studied by sociology or religion, and neural correlates investigated by neurotheology. In any case, both the legitimacy (in a strictly logical sense) and the validity of such attempts need to be very well analyzed before they can be fully understood to be appropriate or not for this type of investigation. In neurotheology, the task is trying to show the correlation between spiritual or religious experiences, starting from an individual level but possibly accounting for interpersonal and social levels as well, via neural structures, brain areas and processes such as the activity of mirror neurons, limbic system and cortical/subcortical structures, and the activity of the brain. This approach has received a lot of criticism from many proponents of both the psychological and the sociological investigation of religion, as the claim of a causal relation between the two (perhaps many more) sides of the spectrum, that is, spiritual-religious experience and neural correlates, appears to be not very well grounded in observational, evidence-based or theoretical-analytical terms. Of course, the same can be said about certain aspects of psychology, for example, the very concept of 'projection' (Tononi 2012). Although the term 'neuroscience' has been used in multiple contexts and with very different meanings and possible interpretations, the word has been first used by Huxley and made popular in both (popular as well as academic) philosophical-theological and (neuro)scientific areas by authors such as Drewermann, Hardy, McKinney and Newberg. Again, the focus of these works is on the connection between what can be experienced in terms of internal, interior, inner perception and the biological observation of specific mechanism of action preceding, following or happening at the same time. Another problem here is whether a conceptualization of time in chronological sense would make sense, but given the laboratory-based requirements for this type of studies, we will have to deal with the theoretical consequences of such approach. The application of psychometrics encounters the same level of problematic questions, as it requires—aside from the specific theories, technologies and/or techniques of psychological measurement chosen for a certain test or analysis—a level of value judgment or even *belief* in the quality vs. quantity of specific certain steps or increments, and whether the separation and/or distance between these has to be previously (a priori?) determined for the satisfactory (again, possible confirmation bias or methodologically flaw included) observation and follow-up analysis to take place. To be sure, here we are not trying to raise the level of skepticism toward any possibility for finding some results in this type of research studies. On the contrary, we do believe in the validity, in terms of practical—even clinical—application and scientific development of such studies. The concern is whether the results thereby obtained can provide more insight to the understanding of the 'whole behind the experience' as opposed to be limited to the psychological or neurological processes, mechanisms or biological, chemical-electric activities investigated.

To begin with an appropriate analysis of what neurotheology and psychology of religion can and should do (not in moral terms but from the theoretical and logical trifold perspective of ontological evidence vs. existence vs. essence), we can list those areas that have been more prone to be investigated under the lens of such disciplines. For instance, researchers have focused their investigative attention on prayer, meditation, mindfulness, ecstatic trance, altered vs. alternate vs. alternative states of consciousness, spiritual-religious (sense of) awe and/or integrity and oneness, wellbeing, whole-being with the divine and/or the universe (investigations on multiverses and parallel universes included), nirvana-like states, increased bright/light perception and enlightenment, and many other aspects. Furthermore, many studies such as the ones presented by Beauregard, Jansen and Van Lommel focused on the effects of such experiences on the very life—in psychological, sociological, neurological as well as philosophical, religious, spiritual terms—of the people that either were the subject of a study or have been affected themselves by phenomena such as near-death experiences or out-of-body experiences. From this perspective, we could almost say that the results are even more important than the cause. In other words, whether the states of consciousness experienced by these subjects were 'original' in the sense of *selbstständig* and autonomous, or (by)product of neural activity, their effect completely changed the worldview of such subjects. Of course, we would certainly argue here—and this is also a matter of personal opinion, taste and preference—that we would rather deal with the possible negative consequences of a disappointing evidential truth rather than live a (at least on the surface) 'happy life' built on/of misguided or entirely false assumptions. Part of our investigation throughout this volume is about the fact that we don't necessarily have to choose between 'misguided' or 'gullible', 'naïve' happiness and 'sad but true' evidence. In this regard, studies on happiness at the intersection between spirituality and religion—especially the latter, because of its intrinsic nature of *legamen*—have indicated a very strong correlation between happiness and good health (Tomasi 2016). More specifically, beside the specificities of apologetics discourse, the underlying factor is that we can see benefits in terms of general amelioration of physical health through 'improvements to spiritual health'. To be clear, these improvements appear to happen to the single individual as well as to and through communities, and that is where the religious factor presents its greatest strengths (Fu et al. 2011). Furthermore, in the analysis by Byron (2011) on the contagious aspect of happiness, positive results were found even after adjusting for genetics, obesity, and tobacco and alcohol use. An even more interesting aspect in this 'spreading of happiness' is that "good feelings continue to move from person to person" (Byron 2011), thereby presenting a model very similar to the one used for airborne droplets, as in the case of a flu. A 'healthy spirit' also provides strength for physical health and fosters a positive cycle in which we find and are in turn motivated to find higher meaning(s) in life. In this regard, the study by Heintzelman and King (2014) focusing on social exclusion, positive mood, and environmental pattern and coherence as main elements provided further connection between a well-defined sense of meaning and health in multiple areas. The concept of an 'existential vacuum' is thereby analyzed from the perspective of the social absence of such meaning, which is defined as purpose and significance in a motivational/existential sense, but it also requires to be understood as 'making sense' by the person experiencing it across space and time and thus with patterns of regularity, predictability or reliable connections. Among the most famous studies published in regard to these considerations, we can think of the experimental work by Michael Persinger and the 'God helmet'. In this well-known experimental example of neurostimulation, more specifically of Low field magnetic stimulation (LFMS), the researchers Stanley Koren and Michael Persinger attempted to analyze and measure the effects of such stimulation, especially in the temporal area, on creativity, religious experience and temporal lobes. Regardless of the specific issues and debates around the replicability, accuracy in terms of double-blindness, and subjective vs. objective responses and bias or suggestibility, these experiments are very valuable from the theoretical analysis of the issue at hand, namely, the possible connection, and possibly causal terms, between an external, artificial stimulation of the brain and the resulting spiritual or religious experience. Of course, the problem here is both ontological as well as methodological, as it involves:

(a) A previous/prior agreement on what we can scientifically define as experience (mechanism, process, feeling, emotion, thought, etc.)
(b) A possible methodological confirmation bias regarding not only the scientific techniques, technologies/instruments used, but also on the very location of such investigation.

In regard to these aspects, we could quote the famous story in which a person who lost the house key was looking for the missing item in the kitchen. Albeit the person knew that he had dropped it on the sofa in the living room, he said that the kitchen was a much better place for his investigation as the light was "far brighter than the one in the living room". A similar joke tells of a scientist who wanted to measure the hearing abilities of an eight-legged spider. The investigator said to the spider "jump!" and the spider, seemingly following the order, proceeded to jump. The researcher asked the question eighth times, each time surgically removing one leg from the body of the spider. At the end of the experiment, where no legs were left attached to the spider's body, the scientist commanded one last time: "jump!" but the spider remained still. The scientist therefore concluded: "The experiment clearly suggests that the removal of all limbs from an eight-legged spider will impair the hearing ability of such spider."

Our intention here is certainly not to undermine the scientific rigor of experiments in this area, but to shed some light on the underlying theoretical issues that arise when attempting to combine what appears to be either two separate fields or two different aspects of a whole. Aside from the potential elements of dualism in this view, it is certainly beneficial to continue our scientific efforts in order to capture the 'how' of such phenomena. Science has certainly provided us with a better understanding of the psychopharmacology of mystical experiences, as we now can use specific chemical elements/synthetized components to elicit certain responses that are akin to spiritual perceptions and monitor their effect on subjects, be these effects deeply connected to the subjectivity of the individual experiencing them. Studies on ketamine, dimethyltryptamine, Lysergic acid diethylamide (LSD), psilocybin, mescaline, for instance, have provided a very precise scientific understanding of what effects are produced by the use of chemically synthetized drugs or elements of spiritual, mystical, shamanic traditions such as ayahuasca, psilocybe mushrooms and peyote. Chemical reactions are also at the center of normal or abnormal (in the psychological sense, or as outlier in comparison to a mean, distant from baseline) neural functions. Thus, investigations on the substrata of certain traits of psychiatric disorders, such as psychotic or manic episodes, delusions or hallucinations exhibiting grandiosity or hyperreligiosity, can provide new insight on associations between these states and, for instance, cases of temporal lobe epilepsy (see the studies by Norman Geschwind or Vilayanur Ramachandran). On the opposite (from a theoretical point of view) side of the spectrum, we find the research by Andrew Newberg and Mario Beauregard on the effect/affect of spiritual and religious practices on brain activity and function, thereby providing support to mindfulness-based interventions as the ones found in behavioral medicine, motivational and positive psychology or mind-body medicine. Among the most important brain areas and neurological functions associated with this type of research, we find the limbic system and the cortical/subcortical structures, the frontal and parietal lobes. More in detail, this type of experiences often result in a (re)activation or increase of processes at the level of nucleus accumbens, frontal attentional regions, ventromedial prefrontal cortex, as well visual processing areas. In regard to the study of the links between areas usually covered in neurotheology, such as spiritual and religious practices in connection to neural activity, and psychiatry, especially regarding mental health disorders, we can refer to our previous analysis in *Medical Philosophy* (Tomasi 2016), especially mentioning that over 450 studies investigating the links between religious and spiritual factors in depression were published in the last 60 years (Bonelli et al. 2011). The meta-analysis indicated over 60% less depression and faster remission from it in those more religious or spiritual, or a reduction in depression severity in response to a religious or spiritual intervention, and "of the 178 most methodologically rigorous studies, 119 (67%) find inverse relationships between R/S and depression" (ibid.). However, among the most recent studies, the University of London created a survey[9] (nearly 10,000 people from Great Britain, Chile, Estonia, the Netherlands, Portugal, Slovenia

[9] Leurent, B., Nazareth, I., Bellón-Saameño, j., Geerlings, M.I., Maaroos, H., Saldivia, S., Švab, I., Torres-González, F., Xavier, M., and Kin, M. 2013. *Spiritual and religious beliefs as*

and Spain) focused on the analysis of the ability of family doctors to predict depression from their patients' worldview, which appeared to yield opposite result. The data in this study indicated that 10.3% of religious people had at least one severe episode of depression during the year; 10.5% of people with a spiritual worldview fell into deep depression and only 7% for secularists, with $p < 0.001$. Furthermore, it appeared that the more people viewed themselves as strongly religious or to have a strong religious faith, the greater the risk of severe depression. Although the results varied depending on the country, the study had found a very strong link between depression and religiosity. We should point out that, based on the results and the very focus of this research, "the notion that religious and spiritual life views enhance psychological well-being was not supported" (Ibid. 2016). Of course, a direct causal effect between religion and depression cannot be inferred as no further analysis post-data collection or within the study design was able to shed light on this aspect. More specifically, it is not clear whether people with depression are more prone to become more religious or at least find some relief in spiritual practices, or if those practices are the cause of depression. In the vast majority of studies, everything else being equal, religious or spiritual practices and beliefs are related to less depression, especially in the context of life stress, it is important to notice that our health is strongly influenced by the interaction between behavioral, cultural, developmental, environmental, genetic, social, etc., factors.

References and Further Readings[10]

Bibliography

Benson, H. 1997. *Timeless healing*. New York: Fireside.

Bonelli, S.B., R. Powell, P.J. Thompson, M. Yogarajah, N.K. Focke, J. Stretton, and M.J. Koepp. 2011. Hippocampal activation correlates with visual confrontation naming: fMRI findings in controls and patients with temporal lobe epilepsy. *Epilepsy Research* 95 (3): 246–254. https://doi.org/10.1016/j.eplepsyres.2011.04.007.

risk factors for the onset of major depression: an international cohort study. Psychological Medicine. London, UK: Cambridge University Press.

[10] Disclaimer and Copyright: "Critical Neuroscience and Philosophy. A scientific re-examination of the mind-body problem" © David Tomasi 2019 for the entire text and all images and tables. The reproduction or copy of the content of this work, in whole or in part, is strictly prohibited.

Brown, W.S., N. Murphy, and H.N. Maloney, eds. 1998. *Whatever happened of the soul? Scientific and theological portraits of human nature*. Minneapolis: Fortress Press.

Byron, P.B., ed. 2011. *The science of emotion*. Boston: Harvard Medicine.

Engeström, Y., and D. Middleton, eds. 1996. *Cognition and communication at work*. New York: Cambridge University Press.

Ewert, J.-P. 1980. *Neuroethology: An introduction to the neurophysiological fundamentals of behaviour*. New York: Springer-Verlag.

Fu, J., et al. 2011. Sympathetic activity controls fat-induced oleoylethanolamide signaling in small intestine. *Journal of Neuroscience* 31 (15): 5730–5736.

Gadamer, H.-G. 1993. *Über die Verborgenheit der Gesundheit*. Frankfurt a. M.: Suhrkamp.

Giddens, A. 1987. *Social theory and modern sociology*. Cambridge: Polity Press.

Grawe, K. 1998. *Psychologische Therapie*. Göttingen, D: Hogrefe.

———. 2004. *Neuropsychotherapie*. Göttingen: Hogrefe.

Haynes, C. 2016. Identity, transcendence and the true self: Insights from psychology and contemplative spirituality. *HTS Theological Studies* 72 (4): 1–9.

Heintzelman, S.J., and L.A. King. 2014. Life is pretty meaningful. *American Psychologist* 69 (561–574). Washington, DC: APA, American Psychological Association.

Klemm, W.R. 2010. Free will debates: Simple experiments are not so simple. *Advances in Cognitive Psychology* 6: 47–65.

Kornhuber, H.H., and L. Deecke. 1990. Readiness for movement – The Bereitschaftspotential-story. *Current Contents Life Sciences* 33 (4): 14.

———. 2012. *The will and its brain: An appraisal of reasoned free will*. Lanham: University Press of America.

Levin, J.S. 1994. Religion and health: Is there an association, is it valid, and is it causal? *Social Science and Medicine* 38 (11): 1475–1482.

Libet, B., Elwood W. Wright Jr., B. Feinstein, and D.K. Pearl. 1979. Subjective referral of the timing for a conscious sensory experience – A functional role for the somatosensory specific projection system in man. *Brain* 102: 193–224. https://doi.org/10.1093/brain/102.1.193.

———. 1996. How religion influences morbidity and health: Reflections on natural history, salutogenesis and host resistance. *Social Science Medicine* 43 (5): 849–864.

Mujkanović, L. 2016. *Nursing research studies*. Ethics, community, and healthcare. Hooksett: SNHU.

Murphy, N., and W.S. Brown. 2007. *Did my neurons make me do it?: Philosophical and neurobiological perspectives on moral responsibility and free will*. Oxford: Oxford University Press.

Pessoa, L. 2008. On the relationship between emotion and cognition. *Nature Reviews Neuroscience* 9: 148–158. https://doi.org/10.1038/nrn2317.

Plantinga, A. 2011. *Where the conflict really lies: Science, religion, and naturalism*. Oxford: Oxford University Press.
———. 2015. *Knowledge and Christian belief*. Grand Rapids: Eerdmans.
Rossouw, P.J. 2014. Neuropsychotherapy – An integrated theoretical model. In *Neuropsychotherapy. Theoretical un-derpinnings and clinical applications*, ed. P.J. Rossouw, 43–72. Sydney: Mediros.
Tomasi, D.L. 2016. *Medical philosophy. Philosophical analysis of patient self-perception in diagnostics and therapy*. New York: Ibidem Verlag/Columbia University Press.
Tononi, G. 2012. *PHI: A voyage from the brain to the soul*. New York: Pantheon Books.
Verene, D.P. 1997. *Philosophy and the return to self-knowledge*. New Haven: Yale University Press.
Waterman, A.S. 2013. The humanistic psychology-positive psychology divide. Contrasts in philosophical foundations. *American Psychologist* 68 (3): 123–196. Washington, DC: APA.
Widerquist, J. 1992. The spirituality of Florence nightingale. *Nursing Research* 41 (1): 49–55.

Further Reading

Sartori, G., and C.A. Defanti. 2012. *Le neuroscienze ed il Mistero del Libero Arbitrio*. Bergamo: Ordine dei Medici ed Odontoiatri della Provincia di Bergamo. Available at: https://www.youtube.com/watch?v=KWRkpU7bmTg

CHAPTER 4

Medicine on, of and off the Brain

4.1 Disease and Disorder, Illness and Issue

How can we be sure that someone is sick? There are very few fields in which evidence-based science is not as applicable in clinical and research settings as it is in medicine. We also discussed the possible answers to this question in *Medical Philosophy* (Tomasi 2016), but we need to:

(a) Present the core concept in this context, as it is fundamental to keep in mind the theoretical background to the scientific method used in medicine, as well as the parameters used to define terms such as disease and disorder, illness and issue.
(b) Understand how such contributions from medical philosophy apply to the mind–body problem, especially in regard to neuroscience.

In the title of this chapter, 'Medicine on, of and off the Brain', we attempt to evidence the ways of interaction between medical science, art and practice and neural activity. More specifically, by 'on' we mean "the medical opinion (thus comprising diagnostic and prognostic aspects) on matters related to brain activity, including function and dysfunction." With 'of' we indicate the specific subfields or areas of medicine that deal with the nervous system, especially the CNS and more precisely with the brain. Finally, with 'off the brain' we want to stress the importance of those areas of scientific investigation as integral part of modern medicine that fail (in

D. L. Tomasi, *Critical Neuroscience and Philosophy*,
https://doi.org/10.1007/978-3-030-35354-4_4

the sense of general application of the scientific method) to account or appropriately describe certain phenomena using *only* medical science. The main effects of this analysis are found in areas related to both neuroscience and psychiatry. This is precisely why in this chapter we will discuss Clinical and Medical Neuroscience, Neurocritical care or Neurointensive care, Neurosurgery and Neurology, including Behavioral Neurology, Paleoneurology, Neurophysiology, Psychoneuroimmunology, Psychobiology and Psychopharmacology in this order. In psychiatry, two names play a very special role in this outcome: Franco Basaglia and Thomas Szasz. The latter is famously known for the critical and controversial stance against classical psychiatry and the 'myth of mental illness' with further developments into the antipsychiatry movement, especially in the United States. Basaglia instead was a true physician-philosopher in the highest sense of Philosophy of Medicine *and* Medical Philosophy. The Italian psychiatrist, neurologist, theoretician and socio-political activist is rightly considered the founding father of the modern conception of psychiatry. Beyond the vast array of accomplishments in medicine, public policy and politics, Basaglia is also the main author (together with the psychiatrist, academician and politician Bruno Orsini) of the *Legge Basaglia*, (also called Legge 180). This psychiatric system reform law represents the Italian Mental Health Act of 1978, which (a) contained specific directives, guidelines and regulations for the closing down of *all* psychiatric hospitals and (b) promoted the replacement of such hospitals with a vast and interconnected system of community-based and acute inpatient care services. Of note, the *Legge Basaglia* still represents the only mental health law of this kind. Italy is in fact the *only* country in the world where traditional psychiatric hospitals are outside the law (Barbui and Tansella 2008). To be sure, Basaglia constructed his views and their direct application in the context of healthcare management and policy making not only on scientific research in the fields of neurology and psychiatry, but also on the basis of a very deep philosophical analysis of the patient as a whole (human being), with all the complex system of issues related to self-perception, self-image, consciousness and awareness, diseases, disorders and challenges. Basaglia fully integrated the *Daseinsanalyse* of Heidegger, but also of Hegel, Feuerbach and Jaspers, especially focusing on the conflict between authentic existence and inauthentic existence. Basaglia had the opportunity to verify the application of this *modus cognendi et operandi* at the Lunatic Asylum of Gorizia-Gorica-Guriza-Görz, a truly *Mitteleuropean* city at the intersection of Italian, Austro-

Hungarian/German, Furlan, and Slavic/Slovenian cultures and a very fertile ground for philosophical and linguistic-communicative investigations. In this context, Basaglia also related to phenomenology and existentialism, particularly in regard to Ludwig Binswanger, Eugène Minkowski, Maurice Merleau-Ponty, Jean-Paul Sartre, but also Giorgio Antonucci, Michel Foucault Anna-Teresa Tymieniecka, Erving Goffman and Frantz Fanon.

The existential component of the appreciation of psychiatric disorders and manifestations of the psyche in general are also representatives of the "integrated and fragmented nature of consciousness" (Tononi 2015) as analyzed by Giulio Tononi and Gerald Edelman, but also (via a multifaceted analysis which covers areas far beyond psychiatry and neuroscience) Ervin Laszlo, Aurelio Peccei, Rick Strassman and Harald Walach. In this regard, philosophy helps us better frame what benefits, in terms of diagnostic classification and targeted clinical interventions, can originate from an accurate analysis of the combinations 'Disease and Disorder, Illness and Issue'. To provide a specific example in this context, we can think of a very complex diagnosis with uncertain etiology such as panalgesia, more commonly known outside of Europe as fibromyalgia, first isolated-classified by the Italian physician Federigo Sicuteri. This condition is characterized by an increase in muscular tension and generalized chronic pain caused by what has been identified as a multifactorial rheumatic idiopathic syndrome often associated with multiple symptomatologies involving disturbances of sensitivity, mood, sleeping pattern, anxiety and depression (Stahl 2008). We will further discuss the definition of fibromyalgia, and we would like to begin with a historical reference. Florence Nightingale is widely known as the founding mother of modern nursing. She was also a researcher, a statistician, and advocate for human rights and a social reformer. Although it is still very difficult to assess the extent of her role as well as her clinical and social achievements in the Crimean war, many historical accounts depicts a clear picture of her progressively complicated health condition (Young 1995). In particular, many have argued that her increasing depressive state might be linkable to conditions such as brucellosis and spondylitis (Bostridge 2008). In the context of the relation between neuroscientific perspectives and medicine, we will focus on the most complicated condition—in the sense of etiology and pathophysiology, as well as in diagnosis and prognosis—of fibromyalgia as potential underlying medical condition Florence Nightingale might have been suffering from (Keddy 2008; Fowler 2013). According to McCance and Huether (2014, p. 1579), fibromyalgia is a "chronic musculoskeletal syndrome characterized by widespread joint and muscle pain, fatigue, and tender points" with a wide symptomatology in multiple body systems, which includes muscle and joint stiffness and "generalized, localized or diffuse pain, tingling and numbness" (Simms and Goldenberg 1988), sleeping pattern disturbances linked to compromised circadian activity in the endocrine system, as well as disturbances in cognition, memory and mood, affecting depression (between 25% and 60% in patient affected by FM), anxiety, and posttraumatic stress disorder (PTSD), which are also discussed as possible precondition or causal effectors for fibromyalgia itself (Schweinhardt et al. 2008). Florence Nightingale was 34 years old when she was deployed to Scutari during the Crimean war and spent over 6 months taking care of the sick and wounded in horrible hygienic conditions, with lack of ventilation and defective sewers, before the British Government sent out the Sanitary Commission to help her. Although Nightingale was able to visibly improve the situation and reduce the death rate of soldiers by 2%, dealing with overcrowding, constant stressors, fatigue and other medical conditions appeared to take a toll on her psychological well-being and immune system (Small 1998). She spent 21 months taking care of approximately 18,000 patients, providing direct clinical care and constantly reassessing, readjusting and improving the quality of care by changing bandages (previously simple rags clotted with blood), food intake, medical supplies and other clinical tools, as well as bathing the wounded, changing linens and helping the wounded with physiotherapeutic strategies (Huntley et al. 2014). Her work as a nurse and as a manager was very intense, psychologically and emotionally, as well as physically. Based on many accounts, including her own writings (McDonald et al. 2015), we can argue that Florence Nightingale might have been developing a history of medical traits very similar to the symptomatology often found in fibromyalgia, more specifically sharp pain in the neck area, shoulders, lower back and hips, with possible myofascial pain syndrome, or MPS. Nightingale also reported vision problems, which became dramatically challenging 30 years after her experience in Crimea, although the early onset was reported during the war, according to the accounts (Genschorek 1990). These issues were followed by a rapid memory loss in 1896 and multiple physical problems, leading her to spend most of her time bedridden (Genschorek 1990). These aspects, together with the more physical symptoms of generalized pain are especially important in our analysis of Nightingale's medical conditions during her work in the Crimean war, as a "history of widespread pain *lasting more than three months* affecting all four quadrants of the body" is a major criteria, together with "generalized pain and 18 designated possible tender points," for the diagnosis of fibromyalgia, according to the American College of Rheumatology (AMR 2015).

The case of Florence Nightingale serves as a perfect scenario, although technically not a true *clinical case* given its history and the lack of specific diagnostic documents, for a thorough investigation of the connections between pain perception, mental states and physical well-being. In the context of a critical investigation of psychophysical components of health, we could imagine a possible treatment plan for the condition, by focusing for instance on immediate pain reduction and general relaxation, both presenting strong effectors in terms of neurological underpinnings. As pain reduction and relaxation are directly linked to the patient's perception of pain, self-assessment and pain threshold, these two therapeutic goals are deeply intertwined with the 'bigger picture' of more advanced, future planning, thus including (a) a general psychological assessment focus on improving psychoneuroimmunology and (b) lifestyle changes and coping skills, physical exercise in particular. As expected, in order to address general relaxation, deep breathing, meditation and mindfulness exercises are recommended. For pain reduction, the physician might recommend supplements, pain relievers and anti-seizure drugs. In these categories we could certainly list vitamin D, CNS-active medications and opioids, as well as anti-inflammatories (Müller et al. 2007). Medication prescription will follow an appropriate clinical evaluation by a specialized physician or rheumatologist. In order to better plan a specialized doctor's appointment, it is important that the patient has educated himself–herself on his/her medical history and symptomatology, as well as family medical history. Furthermore, a detailed list of all the medication currently prescribed or taken in the past—including herbal and/or dietary supplements—will give a better clinical picture, especially because of their direct effects on neurobiological areas as well as due to possible placebo effects. To work on long-term goals it is also important that the patient thinks about potential questions for the physician, in order to gain knowledge and be proactive in the development of effective strategies and a positive attitude toward the healing process. Although the International Classification of Diseases (ICD-10) lists fibromyalgia as a "Disease of the musculoskeletal system and connective tissue" and a "functional somatic syndrome" (Häuser et al. 2009), the multilayered etiology of fibromyalgia is fertile ground for a more omnicomprehensive and balanced approach which includes a psychoneuroimmunological evaluation to support the patient physically, as well as psychologically/emotionally (Mujkanović 2016). As fibromyalgia is observed to reduce pain threshold and tolerance for extended exercise and physical activity, a well-balanced daily schedule will need to be assessed to provide enough stability, routine and motivation to the patient, thereby helping the physician monitor the improvement or difficulties encountered by the patient in his/her recovery, or at least management of the disorder. To plan a long-term treatment, we need to rely on the statistical data found on widespread muscle tension and pain, headaches and migraines, which is present in 50% of cases affected by fibromyalgia; chronic fatigue syndrome, with 70% of the cases; widespread muscle tension and pain, headaches and migraines (50%); irritable bowel syndrome and temporomandibular joint disorder (between 50% and 80%). Finally, to address our specific case (Florence Nightingale) we need to keep in mind that polymorphism and other possible genetic factors and internal/external clinical environmental factors such as viral illnesses, HIV and other infections, Lyme and rheumatic diseases, as well as medications' side-effects might be connected to fibromyalgia, and that a misdiagnosis of the disorder might occur in 75% of the cases (Clauw et al. 2011).

To assess fibromyalgia, comprehensive clinical and diagnostic tools should include a complete blood test with complete blood count and erythrocyte sedimentation rate (Mujkanović 2016). Recommended are also thyroid function tests. In any case, given the complexity of the disorder, it is important to evaluate the specific scenario of each individual patient. In this regard, Müller et al. (2007) identify four typologies of fibromyalgia, with related best clinical strategies for management and treatment:

1. Extreme sensitivity to pain but without associated psychiatric conditions (5-HT3 receptor blockers)
2. Fibromyalgia and comorbid, pain-related depression (antidepressants)
3. Depression with concomitant fibromyalgia syndrome (antidepressants)
4. Fibromyalgia due to somatization (psychotherapy)

Furthermore, clinical assessments based solely on the (18) tender points have been challenged by the American College of Rheumatology (2015) and replaced by the Symptom Severity Scale (SSS) and the Widespread pain Index or WPI, more specifically WPI $\geq$ 7 and SS $\geq$ 5 or WPI 3–6 and

SS ≥ 9 when similar levels are found in a period exceeding 3 months for a diagnosis of fibromyalgia, when other non-related symptoms have been ruled out. According to the American Fibromyalgia Syndrome Association (2014), "fibromyalgia varies from one patient to another, but the multiple symptoms it causes are often intertwined." An accurate diagnosis of fibromyalgia in the case of a deceased individual is a very hard task, especially when the historical and scientific data are collected in ways that do not often match evidence-based medicine. However, relying, at least for the sake of historical accuracy, on lower levels in the hierarchy of evidence can still provide a great insight on the pain suffered by Florence Nightingale. If we had the chance to care for her in modern times, we would have to make sure that our treatment plan would be successful, based on all the premises listed above. For instance, an accurate analysis of the clinical manifestations of psychophysical stressors onto the HPA axis might give us more information on the development of the disorders from Nightingale's experience in Crimea, with a special reference to triggers, infections and physical or emotional traumas and the related progressive sensitization and lowering of the pain threshold. The analysis of a possible diagnosis of fibromyalgia in the case of Florence Nightingale can therefore be used to further expand our knowledge of the pathophysiology of the disorder itself, but also to shed more light on the role of everyday stressors in the nursing profession. What we have said about panalgesia/fibromyalgia can certainly be used to (a) better understand the application of neuroscience to the discovery of neural underpinnings to psychophysical wellbeing, starting from pain-touch pathways, all the way to the core of the limbic system, and (b) to better treat patients in clinical settings as well as to better help clinicians in their profession. This means to understand (a) the long-term effects of physical, emotional and psychological stressors healthcare professionals encounter as part of their profession and (b) prevent those triggers which can diminish the clinical efficacy of a therapeutic intervention or strategy of care and have potential damaging factors on the general well-being of clinicians. In this context, many studies have provided very good results for pain management for this very difficult condition. In the single-blind, randomized trial by Wang et al. (2010), the researchers wanted to verify whether the application of Yang-style T'ai Chi Chuan would yield better results in the treatment of fibromyalgia in comparison to the combination of wellness education and stretching. The results clearly indicate clinically important improvements with T'ai Chi, with no adverse events or side effects observed. In particular, Of the 66 randomly assigned patients, the 33 in the tai chi group rated far better in the FIQ total score and quality of life in comparison to the control group. There is possible criticism to this study, especially due to the relatively low number of subjects involved—albeit justified by the very prevalence of fibromyalgia in the general population—and the short time frame in which the research has been conducted, a 12-week period. In the research by Busse et al. (2013) the focus is on the social and economic components of the possible cause for the development of fibromyalgia, and the structure of the study, that is, a protocol for systematic review and network analysis. The tested hypothesis aims to define the correlational weight of true effectiveness of specific treatments and the connection/possible correlation in statistical terms with substantial socioeconomic loss. The meta-analysis comprised studies published in CINAHL, EMBASE, MEDLINE, AMED, HealthSTAR, PsychINFO, PapersFirst, ProceedingsFirst and the Cochrane Central Registry of Controlled Trials. To test the principle of falsifiability of therapeutic effectiveness, the researchers constructed a random effects model within the Bayesian framework using Markov chain Monte Carlo methods, in combination to the analytical review of the data in each study. Moving on to the systematic review by Li et al. (2014), massage therapy is administered for $\geq$ 5 weeks in the nine randomized controlled trials, thereby allowing the researchers to identify measures such as pain, anxiety, depression and sleep disturbance, which are at the center of our discussion on the possible effectors in the most successful therapies for fibromyalgia, including the examination of possible fight-flight and emotional overactivation and arousal, as observed in the amygdalae (Fig. 4.1). In this study, 404 patients were subjected to massage therapy modalities, after which a general improvement on all the areas discussed above was found. Therefore, although more studies are needed, this meta-analysis clearly indicates that we should include massage therapy as a viable treatment in the multidisciplinary management of Fibromyalgia. The connection between mind and body is explored in the study by Theadom et al. (2015); more specifically the way mind body and behavior mutually influence one another in maintaining well-being and health. The research is an interesting analysis of psycho-physical interventions for the treatment of fibromyalgia, aiming at identifying the effectiveness of mind–body therapies such as psychological therapies, biofeedback, mindfulness, movement therapies and relaxation strategies, in comparison to standard care and attention placebo control groups for adults with fibromyalgia, post-intervention and at three- and six-month follow-up. Among the positive aspects of this

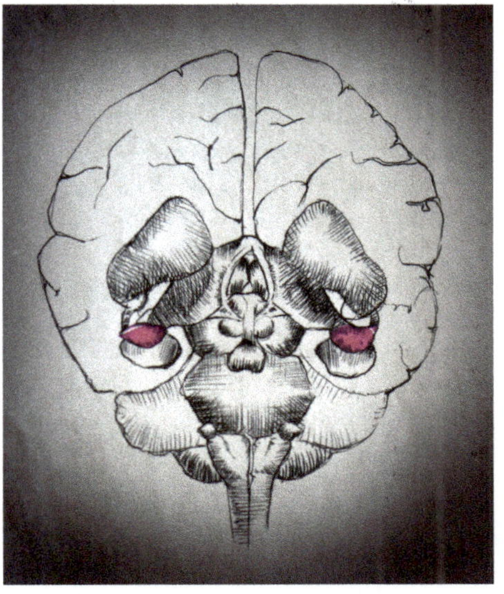

Fig. 4.1 The amygdalae, as primary area of emotional, mnemonic and decision-making processing in the limbic system

meta-analysis, we can list the high number of participants (4234, predominantly female) and trials (61, randomized controlled). With the single-blind, randomized controlled trial by Toussaint et al. (2012) of the 44 patients randomly assigned who completed baseline assessments, 21 patients completed the study, thus receiving amygdala retraining—described in this study as a novel mind–body approach—along with standard care or standard care alone. This study focuses not only on patients affected by fibromyalgia, but also on chronic fatigue—which, in itself is also a widespread defining condition of FM—to demonstrate statistically relevant improvements in the areas of physical health, energy, pain, symptom distress, and fatigue. Of note, the results were based on a validated self-report questionnaire which yielded positive results of the amygdala retraining (the group received an additional 2.5-hour training and instructional videos) course in comparison with the standard of care (1.5-day multidisciplinary program). Finally, Bernardy et al. (2011) observed 239 subjects to address the efficacy of hypnosis and guided imagery in the treatment of fibromyalgia (Mujkanović 2016). Although hypnosis is considered by some among the standard of care within psychotherapeutic, especially psychoanalytic, strategies, this meta-analysis supports a further investigation of non-invasive techniques within the spectrum of possible interventions targeting fibromyalgia.

Among the major weaknesses and problems that this systematic review revealed (as with many other systematic reviews in general), we can identify issues with length and/or time frame in which the single or multiple study has been conducted, as well as the relatively low total number of subjects examined. Certainly, as our analysis covers multiple disciplines we cannot infer that the same patterns can and should be recognized as valid across the studies examined. However, with a fair degree of accuracy we can point out to major positive trends in terms of respectable sources and publishers using sound scientific method involving both qualitative and quantitative research, and a strong indication suggesting the empirical validity of mind–body therapeutic modalities. More specifically, in this regard we briefly examined single studies, meta-reviews and databases published in the *New England Journal of Medicine*, CINAHL, EMBASE, MEDLINE, AMED, HealthSTAR, PsychINFO, PapersFirst, ProceedingsFirst and the Cochrane Central Registry of Controlled Trials. Our analysis clearly suggests that the state of science in regard to the management of fibromyalgia is indicative of higher clinical success in integrative therapies focusing on mind–body connection and psychophysiological stimulus-response mechanisms, in particular meditation and mindfulness-based techniques, massage therapy, gentle movement and T'ai Chi Chuan, relaxation and breathing exercises, guided imagery and hypnosis. The clinical data presented within the quantitative research studies examined indicated low cortisol production, improved and more balanced heart rate, appropriate oxygenation levels (Busse et al. 2013), and general positive response on a neurological level, especially in the areas of procedural processes in the prefrontal cortex and the responses in the parietal lobe (Bernardy et al. 2011). In the qualitative research studies, the responses given via direct questionnaires to the subject were also indicative on a general increase in psychophysiological well-being, albeit partially justified—which doesn't undermine the validity of the examination—by a more appropriate and functional self-perception, in terms of pain management. This is a core concept in the management of fibromyalgia, as the subjects affected by this very complex disorder tend to have a very low threshold for pain in general, in combination with a high tolerance for generalized pain. Examining the effects of natural methods on the promotion and maintenance of health and well-being in our patients is a fundamental responsibility for the practicing medical professional. Healthcare providers often find themselves dealing with a vast array of problems spanning from purely physical issues, to psychological, social and spiritual components. The relation between history, theory, practice and research in medicine and neuroscience from the perspective of integrative approaches can be an important channel for improved therapeutic modalities and the general amelioration of our clinical interactions with our patients. The data collected clearly indicate that a multidisciplinary approach, combining traditional and western medicine with integrative modalities is beneficial in the treatment of the main symptoms of fibromyalgia, as well as in the management of the pain associated with this disorder. More specifically, those practices which focused on the intersection and interaction of mind–body therapy, such as meditation and mindfulness-based techniques, massage therapy, gentle movement and T'ai Chi Chuan, relaxation and breathing exercises, guided imagery and hypnosis have provided the best therapeutic strategy. Furthermore, the results of this meta-analysis provide strong evidence for possible changes in patient care and research. In particular, relational-focused guidelines to ameliorate the general pain management and improve patient-provider relationship have been at the center of funding opportunities promoted by the US National Institute of Health and the National Center for Complementary and Integrative Health. From the clinical perspective, direct care will benefit from new ways to measure patient-reported outcomes, such as pain, fatigue, physical functioning, emotional distress and social role participation as well as their vital role in the maintenance of appropriate interactions to determine physiological differences and similarities from the perspective of generalized chronic pain. Given the aforementioned result data, we recommend more evidence-based, possibly double-blind studies with an extensive cohort/case-control group. In this regard, a very good resource is the meta-analysis and multiple study examinations presented in the National Center for Complementary and Integrative Health at the US National Institutes of Health (NIH). Furthermore, given the very nature of both the symptomatologies (especially due to the connection between mental health and psychosomatic manifestations) and treatment modalities hereby discussed, we would like to stress another fundamental aspect of 'Medicine on, of and off the Brain'. A parallel we would like to draw in this context is between the concept of (a) mental disorder and (a) sin. Alvin Plantinga (2015, p. 50) pointed out in numerous occasions that, following the Christian perspective, especially the 'A/C model' there might be multiple levels of interpretations of a disorder, far beyond the biological, developmental, behavioral, social or environmental areas usually covered by the

Multiaxial Assessment used in psychiatry, according to the DSM model. Plantinga mentions the possibility that original sin (in the orthodox conception of the term) might be responsible for a deficit which we would call *affective disorder*, not cognitive disorder, due to the damage operated on the *Sensus Divinitatis*. This concept brings us to Isaac Newton and his definition of time and space as *sensorium Dei*, as an integral part of the philosophical basis to the scientific debate on time during the Industrial Revolution, and defined eternally and immutably, always the same and the same-to-itself, in its own image. This has a much deeper and more practical applicability in iatrological terms, as it clearly represents a theoretical understanding of the course of a disease. To be more specific, as Newton defined the true motion of a body to be its motion through absolute space, time and space are not true, genuine substances. Therefore, time is completely distinct from the world-space, and it passes uniformly regardless of the succession of (historical) events in the world (Rynasiewicz 2012), giving a different interpretation to the hereby presented notion of an absolute space and absolute time. These 'absolute concepts of absolute' are deeply connected to subjective worldviews. When working with complex—general medical and psychiatric—symptomatologies, professionals need to understand the patient's values, and relate them to the patient's current presentation and diagnosis, especially in relation to philosophical concepts such as the *preservation of the self* (Bunge 2013). Patient's privacy, confidentiality, autonomy, beneficence and nonmaleficence need to be analyzed under the lenses of principlism and universalism, in relation to the wide range of ethical framework such as Kantian and deontological ethics all the way to virtue ethics, care ethics, bioethics and casuistry (McGonigle and Mastrian 2015). These aspects have a practical and immediate application in the patient-provider interaction. In their care, providers need to discuss with the patient the procedures or practices which will be used to minimize their susceptibility to undue influences and unnecessary risk (physical, psychological, etc.). Moreover, healthcare team members need to monitor the specificity of each patient to account for Inclusion/ Exclusion Criteria of care (thus incorporating medical, psychological, and pharmacological interventions), also from the perspective of healthcare coverage: eligibility and ineligibility criteria should be specific and analyzed with reference to the Centers for Medicare and Medicaid Services (CMS), the Internal Review Boards (IRBs) and the broader US healthcare delivery system, in particular the Department of Health and Human Services (DHHS). In order to guarantee the best possible care, each of these interventions involves a constant monitoring of patient condition/presentation with reference to the patient medical history (with further consideration especially under the lens of the Diagnostic and Statistical Manual, as in the case of the Multiaxial Diagnostic Impression in Psychiatry). Thus, medical professionals need to be able to access patient information and share such data with other clinicians within the treatment team and also be able to understand and maintain confidentiality measures according to the Health Insurance Portability and Accountability Act (HIPAA) standards. Medical professionals need to be able to discuss risks and benefits of care with the patient and illustrate the possible alternatives in terms of therapeutic intervention or non-intervention. This includes physical, psychological, social, legal or other risks. Moreover, nurses in particular are responsible for sharing with the patient the planned procedures—previously discussed with the treatment team—for protecting against or minimizing potential risks and assess their likely effectiveness, as well as the probability that given risks may occur, including their severity and potential reversibility. In addition, routine procedures such as additional diagnostic/follow-up tests should be identified. Finally the clinical decisions from physician to other team members in terms of therapeutic intervention versus non-intervention should also be part of the conversation, as the full disclosure of such important pieces of information should be shared by all the members of the treatment team for that particular patient, and include the patient in every decision-making aspect. Certainly, the biggest challenge is delivering enough information to enable the patient to understand the choice, given the specific presentation (in diagnostic terms) of the subject (especially when dealing with mental health disorders) and making sure that the best treatment option is given the appropriate clinical emphasis. In this sense, medical professionals need to understand that "HIPAA seeks to *balance* protecting the privacy of patients' health information and assuring that this information is available to those who need it to provide health care, payment for care, and for other important purposes" (Mujkanović 2016).

In this chapter we have discussed many aspects related to medical philosophy, medical research and medical practice. This discussion won't be complete if we didn't mention, at least in part, patient's rights, especially in connection to safety and privacy. To be sure, this study focuses primarily on neuroscience and the mind–body problem; thus even this discussion is filtered through this viewpoint. In other words, our intention is not to present every philosophical perspective on medical ethics, but provide at least a theoretical basis for the promotion and maintenance of patient's rights, given that the very techniques and technologies used nowadays in neuroscience, most specifically in relation to neuroimaging and the patient's personal-clinical information thereby collected, are connected to the broader technological advancement of medical electronics. Electronic Health Records or EHR in particular, have been progressively and successfully implemented in many countries with a general increase in quality and availability of information to and from patients. To be sure, speed of access is not necessarily a quality per se; however, when there is an immediate need for medical information—as in the case of vulnerable population such as cognitively impaired and elderly persons, traumatized, terminally ill and comatose patients, children and minors, and so on—EHR has been proven a great resource. We should pay special attention to the relation between this type of technology and health literacy in this context. More specifically, the International Ethical Guidelines for Biomedical Research Involving Human Subjects of the Council for International Organizations of Medical Sciences (CIOMS) in collaboration with the World Health Organization (WHO) defines vulnerable persons as "[...] those who are relatively (or absolutely) incapable of protecting their own interests. [...] They may have insufficient power, intelligence, education, resources, strength, or other needed attributes to protect their own interests" (Council for International Organizations of Medical Sciences 2002). The patient's ability to access and understand basic health information and clinical services, including medical-surgical, therapeutic, pharmacological interventions, scheduling, tests, follow-up visits and appointments, and HIPAA rules and regulations and consent forms, is fundamental. To foster accessibility, policy makers and healthcare providers play a very special role in fostering collaboration in the use of Electronic Health Records. Certainly, data alone are not a good indicator of true understanding of the information presented. Therefore, members of the Healthcare team should also:

(a) Help their patient understand the specific aspects of the care they receive, including risks and benefits, typology (and/or dosage and administration) of treatment, therapy, medication, visits, diagnostic tests and laboratory results.
(b) Guide the patients in their personal efforts to obtain more information about their condition, symptomatology and related treatment from the perspective of the information they can find on the world

wide web. They can play an important role in presenting and comparing different websites and other internet resources based on quality, credibility and hierarchy of evidence.

(c) Understand the patient's perspective, both from the medical as well as from the personal perspective, by learning how to accurately, precisely and appropriately interpret the patient's clinical-therapeutic preferences. Integrating classes and continuing education opportunities and certifications in Narrative Medicine could in this sense increase the therapeutic effectiveness of the provider.

In other words, we should be aware of the fact that technology's main purpose is quite similar to ours, as healthcare providers: to serve the patient. Therefore, if technology is a source of problem to the patient (because it is not easily accessible, is hard to understand or due to being unfamiliar with computer-based systems, etc.), we should make changes to the technology itself to make it more user friendly. Furthermore, technology should not replace medical knowledge—that is, every team member should have the technical and scientific skills to understand, check and verify the entries on EHR, especially in regard to medications. If these important requirements are not met, we would exchange one problem for another. We would avoid dealing with (for instance) the unreadable handwriting of a colleague's prescription, but we would not have the ability to spot possible errors in the electronic version of the information provided (Friedman 2013). Health Information Technology represents a broad variety of strategies, interventions, and systems aimed at supporting both the clinician and the patient in an increasingly complex health environment. Telehealth is among the most innovative aspects of this shift in paradigm; the question we need to ask ourselves is whether this type of healthcare delivery system and service based on telecommunication is truly beneficial for both sides of the therapeutic spectrum. From the perspective of technology, a rapid shift occurred through the development of informatics and digital programming starting from postal and telegraphic services, to telephones, videophones, side-specific computers and home computers and finally intranet and internet. Although computerized technology has found an unprecedented increase in efficiency and complexity in the last 10–20 years, the core concepts and application of telemedicine was already developed in the Renaissance, the Middle Ages and in Antiquity, so we cannot argue that it is a truly innovative system, at least from the theoretical point of view (Nutton 2009).

A similar analysis can be applied to telemedicine. To be sure, this technology only works if the human (and humane) presence in the direct delivery of care is present. In other words, the clinician is first and foremost responsible for the appropriate and effective application of such system to the benefit of the patient and of the broader communication with the healthcare team. As an integral part of this team, nurses have added abilities to help the system by focusing on the specific clinical aspects of care—including, but not limited to treatment plan, monitoring and evaluating medical outcomes, assessing possible and present risks, and evaluating clinical workflow—as well as the broader range of safety and security perspectives, such as sharing and delivery mechanisms, signatures and encryption informatics, and biometrics (Mathews and McCormick 2007). Moreover, there are several aspects of telemedicine which improve the general delivery of care to the patient: Telehomecare helps monitor and deliver clinical care, by reaching patients in their homes; telemedicine fosters clinical collaboration and professional autonomy *within* shared responsibility; teletrauma can reach patient even in the most distant and rural areas, thereby bridging the gap in geographic and possibly ethnocultural disparities; e-intensive care units further promote the remote monitoring of critical ill patients' care; telemental health combines the above-listed aspects to improve clinical care in specific vulnerable populations in vulnerable areas such as the aforementioned rural areas, as well as other areas in which access to healthcare might be difficult (war zones, prisons, etc.). Telemental health has been particularly developed in the healthcare system of the US state of Vermont, where a shared effort between the governmental institutions, the University of Vermont and the local hospitals have provided new and innovative developments to support and improve the care to mental health patients throughout the state (Rabinowitz 2015). Given all these premises, we could get the impression that telemedicine in its various and diverse subdefinitions and applications can be a definitive positive shift in the system, directly benefiting the patient. However, in our judgment we might actually be misled by the system itself. If Telemedicine is expected to improve healthcare outcomes, reduce health and geocultural disparities within the population and/or minorities, improve and help develop data infrastructure, collection, quality assessment and delivery, as well as balance patients' interests in protection of their privacy, health, and safety (Mathews and McCormick 2007), we would expect a lesser benefit *in the absence of* Telemedicine. More in detail, there is a disconnect in causal terms between the demands of healthcare and the needs of both business-oriented technological developments and insurance policies. For instance, we can address the supposed increase in cost due to the providers' accountability demanded by patients (often, quite unfortunately—and only in the United States—referred to as 'consumers' or 'clients') or due to medication-related policies. A much easier path rather than an expected increase in cost due to the very development of computer-based technology (at the core of modern telemedicine) is universal, free healthcare. The very claim that increasing healthcare costs result in a decrease in an uninsured population is both a logical fallacy and an oxymoron, in that there won't be any need for private insurances if everything would be covered by the state/country (National Audit Office, London 2003). Aside from the vast literature on these aspects, the strong evidence-based proof is given by the way other countries, especially Europe (some argue with a possible exception of Germany and the United Kingdom), successfully keep healthcare costs down, while at the same time provide the same level (most often a higher level) of both direct clinical care, research and education (World Health Organization, *The world health report* 2010). In this sense, even the need of standardization of healthcare informatics, technologies and programs/systems, albeit very positive in terms of allowing broader access, could be utilized by the same capitalist-oriented policy making strategies in order to foster profit, not necessarily to serve the patient or the provider (unless by 'provider' we include certain pharmacological or medical corporations).

Furthermore, many aspects related to the possible negative effects of social media on our healthcare system in terms of patient's safety should also be analyzed under this socio-economical perspective. By no means should we change our policies to move the focus away from the patients—in this sense, patient's safety and privacy should be always protected. However, there are at least two examples of areas of patient safety in which changes to the system itself could be beneficial. To be sure, due to the very complexity of Healthcare Informatics, the more we strive for accessibility of data and safeguarding the privacy of such data, the more we will find ourselves struggling to balance these two opposites, trying to come up with more and more technological devices to prevent the possible theft of such data. Telemental health is, for instance, a great tool to provide clinical help and support to patient. It is of no use when we are still struggling with social stigma toward definitions-diagnosis of a specific mental health disorder. In other words, instead of spending yet even more time, resources and funds to keep potential theft of private data in relation to these aspects, we should make sure that society understands (and clinicians can be instrumental in this sense) what it means, for instance, to be suffering from depression. Our point of view is that we should not overemphasize the importance that 'nobody will find out about that person's depression', because (in our example) depression *itself* won't be something to be ashamed of. In this sense, a much bigger shift in paradigm and culture than a technological development is needed. A second example could be the theft of person/identity-sensitive data such as social security number (Forsyth 2011). Once more, in countries where a Universal and free healthcare system provides services to everybody, regardless of gender, religion, color, socioeconomic status (even citizenship) and so on, the very social security number is shared—since everybody gets the same services, it is simply pointless to steal it.

A final aspect we would like to address as part of the section 'Disease and Disorder, Illness and Issue' is a very brief discussion on the connection between the information gathered from scientific studies in medical science and neuroscience and the accessibility of this type of information from the general public. We already discussed these *translational* aspects in a previous work (Tomasi 2016); in this section we would simply like to offer some insight on where to find reliable sources in this regard, especially in connection with the risks of 'self-diagnosis' we often encounter when scientific data are directly presented, without appropriate theoretical investigation, from the observational level of science to the level of prescribing advice. In this context, we will give two examples of medical advice websites serving the English-speaking public, especially in the United States of America: *Web MD* (http://www.webmd.com), together with the *Mayo Clinic* website (http://www.mayoclinic.org/), in particular, are among the most used sources of information on the internet for this purpose. The website present several positive aspects, which benefit not only the quality and quantity of information provided, but also the easiness in accessing such information, from the patient's perspective. WebMD was founded in 1996 by Jim Clark and Pavan Nigam under the name of 'Healthscape'. The name was later changed to 'Healtheon', and the corporation acquired WebMD in 1999 from Robert Draughon, thus changing the moniker and logo into Healtheon/WebMD. Finally, the name was shortened to WebMD, and in 2001 the corporation acquired Medscape from *MedicaLogic*. The website clearly displays author (including Chief Medical Editor Michael Smith, MD), credentials and date of last modification. Confidentiality of data is guaranteed for the users who decide to (a) use the website as guest, (b) sign in and/or (c) subscribe to the newsletter. Aside from being considered authoritative and reliable source by many authors, physicians and reviewers (Szalavitz 2017), Web MD was accredited by the Utilization Review Accreditation Commission (URAC 2001). Some criticism has indeed arisen not in regard to the quality of information presented on the website, but on the funding and the alleged profit-driven publishing strategies (Ail and Venkatesan 2013), although some of these sources come from within the broader criticism from proponents of alternative/complementary/integrative therapies to the established mainstream medicine in the United States (Mercola 2010). Certainly, the issues addressed in articles such as the well-known 'A Prescription for Fear' (Heffernan 2011) help evaluate the authority of websites originating in nonprofit groups (such as MayoClinic) versus the ones originating in corporations (such as WebMD) on the basis of the quality and accessibility of information available to patients (often referred, also—unfortunately—in clinical settings as 'customers'). Therefore, the final positive judgment of WebMD is related to the core medical values and clarity of information provided. As advice for further analysis of valid and respected sources of medical information, we would like to refer to international websites in languages other than English, especially German, Italian, Finnish, Norwegian, French, Russian and Spanish. Of note, a broad range of information derived from Scientific Research published in these languages is monitored and translated into English by authoritative publications such as *The Lancet* and *The New England Journal of Medicine*. Keeping the patient at the center of our care should be the most important goal of any medical or medically-related profession, including the clinical aspects of neuroscience. In the context of neuro-intensive nursing/care nursery, the question is what can researchers and healthcare providers do to ensure and protect patient safety? There are certainly specific strategies we can follow, in order to identify and reduce medical errors. According to Zhang, Patel, Johnson and Shortliffe (2004), a cognitive taxonomy of errors can "(1) categorize major types of medical errors along cognitive dimensions, (2) associate each type of medical errors to a specific underlying cognitive mechanism, (3) describe how and explain why a specific error occurs, and (4) generate intervention strategies for each type of error" (Zhang et al. 2004, p. 202). This approach can help neuroscientists, medical researchers, and healthcare providers in general understand the cognitive mechanisms of such errors. Furthermore, it provides guidelines and structuralization of intervention, focused on both prevention and aftermath. This certainly includes pragmatic and ethical aspects connected to error reporting systems and solution fostering systems. In order to achieve this goal, we must also analyze the specific categories of errors. As Riley, Davis, Miller, Sweet and Hansen (2010) suggest, we can focus in particular on active and adverse errors, errors of omission and commission, errors of execution, iatrogenic and latent injury, near miss, and finally sentinel events. From the perspective of Electronic Medical Records (EMR) we could argue that, aside from the great advantages this technology has brought to healthcare in general, it has also generated errors related to the very nature of such technology. EMR has certainly improved a wide range of effectors in clinical care, including time management, organizational skills, and patient safety, by allowing a direct implementation of the flow of information pertaining to a single patient, and also foster a more precise categorization of patient groups by typology and characteristics. In fact, the data collected from patients and available to nurses, physicians and other staff members can be continuously monitored, implemented and updated. This in turn benefits the prevention and correction of errors, thereby improving a more ethically sound report system, especially since the information is available to the healthcare team as a whole (Mujkanović 2016). In particular, new structure of communicating possible errors arises, by connecting healthcare staff via electronic information technology, including, but not limited to, intranet and internet access portals, internal/hospital/university-based (encrypted) e-mail systems, digital safety net plans and reporting systems. This allows for more precise, direct and safeguarded flow of information, for instance, between a nurse who noticed a certain error and the direct supervisor to whom the error needs to be reported. Since 2009, the Patient Protection and Affordable Care Act (PPACA) pushed the US healthcare system toward an increased utilization of information technology, by elucidating important positive effectors such as:

- Increased and improved user-centered design in healthcare (from both the perspective of the patient as well as the provider)
- Addressing workplace safety
- Anticipating unexpected outcomes
- Avoiding reliance on provider's memory
- Improving access to healthcare
- Directly involving patients in the care provided and administered to them, thereby contributing to future amelioration of clinical framework

However, EMR is based on a technological infrastructure that can also allow for important threats to patient safety, especially in the context of privacy, and especially in vulnerable populations such as children and minors, mental health patients, cognitively Impaired Persons, traumatized and Comatose Patients, terminally Ill Patients, elderly persons and minorities. As Samuel Lustgarten (2016) points out, electronic-informatics and internet/intranet-based technologies contain:

- Risks from individuals and collective actors, including stealing data from digital storage areas
- Risks from corporations, including shared profiles and systems of data collection and analysis
- Possible risks from the government. Regarding the latter, the author points out to the breach in confidentiality, privacy and safety within email exchanges in public universities. These aspects have a direct link to the security policies in University Hospitals, and healthcare facilities in general, as the sharing technologies allowing direct access to Electronic Medical Records could potentially cause harm to patients.

Strictly following HIPAA guidelines can help prevent this type of issues. At the same time, healthcare needs to further develop both the technology and the ethical understanding of EMR technologies to help providers continue monitoring and evaluating specific clinical-medical information, for instance, the 'The Five Rights of Medication Administration': the right patient, the right drug, the right dose, the right route and the right time.

4.2 To Treat, to Heal, to Cure

4.2.1 Clinical and Medical Neuroscience

Clinical and medical neuroscience focuses on the processes, mechanisms, and diagnostic, prognostic and therapeutic aspects related to the function or malfunction of the nervous system. Therefore, this field serves as a broad combination of activities at the intersection of (especially clinical and health) psychology, neurology, psychiatry, general neuroscience and general medicine, nursing, rehabilitation medicine, biology and technology, especially medical biotechnology and artificial intelligence, particularly in the context of neurogenesis (Fig. 4.2). Furthermore, other sub- and

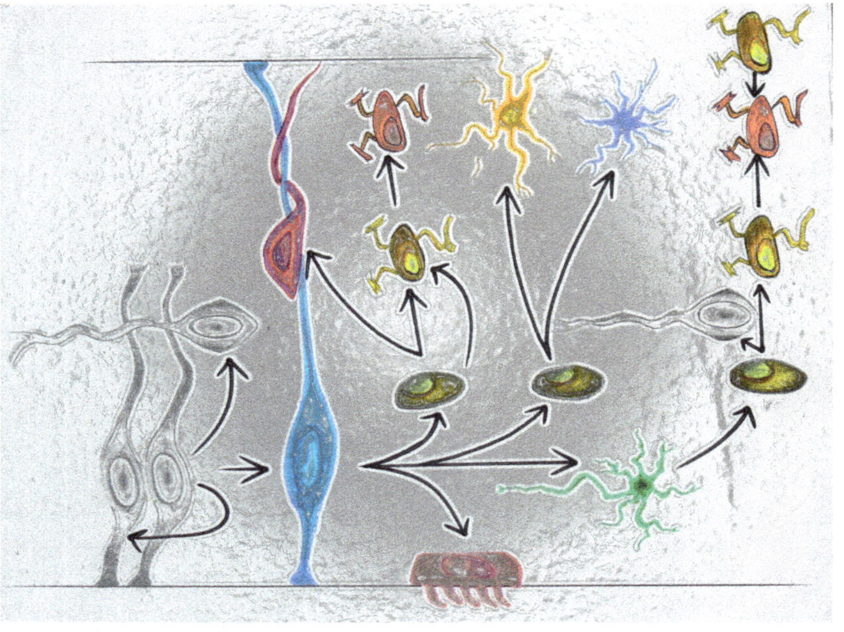

Fig. 4.2 The process of neurogenesis as a reinterpretation of the model. In the bottom left ventricular zone, we observe the neural stem cell (neuroepithelial cell) and the neuron (silver), followed by the radial glial cell and the neuron (in blue and bright red, respectively), followed by type 3 cells (dark yellow), ependymal cell (gray/pink), subventricular-zone astrocyte (type 2 cell, in green) and oligodendrocyte precursor cell (in bright orange/yellow), all in the subventricular zone. In the white matter we observe oligodendrocytes (light orange) and ALDg1L^{+} GFAP^{+} (yellow)

related branches of medicine, including neuroradiology, neuropathology and anesthesiology, share theoretical frameworks as well as research methods, data analysis and clinical interventions with clinical and medical neuroscience. Studies in this field cover basic neurobiology, statistical-epidemiological analysis in relation to disorders and disease, developmental aspect of neurology and psychology, with special foci on neuropathophysiology, functional human neuroanatomy, but also including foundational elements of developmental neuropsychology. In this sense, research on areas such as neurogenesis and neuroplasticity, genetic transmission, and the activity of mirror neurons is of fundamental importance.

From the clinical perspective, this field promotes a deeper understanding of disorders and diseases such as addiction, Alzheimer's disease, amyotrophic lateral sclerosis, anxiety disorders, attention deficits and ADHD, autism spectrum disorder (ASD), many other psychiatric disorders including bipolar disorder, BPD, MDD, OCD, SAD and schizophrenia, but also brain tumors, cataplexy, dyslexia, Down syndrome, epilepsy, Huntington's Disease, multiple sclerosis, narcolepsy, neurological AIDS, neurological trauma, Parkinson's disease, stroke and Tourette syndrome. Technologies and techniques used in this setting make great use of MEG/EEG, MRT, TMS and tDCS/tACS with advanced analysis techniques. Among the many scopes of investigation in clinical and medical neuroscience, we find developing new clinical interventions for neurological and psychiatric disorders. Such interventions focus on the amelioration of medical, especially psychiatric, diagnosis, as well as developing patient-centered approaches, and multidisciplinary treatment teams. Furthermore, these changes link perspectives developed within translational science to new treatment theories and applications for (especially neuropathic) pain, speech/person recognition (including associative visual agnosia and category-specific visual object agnosia such as prosopagnosia), as well as identifying developmental risk factors in relation to biomarkers and genetic transmission, and improving treatment of addiction and lifestyle/behavioral problems. These include sleep patterns and multiaxial effectors on health and wellbeing, preventing and treating disorders of immunity or inflammation (in relation to psychoneuroimmunology), cerebrovascular disease, and metabolic and mitochondrial disorders, promoting further study of anesthesia, and (better) understanding of mechanisms and processes of neurogenesis and neurodegeneration.

4.2.2 Neurocritical Care or Neurointensive Care

In a study on critical neuroscience, the term 'critical' conveys a multilayered interpretation. In the context of neurocritical care or neurointensive care, the focus is on medical interventions targeting life-threatening diseases of the nervous system from the emergency department and intensive care unit through the operating room and post-surgical period. As with previous subfields of both medicine and neuroscience, a multidisciplinary approach is used to treat diseases such as trauma-related medical problems, (major acute) strokes, seizures and epilepsy; specific, local or non-localized intracranial edema or encephalitis; intracerebral and subarachnoid hemorrhages; meningitis; brain tumors, as well as to prevent or treat myasthenic crisis, spine/secondary neurological injuries and non-neurological/systemic issues, immune system-related problems such as Guillain-Barre syndrome and possible medical complications with the collaboration of treatment teams in anesthesiology, emergency medicine, neurology and neurosurgery. Given the nature of this discipline, efforts in this area cover technical and technological advancements, scientific discoveries, social policy and public health, as well as patient–provider relations, especially in the case of direct clinical care (in particular medicine and nursing, especially neuro-intensive nursing/care nursery as we have previously seen) and long-term and palliative care.

Many of the philosophical, social and legal debates around the individual's functionality and ability to think and perform certain tasks following a medical or mental problem focus on both etiology of the condition and the individual's personal action, that is, responsibility in terms of actions to take to prevent or treat the problem. The etiology of epilepsy is still unknown, although there have been many speculations based on scientific data and peer-reviewed research which appear to indicate a probable neurological cause, oftentimes following other medical problems located in the cerebral areas such as brain injury, tumors, infections, and stroke, as well as possible genetic factors, including birth defects. Aside from the more historical-spiritual/spiritist speculations (in any case absolutely confuted by modern science, in part due to the very scope and areas of investigation of evidence-based medicine) on possible demonic possession of once-called 'Sacred Disease', the issue of personal responsibility (leaving aside the free will problem for a moment) seems to have a clear answer: the person affected by epileptic seizure is not at fault (and we will further explore the complexity of personal responsibility, especially in the context of free will and lateralization) for either eliciting the problem or failing to prevent it; thus we can benefit from more research studies in those critical areas of medicine that deal with neural activity. In particular, neurocritical care or neurointensive care utilizes many theoretical and technological approaches to investigate these aspects. Studying the normal/regular/baseline function and the abnormal/mal-function of the nervous system, researchers develop cognitive, computational and developmental models of (expected) biochemical response or behavior in order to ameliorate the level of fidelity of artificial neural models and neurotechnological devices to mimic the natural neural networks. Focusing on complex aspects such as channel kinetics, synaptic transmission from the perspective of specific frameworks such as that of dynamic system theory, tri-dimensional network templates are created to combine theoretical investigations coded in mathematical equations to the actual (observed) behavior elicited. This approach can also be expanded via the modulatory application of brain-computer interfaces and virtual reality technologies to treat mental health disorders such as OCD as well as neurological disorders such as Parkinson's disease. Although technological advancements in this area represent an interesting development of applied medical science, standard neural interfaces are used by engineers as a solid base upon which are determined the structure and function of the artificial neural renderings. More specifically, research studies focus on the regeneration of damaged or lost neural tissue via the engineered versions of the mechanical properties of the nervous system itself. Creating artificial electronic circuits mimicking the tissue is a very important part of neural-interfaced structures that replicate the electro-chemical activity in the body and account for the possible challenges and rejection of artificial materials by the body. In this context, further research developments cover microelectrodes and optical neural interfaces with fiber optics, as well as complete microsystems to collect and modulate multiple signals and deliver them to the neural tissue. Researchers can therefore study the properties of these chemical, electrical, magnetic and optical signals, including production, transmission and (artificial vs. natural) delivery modalities, that is, aperture, amplitude, length, shape and population of spikes/action potentials. Focusing on both neuromodulation and neuroregeneration, the cutting-edge scientific research contributed to regulating the electrochemical activity to correct and re-balance the neural functions of specific brain areas without negatively affecting neighboring regions and/or causing undesirable side effects such as alteration in visuo-spatial perception, psychomotor agitation and tremor, localized or generalized pain, and many others.

Furthermore, enhancing the naturally occurring neurogenesis using the above-discussed techniques helps address problems of partial/limited function or relative/absolute malfunction by using artificial-engineered tissues (as in spinal cord tissue) or neural prostheses allowing for stimulation, regeneration, reactivation of damaged areas and recording neural activity to foster adjusted signal neurotransmission or stimulation—for instance, Functional electrical stimulation or FES to restore motor processes—through artificial sensors interfaced with natural neural tissue or through the interaction between extracellular matrix protein and neural stem cells. Of course, narrowing down to pre-op and post-op elements of neurointensive care and management, we find ourselves in the realm of neuroengineering, which we already discussed in Chap. 1.

4.2.3 *Neurosurgery and Neurology*

As the medical aspects of the direct application of neuroscience for the clinical treatment of patients are discussed throughout this volume from a theoretical framework, we would like to provide a very brief definition of the specialties neurosurgery and neurology. As a sub-branch of medicine, neurology involves the clinical study of the nervous system, more specifically investigating medical conditions, disorders and diseases involving the central and peripheral nervous system.

The main difference between neurology and neurosurgery is in fact the focus on diagnosis, prevention, prognosis and treatment of such disorders of the latter. In this context, the surgical methods and follow-up rehabilitation techniques are an integral part of neurosurgery, albeit based on the scientific investigations developed in neurology. A related specialty to neuroscience, neurology is therefore also a closer discipline to both neuropsychiatry and neuropsychology, with specific sets of clinical interests on Agnosia, Agraphesthesia, Agraphia, Alexia, Amnesia, Anosognosia, Aphasia (especially thanks to the studies by Broca and Wernicke), Apraxia, Aprosodia, ADHD and ASD, but also disorders such as Alzheimer's, Dementia, Dyslexia, Epilepsy, Psychosis, Stroke and TBI. Aside from the contributions of Alois Alzheimer, Karl Lashley, Arnold Pick and Norman Geschwind, other researchers such as Antonio Damasio contributed to further develop the discipline into behavioral neurology. Of course, among the most important research areas in these fields we need to mention neuroendoscopy and spinal surgery as more specialized subspecialties for neurological treatment, the surgical removal of tumors in the brain, the localization of new structures such as lymphatic vessels, the accuracy of techniques with frameless stereotaxy, and the beneficial effects of a more precise understanding of the neuroplastic activity to promote healing and growth processes.

4.2.4 *Paleoneurology*

A further combination of classical neurology, anthropology, evolutionary biology, and paleontology is represented by the field of paleoneurology. Historically, parallels between the external features of specific animal (especially human, as in Fig. 4.3) brains and skulls have been drawn to infer certain characteristics and traits, from the very early stages of precursors of modern psychiatry and psychology such as physiognomy, phrenology to very controversial claims of racial-based divide (read: racial superiority vs. inferiority) as pseudoscientific basis for practices such as the ones found in the eugenic movement. By sterilization, the medical community and the social-legal debate usually indicates the process and the medical technique in which the (human) reproductive organs are altered in order to terminate the ability to procreate, for example, in the strict biological sense, to produce offspring. A broader definition might focus on the following inability to reproduce, thus representing a method of birth control. More specifically, sterilization processes, which included surgical, pharmacological, and transluminal methods, can be voluntary or involuntary, therefore producing possible and unwanted (side) effects on the physical, psychological, familial-cultural, and social levels (Gillespie 2003). Sterilization can be used for therapeutic purposes, especially when the health of the patient is in danger (Fremgen 2012). However, aside from the specific medical problems that might affect the patient, the risk factors for related or other clinical issues might affect the possible off-

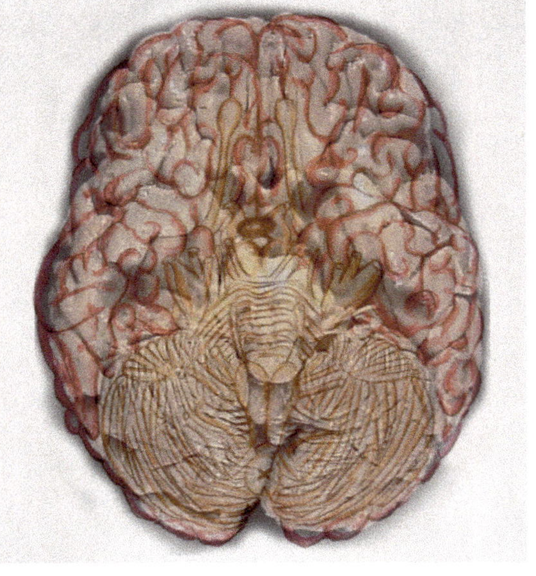

Fig. 4.3 A ventral, semitransparent and superimposed view of the brain showing inferior olives, flocculi, middle cerebellar peduncles, optic tract, nerve, and chiasm, olfactory bulb and corpus callosum, infundibulum, mammillary body, cerebellum, pyramid, and posterior perforated substance

spring. This is the case of sterilization that is 'therapeutic' only in the prospective sense, thus conducted in order to prevent the birth of children with possible genetic defects affecting physical and/or mental health disorders (Zite and Borrero 2011). These considerations, which are rightfully medical as well as philosophical and legal in their nature, are at the center of the ethical issues surrounding sterilization processes and birth control methods in general. To be sure, the focus is on the capability, personal responsibility and decision-making power of a human being or a series of human beings onto another or multiple human beings. The advances in medical sciences in this area come therefore with an enormous 'ethical baggage' that makes investigation, analysis and implementation of the related medical techniques very complex and complicated. Among the most evident cases, we certainly find the eugenic methods implemented especially in England, the United States, Germany, and Sweden to control certain (read: arbitrarily decided on racial/racist pseudoscientific claims on ethical, moral, intellectual, psychological, and physical superiority vs. inferiority) characteristics in human offspring (Tydén 2001; Galton 1883). Thus, the development of paleoneurology happened in multiple stages and steps, oftentimes through conflicting and opposed views of researchers like the early studies by Giambattista Della Porta, Thomas Browne, Franz Gall, Johan Spurzheim, to the research by Georges Cuvier and Étienne Geoffroy St. Hilaire (comparative anatomy), Ludwig and Tilly Edinger (comparative neurology), Louis Corman (father of morphopsychology), Ralph Holloway, Dean Falk, Veronika Kochetkova, Vladimír Blažek (general paleoneurology), Emiliano Bruner (Hominid Paleoneurobiology) as well as Miguel Burgaletac, Manuel Martin-Loechesb (cognitive paleoneurology) and Roger Sperry (neuropsychology). The underlying issue in paleoneurology is the connection between external, physical manifestations and internal, mental/psychological features, and one of the biggest challenges is the impossibility, aside from the insight provided by comparative methods and theoretical frameworks to determine specific functional elements of ancient and/or extinct species. From the perspective of neuroanatomical and structural/functional development, a very interesting aspect of the cranium is the fact that the growth process appears to stem from the development of the brain itself, rather than due to genetically transmitted sequence/guidance. In this context, paleoneurologists study the effect of the brain on the skull, more specifically the pressure or imprint (a process called 'imprintation' or 'endocast') it exerts on the innermost layer of the skull.

This process can also be replicated artificially—as opposed to the naturally occurring sedimentation through the cranial foramina—to support modeling on the basis of deductive/applied approaches and further developed by the implementation of computed tomography and computer algorithms/mathematical equations (to account for the differences between the original brain-skull combination and the endocast-based replica). Among the advantages of CT technologies in this area is the protection of the very structure of the fossils achieved by the non-invasive technology which allows for a very accurate mapping of the underlying structures—including general and area-related variations, cortical asymmetries, sulcal length, consistency, density, and endocranial capacity volume—and the possibility of virtual reconstruction of missing elements. These efforts contribute to a fundamental comparative challenge, that is, the investigation of possible relations and differentiations between animal species through morphometric analysis. The analysis of volume, structure and patterns is especially important for this comparison, as it provides more information on blood supply necessary for brain function (especially from cognitive and computational perspectives, with special regard to the middle meningeal system in humans) and relative lobes size (for differentiation and localization).

This research also comes with challenges such as the ones discussed previously and regarding the translation of data obtained via dioptograph-based measurements in actual-applicable information—from 3-D rendering to 2-D view—for different, extinct or living, species, as well as due to the conflicting points of view on the interpretation of these data. Differences in interpretation generally swing between more or less distance, in terms of anatomical-structural identification between human and non-human animals. For instance, some researchers believe that certain anatomical features are a sign of cerebral organization moving toward a more human pattern (in the case of Ralph Holloway and the endocast of *Australopithecus afarensis*), while others (in this example, Dean Falk) suggest that these patterns, more specifically the position of the depression, are indicators of the lunate sulcus, similarly to what is found in an ape-like sulcal pattern. Of course, some scientists believe that similarities between skulls from different species could provide proof of relations, which could be justifying similarities in cognitive-emotional functions as well. From this perspective, geometry, asymmetry and/or unevenness (petalia) between right and left hemispheres could indicate hemispherical specialization, which in turn could account for specific qualitative/quantitative differences in terms of emotion and behavior, as well as processing 'speed' in terms of computational-cognitive performance.

4.2.5 Neurophysiology

The combination of comparative studies found in paleoneurology and the analytical research in disciplines such as neuroanatomy and neurology contributes to the strength of neurophysiology as a scientific field interested in the functional properties of the nervous system, more specifically of glia, neurons and neural networks. Of course, neurology contributes to a further understanding of disorders from a medical-scientific perspective, while neurophysiology—aside from being, at least academically, not a sub-branch of (clinical) medicine—studies 'the whole spectrum' of neural function, including its physiology. Of course, further differentiation and specialization in areas such as clinical neurophysiology—which is part of medicine in countries like the United States, and a separate field in many European countries—truly bridge the two, artificially separated, perspectives. From the perspective of research tools, neurophysiology uses technologies similarly to the ones utilized by psychophysiology and psychological physiology. Therefore, molecular biology-based technologies are used in combination with multiple imaging techniques such as calcium imaging, electrophysiological recordings (including patch and voltage clamp) but also field potential and single-unit recording technologies, and optogenetics resources.

Of course, these techniques directly depend on the focus of specific areas of investigation within neurophysiology, namely, which system is primarily analyzed, given the main focus on the nervous system. For instance, aspects such as cognition, emotion and computation might be analyzed through comparison between areas, but the connection between (sub) structural anatomy and function will be more evident using electrophysiological as opposed to imaging modalities. Furthermore, if the focus is extended to clinical aspects, then diagnostic techniques might be used in order to provide the best possible testing method for certain disorders. For instance, we will see EEG utilized to monitor brain waves and compare their structure/frequency for the evaluation of possible CNS abnormalities such as seizures. Nerve conduction studies are instead used to investigate the PNS, and, in combination with EMG, to analyze muscles and nerves, while PSG is used for sleep studies. Moreover, combinations of electroencephalography, electromyography and evoked potentials are found in Intraoperative neurophysiological monitoring (IONM), for instance, in somatosensory evoked potentials (SSEPs), transcranial Doppler imaging (TCDI) and transcranial electrical motor evoked potentials (TCeMEP). Although these techniques and tests were developed in recent times, the history of neurophysiology dates back millennia, from the ancient Egyptian studies on medical properties of plants (for instance, the Edwin Smith papyrus) to the medical science found in ancient Greece, especially with Hippocrates or Erasistratus of Chios. These studies contributed to our modern understanding of processes in the nervous system and were the scientific basis upon which theories were built and observations verified by scholars such as Marinus and Al-Zahrawi. During the Middle Ages and the Renaissance, Italy was the center of scientific progress, especially in areas such as medical-clinical investigations. In the context of neurophysiology, aside from the previously mentioned Mondino de' Liuzzi and Leonardo da Vinci, it is important to mention the description of cerebrospinal fluid and syphilis by Niccolò Massa, the research at the Venetian College of Physicians or the very work by Fernelius, who was both the first physician-scientist to describe the spinal canal and to use the term physiology. Other famous names for the history of this discipline include Jason Pratensis, Giulio Cesare Aranzio, Johann Wepfer, Robert Burton, David Hartley and Vincenzo Malacarne. Of course, the old mind–body problem reappears in physiology, and alternate views on monism vs. dualism are part of the development of methods and techniques, from the clinical diagnoses and invention of tests to measure their intensity, to therapies to treat them. Furthermore, the analysis of specific brain areas provided more information on the functional properties of the nervous system, especially the CNS, for instance, the location of processes necessary to emotion, cognition or more specifically-localized activation, as in eye movement, for instance, in the basal ganglia (Fig. 4.4). Malacarne closely analyzed the cerebellum (an analysis fully implemented by Magendie in the nineteenth century), while Galvani observed the function of nerves and their reaction to electric impulses; Vicq-d'Azyr first and von Sömmerring after appropriately described the substantia nigra, while Burdach studied the subiculum and lateral and medial geniculate bodies, Legallois studied the medulla oblongata, and the Schwann cells got their name from the famous German physiologist.

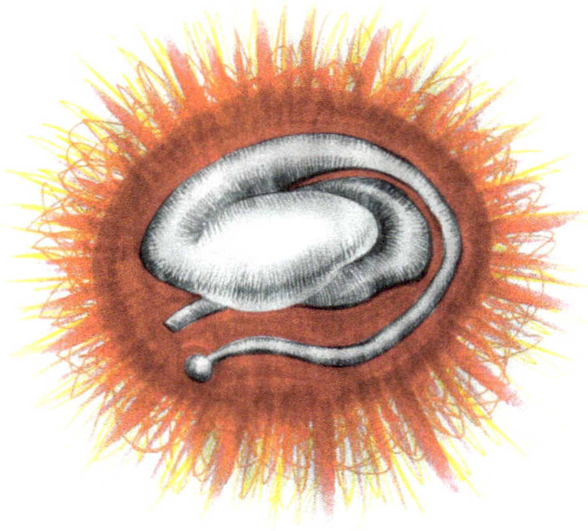

Fig. 4.4 Rendering of the basal ganglia with pallidum, striatum, substantia nigra and subthalamic nucleus

4.2.6 Psychoneuroimmunology, Psychobiology and Psychopharmacology

In a volume addressing critical aspects of the mind-brain-problem, it is fundamentally important to discuss the connection between disciplines involved with the study of the mind-psyche, the study of biology and the study of immunology. As psychobiology is a term often used as a synonym for behavioral neuroscience, we will refer the reader to our analysis of such discipline, as found in Chap. 2, *Between Psyche and Mind*. In any case, as psychobiology "describes the interaction between biological systems and behavior" (British Psychological Society 2015), the focus on the system-based (especially the nervous system) processes determines the mechanic basis of emotions-thoughts, and actions. A further elaboration of this point of view and methodological framework is represented by psycho-neuroimmunology (a term officially used in academia for the first time by the psychologist Robert Ader and the immunologist Nicholas Cohen, while studying conditioning and immunosuppression), which further develops the analysis on the connection between these elements. Psychoneuroimmunology moves the attention onto the etiological and diagnostic aspects of possible medical-mental syndromes and disorders (especially including allergies, hypersensitivities, intolerances, immune deficiency, and autoimmune diseases), and preventive measures to foster effective health strategies and well-being. Moreover, psychoneuroimmunology promotes a more comprehensive view of the effects which complex symptomatologies and processes such as the HPA axis and the stress response on psychophysical well-being. This is evidenced by the studies of Hans Selye on 'General Adaptation Syndrome' and George Solomon on 'psychoimmunology'. Thus, this discipline connects multiple systems, including the nervous, immune and endocrine systems as evidenced by the alternative names it is known by, that is, psychoendoneuroimmunology or psychoneuroendocrinoimmunology. For these reasons, the history of psychoneuroimmunology dates back to both 'holistic' approaches of traditional-ancient medical practices as well as, in the modern sense of evidence-based medical science, the studies by Claude Bernard and Louis Pasteur. In this sense, definitions such as the previously encountered 'pneuma', 'vis vitalis', 'milieu interieur' (internal state) and 'homeostasis', all define the focus and scope of analysis and intervention of psychoneuroimmunology, and define the viewpoint according to which health disorders truly describe disturbances, dislocation, disruption, perturbation of a (natural) order, which creates, monitors and promotes the regular/regulated functions of the body. In the case of the homoeostatic process of course, the final *stasis* in never achieved, as it would imply (mean) instant death, being the intention-tension, an ever functioning 'motion/movement towards balance' without ever fully achieving it. These balancing features are at the center of the connection between mind and body, and it is important to understand that, although some scholars would completely disagree with the existence of a dualistic separation of the two, and some would argue for a monistic perspective according to which only matter (in our case, the body, especially the CNS) exists, from the perspective of clinical interventions saying that mind and body influence one another is at least very useful and effective to promote health.

Psychoneuroimmunology, psychobiology and psychopharmacology are also among the primary strategies to understand empathy. In fact dopamine and serotonin, associated with the emotion center of the brain, which is in turn connected to the brain's reward center, are the primary neurotransmitters that modulate our biologically-based connection to the external world, in terms of praise and reward, but also in relation to stressors and triggers, and human interaction starting with the earliest stages of development. This obviously includes mother-infant connection and the fact that human babies thrive when there is empathy, having the most postnatal neuronal growth of any species (Marci et al. 2007), fundamental in the psychoneural development of attachment, which in itself is fundamental for our survival. From the perspective of neuroscience, several studies clearly indicate that the stress response is also mediated by these connections, as mothers directly impact their babies' cortisol levels by their reaction to internal and external stressors and related hormonal levels. More specifically, these reactions cause different responses in the modulatory ability of their children (Lyons-Ruth et al. 2013). In summary, psychoneuroimmunology, psychobiology and psychopharmacology research studies indicate that our physical health depends on our biology and its related activity on multiple levels, especially in the endocrine, immune and nervous systems. Thus, health depends on the factors that directly influence biology and is influenced by biology, as in our sense of connection-connectivity-connectedness with others. In the context of practical applications of neuroscientific research on mind–body medicine, Candace Pert was able to show that neuropeptides and neurotransmitters directly impact the immune system, and that this system and the endocrine systems are modulated by the entire central nervous system (therefore, not only the brain) and are connected deeper to the processes involved in emotions, cognition, self-awareness, etc. Another important research on neuro-immune interaction came from David Felten, while new perspectives on the Enteric (Intrinsic) Nervous System, the so-called Second Brain and the function of gastroenteric neurotransmitters from the studies conducted by McConalogue, Furness, Gershon, Lißina-Krause, Lychkova, Baldissera, Luczak and Porro. More specifically, connecting multiple systems, and working, for instance, on perspectives of immune alterations producing changes in behavior, and (vice versa) behaviorally induced changes to the immune system, directly help disciplines such as psychopharmacology to study the best medication strategies for the treatment of mental health disorders, fostering an understanding of "Protein binding (how available the medication is to the body), Half-life (how long the medication stays in the body), Polymorphic genes (genes which vary widely from person to person), and Drug-to-drug interactions (how medications affect one another)" (The American Society of Clinical Psychopharmacology 2015).

Of course, scientific efforts in psychopharmacology did not come without controversies, especially regarding the possibility of artificially induced behavioral aspects in subjects, and thereby changing their (self) awareness, perspectives, and, according to some, even conscience-consciousness. In this regard, we refer to the Vichian expression *coscienza non è conoscenza*.[1] This is true not

[1] Especially in regard to epistemological consideration of causal relationship.

only for the earliest stages of psychopharmacological research on psychotropic agents (especially barbiturates and opiates), but far into contemporary analysis and clinical use of antipsychotics and antidepressants (including selective serotonin reuptake inhibitors, serotonin–norepinephrine reuptake inhibitors, monoamine oxidase inhibitors, noradrenergic and specific serotonergic antidepressant, reversible monoamine oxidase A inhibitors, tetracyclic antidepressants tricyclic antidepressants), benzodiazepines and others.

The general umbrella for psychoactive drugs generally includes psychedelics, stimulants, depressants and opioids. More specifically, the term 'drug action' defines the interaction between the involved chemical substance and the relative receptor, while by 'drug effect' the scientific community indicates the whole series of functional changes, both physiological and psychological, in the subject to whom the drug had been administered, as seen in Table 4.1.

A very important thing to note when examining the connection between type and class of psychotropic medications and the corresponding neuroreceptors and neurotransmitters is that the primary factor is biomechanistic in nature. Thus, the very notion according to which there are specific psychiatric medications for specific psychiatric disorders is in itself both tautological and controversial, especially given some phenomenological manifestations related to the observation that "celestial, paradisean, and infernal visions are important components of the experiential spectrum of psychedelic inner journeys, near-death states, mystical experiences, as well as shamanic initiatory crises and other types of 'spiritual emergencies'" (Grof 2010). These factors represent truly fundamental cores of both philosophical investigation, especially in the heideggerian-husserlian sense of "inner experience" and the exploration of the self in the current neuroscientific paradigm. In the context of critical neuroscience *applied* to psychiatry, we can certainly notice how, on one side, "psychiatry is becoming increasingly neurologized" (Choudhury and Slaby 2016, p. 21) and, on the other side, it is victim of a reductionist process which makes it heartless-mindless, to the point that (a) it is at risk of missing its intrinsic 'iatric' aspect, and 'b' it is progressively and artificially, without a truly scientific basis for it, separated from its context. This context is not only (which doesn't mean that is in any way less important) historical or cultural, it is also 'meaning-defining'. In fact, here we are talking about the problem of definition, because *a* definition can be translated into diagnostic label and "greatly expand the potential use of this category, because the symptoms and treatment of many mental disorders include religious and spiritual aspects" (Scotton et al. p. 240). Another very important

Table 4.1 List of the most important neurotransmitters and neuroreceptors subdivided by type/class

Type / Class	Neurotransmitter / Neuroreceptor
○ Cholinergics (acetylcholine receptor agonists) ○ Muscarinic antagonists (acetylcholine receptor antagonists) ○ Nicotinic antagonists (acetylcholine receptor antagonists)	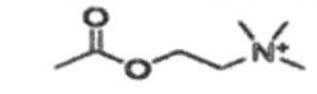 Acetylcholine
○ Adenosine receptor antagonists	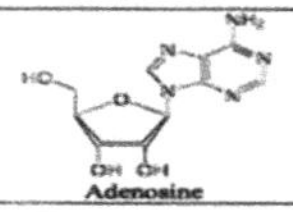 Adenosine
○ Dopamine receptor agonists ○ Dopamine receptor antagonists ○ Dopamine receptor partial agonists ○ Dopamine releasers ○ Dopamine reuptake inhibitors (DRIs)	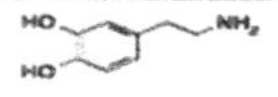 Dopamine
○ GABA receptor agonists ○ GABA receptor antagonists ○ GABA reuptake inhibitors	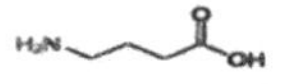 GABA (gamma-Aminobutyric acid)
○ Norepinephrine receptor agonists ○ Norepinephrine receptor antagonists ○ Norepinephrine releasers ○ Norepinephrine reuptake inhibitors	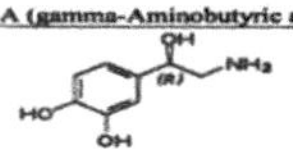 Norepinephrine
○ Serotonin receptor agonists ○ Serotonin receptor antagonists ○ Serotonin releasers ○ Serotonin reuptake inhibitors	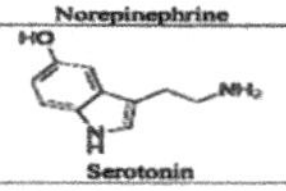 Serotonin
○ AMPA receptor antagonists ○ AMPA receptor positive allosteric modulators	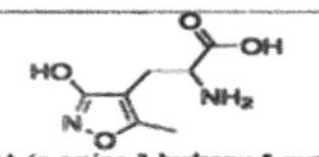 AMPA (α-amino-3-hydroxy-5-methyl-4-isoxazolepropionic acid) receptor
○ Anandamide reuptake inhibitors ○ Cannabinoid receptor agonists ○ Cannabinoid receptor inverse agonists ○ Cannabinoid receptor partial agonists	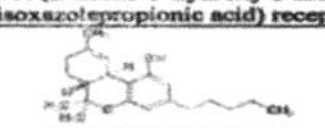 Cannabinoid receptor
○ MAFP, URB597, N-Arachidonylglycine	FAAH (Fatty acid amide hydrolase) enzyme inhibitors
○ Melanocortin receptor agonists	Melanocortin receptor
○ NMDA receptor antagonists	NMDA (N-methyl-D-aspartate) receptor
○ GHB receptor agonists	GHB (γ-hydroxybutyrate) receptor
○ Sigma-1 receptor agonists ○ Sigma-2 receptor agonists	Sigma receptor
○ κ-opioid receptor agonists ○ κ-opioid receptor antagonists ○ μ-opioid receptor agonists ○ μ-opioid receptor inverse agonists ○ μ-opioid receptor partial agonists ○ μ-opioid receptor antagonists	Opioid receptor
○ H_1 histamine receptor antagonists	Histamine receptor
○ Monoamine oxidase inhibitors (MAOIs)	Monoamine oxidase
○ Melatonin receptoragonists ○	Melatonin receptor
○ Imidazoline receptor agonists	Imidazoline receptor
○ Orexin receptor agonists ○ Orexin receptor antagonists	Orexin receptor

point to make in this area is that the aforementioned transcendental experiences transcend the individual and his vs. her 'self' when the individual is 'ready'. This concept is truly related to the neurophysiological processes underlying the phenomena observed in the context of 'Readiness potential' and a vast array of philosophical implications for free will (as we will observe in the following chapters) and to the experiential and experimental observation that "the individual's readiness for inner transformation plays a far more important role than the external stimuli" (Grof 2012). In any case, psychotropic medications at least in the Western culture and medical practice, follow the standardized model offered by the DSM manual, as presented in Table 4.2.

Table 4.2 List of the most commonly prescribed modern psychotropic medications with generic and US brand names

Antidepressants	*Antipsychotics*	*Stimulants*
Selective serotonin reuptake inhibitors, SSRI:	*Typical antipsychotics:*	Adderall (amphetamine, mixed salts)
Celexa (citalopram)	Haldol (haloperidol)	Concerta (methylphenidate, long acting)
Luvox (fluvoxamine)	Haldol Decanoate (long acting injectable)	Cylert (pemoline)
Paxil (paroxetine)	Loxitane (loxapine)	Dexedrine (dextroamphetamine), Dextrostat
Prozac (fluoxetine)	Mellaril (thioridazine)	Dexedrine Spansules (dextoamphetamine, long acting)
Zoloft (sertraline)	Moban (molindone)	Metadate (methylphenidate, long acting), Ritalin SR
Tricyclics:	Navane (thiothixene)	Ritalin (methylphenidate), Methylin
Anafranil (clomipramine)	Prolixin (fluphenazine)	
Asendin (amoxapine)	Serentil (mesorizadine)	
Elavil (amitriptyline)	Stelazine (trifluoperazine)	
Norpramin (desipramine)	Thorazine (chlorpromazine)	
Pamelor (nortriptyline)	Trilafon (perphenazine)	
Sinequan (doxepin)	*Atypical antipsychotics:*	
Surmontil (trimipramine)	Aibilfy (aripiprazole)	
Tofranil (imipramine)	Clozaril (clozapine)	
Vivactil (protriptyline)	Risperdal (risperidone)	
MAOIs:	Seroquel (quetiapine)	
Nardil (phenelezine)	Zyprexa (olanzapine)	
Parnate (tranylcypromine)		
Others:		
Effexor (venlafaxine)		
Desyrel (trazodone)		
Ludiomil (maprotiline)		
Parnate (tranylcypromine)		
Wellbutrin (bupropion)		
Zyban		

(*continued*)

Table 4.2 (continued)

Antidepressants	*Antipsychotics*	*Stimulants*
Mood stabilizers (Antimanic agents)	**Antianxiety drugs**	**Anti-obsessive agents**
Depakene (valproic acid, sodium divalproex)	Ativan (lorazepam)	Anafranil (clomipramine)
Depakote	BuSpar (buspirone)	Luvox (fluvoxamine)
Lamictal (lamotrigine)	Klonopin (clonazepam)	Paxil (paroxetine)
Lithium, (lithium carbonate)	Lexapro (escitalopram)	Prozac (fluoxetine)
Lithobid	Librium (chlordiazepoxide)	Zoloft (sertraline)
Tegretol (carbamazepine)	Serax (oxazepam)	
Topamax (topiramate)	Tranxene (clorazepate)	
	Valium (diazepam)	
	Xanax (alprazolam)	
Anti-panic agents		
Klonopin (clonazepam)		
Paxil (paroxetine)		
Xanax (alprazolam)		
Zoloft (sertraline)		

Certainly, culture plays a fundamental role in determining the indications for medications in every geographical area in the world, and that is true for both modern, laboratory-based pharmacology, as well as for natural remedies, folk medicine and ethnobotany-ethnopharmacology (although, to be fair, the very definition of 'ethno-' is in itself an ethnocentric separation similar to 'complementary and alternative medicine'). This also means that, once again, psychiatric disorders are very rarely individual disorders, and much more inter-individual, social, situated, cultural, societal disorders, and therefore require a further analysis of brain, culture and society, as we will see in the next chapter.

References and Further Readings[2]

Bibliography

Ail, P., and A. Venkatesan. 2013. *WebMD CEO Redmond leaving; company reports narrower loss*. Washington, DC: The Washington Post, May 7. Retrieved from: http://articles.chicagotribune.com/2013-05-07/lifestyle/sns-rt-us-webmd-resultsbre94610c-20130507_1_affordable-care-act-health-plan-revenue

American College of Rheumatology. 2015. *Fibromyalgia*. Atlanta: ACR. Retrieved from: http://www.rheumatology.org/I-Am-A/Patient-Caregiver/Diseases-Conditions/Fibromyalgia

American Fibromyalgia Syndrome Association. 2014. *What is Fibromyalgia?* Tucson: AFSA. Retrieved from: http://www.afsafund.org/fibromyalgia.html

American Society of Clinical Psychopharmacology. 2015. *What is psychopharmacology*. Brentwood: ASCP.

Barbui, C., and M. Tansella. 2008. Thirtieth birthday of the Italian psychiatric reform: Research for identifying its active ingredients is urgently needed. *Journal Epidemiology and Community Health* 62 (12): 1021.

Bernardy, K., N. Füber, P. Klose, et al. 2011. Efficacy of hypnosis/guided imagery in fibromyalgia syndrome – A systematic review and meta-analysis of controlled trials. *BMC Musculoskeletal Disorders* 12: 133.

Bostridge, M. 2008. *Florence nightingale, the woman and her legend*. New York: Viking Press.

British Psychological Society. 2015. *Psychobiology – Where mind meets matter*, The psychologist. Leicester: British Psychological Society.

Bunge, M. 2013. *Medical philosophy. Conceptual issues in medicine*. Singapore: World Scientific Publishing.

Busse, J.W., S. Ebrahim, G. Connell, E.A. Coomes, P. Bruno, K. Malik, and G.H. Guyatt. 2013. Systematic review and network meta-analysis of interventions for fibromyalgia: A protocol. *Systematic Reviews* 218. https://doi.org/10.1186/2046-4053-2-18.

Choudhury, S., and J. Slaby. 2016. *Critical neuroscience: A handbook of the social and cultural contexts of neuroscience*. New York: Wiley.

Clauw, D.J., L.M. Arnold, and B.H. McCarberg. 2011. The science of fibromyalgia. *Mayo Clinic Proceedings* 86 (9): 907–11. Retrieved from: https://www.ncbi.nlm.nih.gov/pmc/articles/PMC3258006/

Council for International Organizations of Medical Sciences. 2002. *International ethical guidelines for biomedical research involving human subjects*. Geneva: CIOMS.

Forsyth, J. 2011. Records of 4.9 million stolen from car in Texas data breach. *Reuters*. Retrieved from: http://www.reuters.com/article/2011/09/29/us-data-breach-texas-idUSTRE78S5JG20110929

Fowler, B. 2013. *Famous faces of fibromyalgia: Florence nightingale*. San Francisco: Healthline.

Fremgen, B. 2012. *Medical law and ethics*. 4th ed. Boston: Pearson.

Friedman, C.P. 2013. What informatics is and isn't. *Journal of the American Medical Informatics Association: JAMIA* 20 (2): 224–226.

Galton, F. 1883. *Inquiries into human faculty and its development*. London: Macmillan. New Edition 2003.

Genschorek, W. 1990. *Schwester Florence Nightingale*. Leipzig: Teubner Verlag.

Gillespie, R. 2003. Childfree and feminine: Understanding the gender identity of voluntary childless women. *Gender and Society* 17 (1): 122–136.

Grof, S. 2010. *Holotropic breathwork: A new approach to self-exploration and therapy*. New York: SUNY Press.

———. 2012. *Healing our deepest wounds: The Holotropic paradigm shift*. Newcastle: Stream of Experience Production.

Häuser, W., W. Erich, M. Herrmann, D.O. Nutzinger, M. Schiltenwolf, and P. Henningsen. 2009. Fibromyalgia syndrome: Classification, diagnosis, and treatment. *Deutsches Ärzteblatt International* 106 (23): 383–391.

Heffernan, V. 2011. *A prescription for fear*. New York: The New York Times. February 4. Retrieved from: http://www.nytimes.com/2011/02/06/magazine/06FOB-Medium-t.html?_r=0

Huntley, D., S. Lawrence, S. Ellis, S. Reeves, F. Inglis, J. Dorn, et al. 2014. *Information and articles about Florence Nightingale, a famous woman in British history*. London: British Heritage.

Keddy, B. 2008. *Fibromyalgia and Florence Nightingale. Women and fibromyalgia. Living with an invisible disease*. Lincoln: iUniverse publishers.

Li, Y.H., F.Y. Wang, C.Q. Feng, et al. 2014. Massage therapy for fibromyalgia: A systematic review and meta-analysis of randomized controlled trials. *PLoS One* 9 (2): e89304.

Lustgarten, S.D. 2016. New threats to client privacy. *Monitor on Psychology* 47 (1): 66–72. Washington, DC: American Psychological Association.

Lyons-Ruth, K., J.F. Bureau, M.A. Easterbrooks, I. Obsuth, K. Hennighausen, and L. Vulliez-Coady. 2013. Parsing the construct of maternal insensitivity: Distinct longitudinal pathways associated with early maternal withdrawal. *Attachment & Human Development* 15 (5–6): 562–582. https://doi.org/10.1080/14616734.2013.841051.

Mathews, I.Z., and C.M. McCormick. 2007. Female and male rats in late adolescence differ from adults in amphetamine-induced locomotor activity, but not in conditioned place preference for amphetamine. *Behavioural Pharmacology* 18: 641–650.

Marci, C.D., J. Ham, E. Moran, and S.P. Orr. 2007. Physiologic correlates of perceived therapist empathy and social-emotional process during psychotherapy. *The Journal of Nervous and Mental Disease* 195 (2): 103–111. https://doi.org/10.1097/01.nmd.0000253731.71025.fc.

McCance, K.L., and S.E. Huether. 2014. *Pathophysiology: The biological basis for disease in adults and children*. 7th ed. St. Louis: Mosby.

McDonald, C.R., M.E. Ahmadi, D.J. Hagler, E.S. Tecoma, V.J. Iragui, L. Gharapetian, A.M. Dale, and E. Halgren. 2015. Diffusion tensor imaging correlates of memory and language impairments in temporal lobe epilepsy. *Neurology* 71 (23): 1869–1876. PMID: 18946001.

McGonigle, D., and K. Mastrian. 2015. *Nursing informatics and the foundation of knowledge*. 3rd ed. Boston: Jones and Bartlett.

Mercola, J. 2010. *BUSTED: This popular "Independent" health website is deceiving you*. www.mercola.com. December 10. Retrieved from: http://articles.mercola.com/sites/articles/archive/2010/12/14/webmd-not-the-independent-health-source-you-expected.aspx

Mujkanović, L. 2016. *Nursing research studies*. Ethics, community, and healthcare. Hooksett: SNHU.

Müller, W., E.M. Schneider, and T. Stratz. 2007. The classification of fibromyalgia syndrome. *Rheumatology International* 27 (11): 1005–1010. https://doi.org/10.1007/s00296-007-0403-9. PMID 17653720.

National Audit Office. 2003. *International health comparisons: A compendium of published information on healthcare systems, the provision of health care and health achievement in 10 countries*. London: National Audit Office.

Nutton, V. 2009. *The Western medical tradition: 800 BC to AD 1800*. Cambridge, MA: Cambridge University Press.

Plantinga, A. 2015. *Knowledge and Christian belief*. Grand Rapids: Eerdmans.

Rabinowitz, T. 2015. *Telemedicine at the University of Vermont Medical Center*. VITL Summit.

Riley, W., S. Davis, K. Miller, R. Sweet, and H. Hansen. 2010. Improved patient safety: Near misses, errors and breaches. *Journal of Nursing Management* 17 (2): 247–255.

Rynasiewicz, R. 2012. Simultaneity, convention, and gauge freedom. *Studies in History and Philosophy of Science Part B: Studies in History and Philosophy of Modern Physics* 43 (2): 90–94.

Schweinhardt P., K.M. Sauro, and M.C. Bushnell. 2008. Fibromyalgia: A disorder of the brain? *Neuroscientist* 14 (5): 415–21. Retrieved from: https://www.ncbi.nlm.nih.gov/pubmed/1827031, https://doi.org/10.1177/107385840731252.

Simms, R.W., and D.L. Goldenberg. 1988. Symptoms mimicking neurologic disorders in fibromyalgia syndrome. *Journal of Rheumatology* 15 (8): 1271–1273. PMID 3184073.

Small, H. 1998. *Florence nightingale, avenging angel*. London: Constable.

Stahl, S.M. 2008. *Stahl's essential psychopharmacology: Neuroscientific basis and practical applications*. Cambridge: Cambridge University Press.

Szalavitz, M. 2017. *Unbroken brain. A revolutionary new way of understanding addiction*. New York: Picador Books.

Theadom, A., M. Cropley, H.E. Smith, et al. 2015. Mind and body therapy for fibromyalgia. *Cochrane Database of Systematic Reviews* 2015 (4): CD001980.

Tomasi, D.L. 2016. *Medical philosophy. Philosophical analysis of patient self-perception in diagnostics and therapy*. New York: Ibidem Verlag/Columbia University Press.

Tononi, G. 2015. *Phi. Un viaggio dal cervello all'anima*, Turin, I: Codice edizioni.

Toussaint, L.L., M.O. Whipple, L.L. Abboud, et al. 2012. A mind-body technique for symptoms related to fibromyalgia and chronic fatigue. *Explore* 8 (2): 92–98.

Tydén, M. 2001. *Från teori till praktik. De svenska steriliseringslagarna 1935–1975.* Stockholm: Acta Universitatis Stockholmiensis.

Utilization Review Accreditation Commission. 2001. *Accreditation status – WebMD health services group, Inc.* www.webmdhealth.com. Retrieved from: https://www.urac.org/directory/CompanyView.aspx?cid=C8700007DBE&aid=3367&Type=Accreditnet

Wang, C., C. Schmid, R. Rones, R. Kalish, J. Yinh, D. Goldenberg, et al. 2010. A randomized trial of tai chi for fibromyalgia. *New England Journal of Medicine* 363 (8): 743–754. https://doi.org/10.1056/NEJMoa0912611.

World Health Organization. 2010. *The world health report: Health systems financing: The path to universal coverage.* Geneva: World Health Organization (WHO).

Young, D.A.B. 1995. Florence Nightingale's fever. *British Medical Journal* 311 (12): 1697–1700.

Zhang, J., V.L. Patel, T.R. Johnson, and E.H. Shortliffe. 2004. A cognitive taxonomy of medical errors. *Journal of Biomedical Informatics* 37 (2004): 193–204.

Zite, N., and S. Borrero. 2011. Female sterilisation in the United States. *The European Journal of Contraception and Reproductive Health Care* 16: 336–340.

CHAPTER 5

Brain, Culture, Society

5.1 Context and Situation

Brain, Culture, Society are terms which we will find multiple times in our analysis, as they are at the center of the nature vs. nurture discussion. 'Community' is a term that has generally had a much deeper significance than 'society' in the sense that it carries forward, in a traditional sense, some values and principles which are closely linked to ethnicity and sense of belonging. Furthermore, culture also shapes societies and their members by way of relating, comparing and juxtaposing a multitude of frameworks, rituals, attitudes, worldviews and many other aspects which ultimately determine our sense of self. Of course, our sense of self is deeply connected with our identity, and when we discuss identity politics, we have to be extremely careful in *identifying* identity, in the sense of 'being identical to itself' (Lat. Idem, Gr. ταὐτότης). This is not sheer philological speculation; it really has to do with the level of theoretical justification we find applied to terms closely linked to identity, as in ethnicity, culture, tribe, family, community and even sexuality, the latter term especially important for descriptors such as sexual orientation, inclination and gender. In regard to the dichotomy heterosexual vs. homosexual, these terms were coined very recently, possibly by Karl-Maria Kertbeny in 1869. In fact, we could argue that the main element defining one's own sexuality had to do more with ritualistic, enacting (thus, 'passive' vs. 'active') and romantic aspects, rather than purely biological or social. In other words, a homosexual orientation was found pretty much everywhere in the Ancient

D. L. Tomasi, *Critical Neuroscience and Philosophy*,
https://doi.org/10.1007/978-3-030-35354-4_5

world, often with an aura of 'Sacred', as in its etymological value of 'separated from the rest, the norm, the common behavior'. It is truly interesting to note that this separation, this difference, was certainly not meant or understood in terms of superior or inferior, but more of 'appropriate for the setting', with all the possible implications that these terms would entail. Historically speaking, one could argue that the Judeo-Christian-Muslim cultural/religious influence played a fundamental role in the modern concept of sexual labeling of this type of human *orientation* (Fone 2000). However, equating modern monotheism (at least for the last 4500 years) or religion in general with homophobia or LGBTQ persecution would be a wrong assumption, as homosexual orientation and behavior were found from the romantic relationship between King David and Jonathan all the way to the Italian Renaissance and beyond (Dynes 1990). We should also remember that the Roman Catholic Church still contains many elements of Roman paganism, a culture in which homosexuality was accepted and open (although someone might argue that there is a wide gap between Roman Catholicism and American Catholicism in this sense), and the very adjective 'Romantic' stems from 'Rome' (Kuefler 2006). As much as there is plenty of historical evidence of homosexual behavior throughout human history, homophobic reactions are also found cross-culturally, pretty much the same way we find discussions of 'value' regarding skin color or religious affiliation. In the opening question, it is mentioned that St. Augustine in the fourth century AD may have been 'black', that is, Afro-Roman, and that it would not have occurred to the Romans to mention his 'ethnicity' based on skin color. To be sure, in this context, we refer to 'tradition' as a series of analyzed, debated and applied guidelines through centuries and millennia—we do not refer to 'traditional' as 'dogmatic'. In other words, the need that someone might experience in defining himself/herself as a separated/different from others on the base of religion, sexual orientation, ethnicity and so on might stem from the need of being accepted within a certain group and thus be defined by comparison to other groups. From a sociological standpoint, addressing the concept defined by Patricia Hill Collins as the "matrix of domination" or "matrix of oppression", we need to better define terms such as freedom of expression, power, control, inclusion and exclusion, discrimination and incrimination, difference and diversity, oppression and suppression, group and individual, personal and public, sex, gender, ethnicity, religion and race. Furthermore, we should evaluate the generalizability of the application of these definitions outside an American cultural environment. In Hill Collins' case, this concept identifies multiple issues related to domination/oppression within a specific sociological paradigm and classification. More specifically, the target of this paradigm was the understanding of the African-American Women's perspective, deeply embedded in black feminist analysis.

Furthermore, in the work by Hill Collins, we understand how there are several layers of deeply intertwined oppressive domains, specifically divided in the structural domain, which serves as superior structuralization of oppression and power; the disciplinary domain, which appears to manage oppression with the ultimate goal (often hidden) to support and sustain this very oppression; the hegemonic domain and the interpersonal domain. These layers represent multiple forms of powers which in turn apply multiple forms of domination on age, ethnicity, gender, race, sexuality and socioeconomic status. Perhaps, among the weakest elements of analysis in Patricia Hill Collins' matrix of domination, there seems to be a little confusion in the understanding and use/usage of the terms. More in detail, Hill Collins argues that:

> [...] These systems and the economic, political, and ideological conditions that support them may not be the most fundamental oppressions, and they certainly affect many more groups than Black women. Other people of color, Jews, the poor white women, and gays and lesbians have all had similar ideological justifications offered for their subordination. All categories of humans labeled Others have been equated to one another, to animals, and to nature. (Hill Collins 1990, p. 231)

Certainly, Hill Collins knows the difference between race, ethnicity, religion, gender, sexuality, and social class and does a wonderful job in describing the social implications of identity politics as a response to increase of sectorialized power and domination of specific subgroups, most specifically a certain majority, over minorities in general. However, we are not entirely sure whether this difference is somewhat forgotten in the attempt of evidencing the common features each of these categories have in the context of suffering a targeted and targeting discrimination within the matrix. Another conscious or subconscious misunderstanding or bias appears in statements such as "additive models of oppression are firmly rooted in the either/or dichotomous thinking of Eurocentric, masculinist thought."[1] The author seems to overlook one of the core issues in this analysis, namely the fact that there is not a 'single Europe' and the

[1] Ibid., p. 230.

fact—rooted in history and evidenced by genetic studies—the assumption of a single and/or single-minded 'European thought' within the context of domination and oppression is simply put, false. Furthermore, even from a purely neuroscientific–biogenetic point of view (which is the central framework of our analysis), we will have to focus more on the possible evidence—at this stage still very controversial—of possible neural mechanisms of differentiation for neural structures more prone to react to sociocultural elements of gender bias. A similar conversation can happen on theoretical grounds, for instance, the difference between Continental and Analytic philosophy, especially when addressing issues of socioeconomic status in relation to the specificity of (assumed and/or presumed) dominant cultures as in 'WASP' terminology (Kearney and Rainwater 1996). In conclusion and beyond these considerations, the greatest contribution of the sociological matrix of domination as active through the aforementioned axes of age, gender, sex, race and so on is showing that each and every one of us might fall prey of a dichotomy between an overlay and overlap of multiple dominant groups while at the same time being a member of multiple subordinate groups. In fact, sociological and psychological elements of feminism are important cross-culturally and even beyond gender issues. More specifically, there are strong ties between community psychology and feminism, in the sense that they "emphasize the importance of context for understanding behavior; both value social justice and diversity; and both adopt a critical stance toward much of traditional psychology". Thus, feminism represents a relatively modern and psychologically rich term to identify a broad range of intellectual, social-activist, and political discussions, debates and actions.

Based on all the premises evidenced above, it is evident how our identity is truly shaped by the identity of the other (person, community, culture and society) in front of me. However, there are certainly very few concepts that so profoundly shape our identity than the sense or thought of death. Our ideas about the ultimate moments of our life truly determine what we think our life is and should be *about*. It determines our meaning and purpose, our role in this world. This is part of a vast discourse we already discussed (Tomasi 2016), but in the context of Critical Neuroscience, we would like to shed some light on the current identity-related aspect of neuroscientific research, Extrasensory Perception (ESP) and Near-Death Experience (NDE). This is truly an incredibly vast field to analyze, and we will reconnect to this discussion at the end of this volume, to focus on 'Sense, Purpose, Meaning' as part of Chap. 6,

Philosophy as Basic Approach Toward Neuroscience. David Eagleman discussed the role of perception and faith (or lack thereof) in the description of a middle ground between traditional theism and activist atheism, which he and Robbie Parrish called Possibilianism. Although, in our opinion, this position is philosophically quite weak, in the sense that it is absolutely needed in the sociopolitical discourse to mediate extremists' views on both sides, it does not present any real philosophical groundbreaking ideas or theory. However, in the context of identity and (self) perception, the analysis by Eagleman can provide interesting elements, especially concerning our place in the world under the lenses of visuospatial perception, time analysis, synesthesia and what the neuroscientist defined as 'Sensory substitution' (Eagleman 2015), a neurotechnique which substitutes sensory inputs/stimuli from sound to touch, thereby feeding information to the cortical areas in the brain via the activation of neurons (Fig. 5.1).

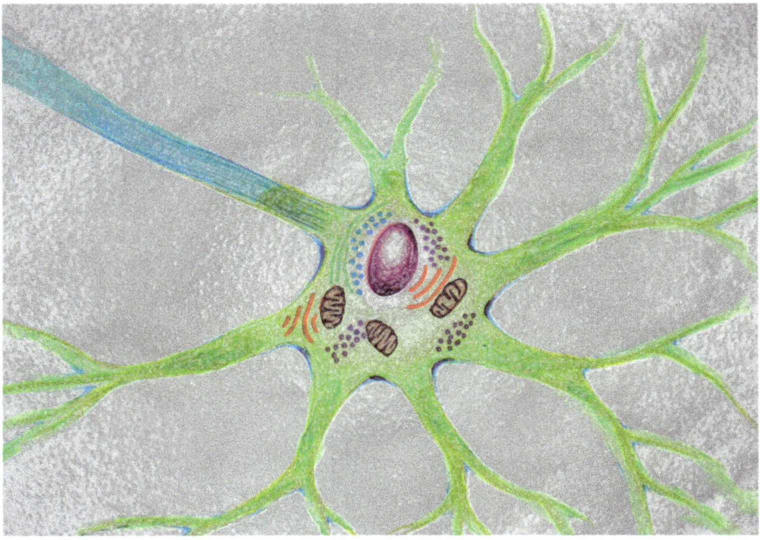

Fig. 5.1 Rendering of a neuron with nucleus (purple), mitochondria (beige), endoplasmic reticulum (light green), dendrites and dendritic branches (green/fluorescent green), microtubules (dark green), sections of the Golgi apparatus (orange/red) and of the Nissl bodies (dark red), polyribosomes (dark purple) and axon (light blue)

5.2 Mother Nature, Father Nurture

5.2.1 Medical Anthropology and Ethnomedicine

The terms ‘Medical Anthropology and Ethnomedicine’ have been used to identify two different aspects and at times two different approaches within the same investigation, namely the analysis of the connection between medical system(s) and the individual. Therefore, the ethnic, cultural and social dimension of the healing art, science and practice is understood as an integral part of the philosophical and psycho-sociological elements of the existence of (human) beings in the world. More specifically, medical anthropology studies the etiology, diagnostic approaches and treatment of disorders, the processes related to the development and maintenance of health, and the impact of healthcare systems. As such, it is certainly a branch of (applied) anthropology. Ethnomedicine, although the term has been at times used interchangeably with (or as a substitute for)[2] medical anthropology, indicates a specifically ethnological approach which analyzes, collects, categorizes and compares oral and written traditions of healing practices and methods on a local and intercultural, intersocial and international levels/layers. An important question in this context is how we can differentiate between traditional, classic, rural and/or folk healing practices, in the same way we separate, for instance, ‘scientific psychology’ from ‘folk psychology’ or even ‘pop psychology’. Answering this question is a central part of the research scope of this discipline, including the very (re)definitions, according to the classical scientific process of experimental trial and error, of ‘golden standards’ of medicine, including the hierarchy of evidence and evidence-based, double-blind, clinical case-control studies. These standards obviously define what can and should be considered ‘real’ medicine from other (complementary, alternative, integrative) types of medicine. Of note, one of the most problematic aspects in regard to these distinctions comes from the very diagnostic criteria used, especially in regard to symptomatological and psycho-physical (given the social-cultural aspects, even spiritual) presentation of the patient-subject. Very interestingly, definitions such as ‘ethnic disorders’ and ‘culture-bound

[2] This is particularly the case of countries such as Italy and Germany, where the discipline had received more attention and scientific development, as opposed to medical anthropology. Of note, terms such as ‘anthropology of health’, ‘anthropology of illness’ and ‘anthropology of medicine’ have also been used historically in this context.

syndromes' (such as the evil eye) have been generally assigned to 'primitive' medical diagnosis by Western medicine and psychology, without even recognizing that evidence-based biomedicine also promotes 'ethnocentric' views on what ought to be considered 'normal' vs. 'abnormal' or 'pathological', and constantly produces 'culture-bound syndromes' especially in regard to psychiatric disorder, which appear to grow 'by the minute' given the ever increasing size of the Diagnostic and Statistical Manual of Mental Disorders (DSM). Of course, the very concept of 'folk' and/or 'pop(ular)' as opposed to 'hospital/science-based' is a (by)product of post-enlightenment, post-industrial revolution, utilitarian, mechanistic-positivist views originating in the scientific-academic communities in Europe and the United States and the work by scholars such as Gonzalo Aguirre Beltrán, Jean Benoist, Claude Bernard, Gilles Bibeau, William Caudill, George Foster, Ronald Frankenberg, Byron Good, Gilbert Lewis, Abram Kardiner, Arthur Kleinman, Robert Levy, Eduardo Menéndez, W.H.R. Rivers, Tullio Seppilli and Andràs Zempleni. This approach also determined the final separation between the methods used by medical science and the religious practices, seen from this historical moment on as forms of 'magic'. Aside from the conceptual weight of this term, a similar attitude is nowadays found in psychological examinations of 'magical thinking', defined as 'cognitive distortion'.

In any case, evidence-based techniques in the modern conception of the term have been used to test the natural-based substances sued for healing purposes in different cultures and ethnic groups. The investigation of these drugs and their application (also in terms of clinical effectiveness) to modern pharmacological research studies and clinical use is the main scope of ethnopharmacy, more specifically the connection between the (cultural, social, ritual-istic) perception of these substances and their practical use. While it is true that related disciplines such as ethnopharmacology, which focuses mostly on the evaluation and application of plant derivatives; ethnopharmaceutics, the study of the traditional methods and techniques of preparation of pharmaceutical forms; and ethnobotany, which analyzes the production and delivery of pharmaceuticals through traditional knowledge are often viewed as separate fields, they all contribute to the multidisciplinary approach of ethnomedicine. As such, these study efforts analyze multiple aspects of the natural remedies/pharmacological interventions and help identify, label and categorize the natural substances (especially ethnotaxonomy) and discuss their application not only in terms of clinical use but also in regard to social policies and public health strategies and legislation.

5.2.2 *Cultural, Cross-cultural and Trans-cultural Psychiatry*

When talking about multidisciplinary fields, words like 'cultural', 'cross-cultural' and 'trans-cultural' play a great role in defining the relatively new approaches in the study of the mind/brain and are therefore applied to disciplines such as neuroscience, psychology and psychiatry. Of course, depending on which term we use, we will expect to study the interaction between (single or multiple, compared) culture(s) and neural activity. The concept of culture is itself a very complex one, related to cult(us) and cultivation (Lat. *colere*) at the individual and community-social sense (as in the Ciceronian 'cultivation of the soul'), which is something we "grow, worship, inhabit (determining and determined, as in the nature-nurture concepts of *habitat* and *habitus*), and are part of" at the same time. In the case of Habermas (1993), there is a first type of culture, which is understood from a general perspective and is not necessarily shared by all members of society, and a second type of culture, which is both common and political, and marked by a shared (and mutual) respect for rights. Thus, the previously discussed investigations of cognition, emotion and computation and the related combinations (human) perception/awareness, understanding/intelligence and behavior/communication are understood in terms of intercausal (mutual) relations, with specific 'underpinnings' in sight. This also implies that the methods and technologies utilized will be derived from the ones used in the fields this discipline attempts to bridge while bringing together different approaches (again, nature vs. nurture) through multiple and multilayered combination of quantitative and qualitative research methods from social and applied natural sciences, including observational, longitudinal studies, single-case and case-control studies, meta-analyses and so on, for instance neuroimaging techniques, genetic testing, as well as surveys and questionnaires. The scope here is again on the origin/production and delivery/transmission of specific cultural-social and genetic-biological (especially neural) traits in human beings and societies. Studies in this area have indicated that cognitive processes can be better understood under the framework on socio-cultural and anthropological analysis, for instance how artificial categories such as East vs. West—interestingly, a juxtaposition deemed true, especially within self-referred 'Western culture'—can be translated into different and often opposed ways of interpreting reality, for instance symbolic vs. diabolic (in the etymological sense) or 'the detail vs. the whole'.

What is interesting about transcultural neuroscience and psychiatry is that these differences are investigated with the goal of identifying neural correlates, that is, in order to see which brain areas are activated either in the sense of fully autonomous-independent activation of neural functions, such as involuntary (muscle) functions, as in the case of the medulla oblongata (Fig. 5.2) or depending on the ethno-cultural framework used, as in the studies by Blizinsky, Cacioppo, Chiao, Domínguez, Farah, Han, Hargittai, Iacoboni, Kirmayer, Nothoff, Seligman and others in the fields of neuroethics, neuroanthropology, cultural and social neuroscience psychology and psychiatry. In the case of cultural, cross-cultural and trans-cultural psychiatry, of course, the main attention is geared toward the social, ethnic and cultural context of mental disorders, and it therefore provides more information not only on the etiological and perceptual aspects of a specific disorder within the above-discussed framework, but

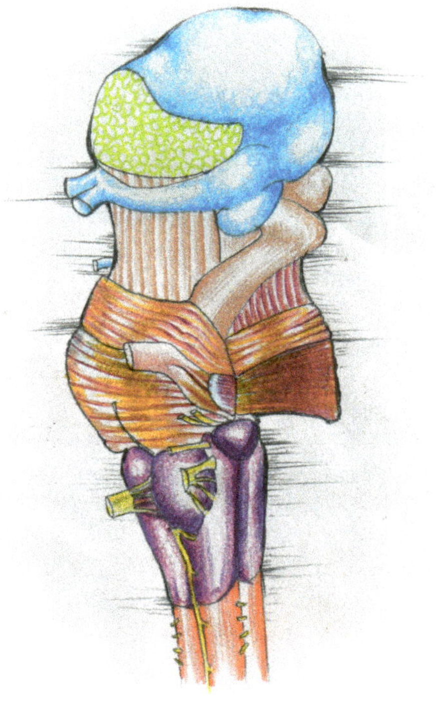

Fig. 5.2 A colored representation of the main elements of the brain stem, with the thalamus (light blue), the edge of cerebrum (green), the midbrain (light orange), the pons (bright orange), the cranial nerves (yellow), the edge of cerebellum (dark red), the medulla oblongata (purple) and the spinal cord (red)

also investigates the ratio/prevalence of such disorder within a single culture, via the comparison of multiple cultures, and what is even more interesting from the perspective of 'eradicated culture'. With this term, we refer to those populations that have been willingly voluntarily or forcibly removed from their society-nation of origin and are living as immigrants, refugees or in a state of exile and are thereby separated from their culture. Aside from anti-psychiatric considerations on the definition of psychiatry as a cultural (by)product, there are several research studies that shed light on the development of mental disorder in a causal relation not only with personal, subject-based abuse but also with ethno-social exploitation, as in the case of colonialism (military-political and cultural), conquest, occupation, war and genocide. Furthermore, the very development of 'DSM-based disorders' such as schizophrenia, schizoaffective disorder but also depression and other mental disorders appear to be linked to the process of alienation-estrangement from the culture of origin, thus increasing the ratio of these disorders within the aforementioned 'eradicated culture'. This is very well understood in psychoanalytic terms, especially from a Jungian perspective of 'collective unconscious' and 'archetype', where the subject can become a 'stranger in a strange-estranged-foreign-fiend land' as in *Entfremdung*. This aspect has obviously strong social and economic outcomes, as in the opposition to *Gattungswesen* in the sense coined by Feuerbach and Marx, but with clinical-anthropological reference to the 'new cross-cultural psychiatry' by Kleinman. Of course, trans-cultural means bridging multiple and diverse cultures, and each culture is made up by its members, so that we need first to understand how each member interacts with one another and then how groups are formed and in turn interact. We refer in this context on this member-subject becoming object, in the Levinasian relation between the 'I' and 'the Other'. There is in fact a movement from myself or the *personal I*, the *I-think*, the *me*, which is observed and observable, toward ('in confronto a') the Other. This is an *incontro-scontro*, an encounter-based experience, which is absolutely founding from the cognitive perspective of the inescapable and existential struggle. Levinas (1947) defines (this) Other as Other-from-me, and yet, it is and represents a/the 'face', another and the same, in the consequential relationship with the (conceptual and spiritual) *image/presentation/epiphany* of 'the (my) Neighbor' (Tomasi 2016). The neighbor is 'the other me', my correlative, my peer. I rely on *him* for support and meaning, thus presenting a divine dimension of the Other, which calls us to (personal) responsibility (toward, in a movement outside ourselves, an e-motion) and service to him who is calling us. This is thus the dimension of God, who does not act directly, but via the face of the Other; a God who/that "has implanted in us a spark of Reason and no matter how much we try to step on and extinguish it, that is impossible" (Dimitrova 2011) and commands/commits us to a search for justice inspired by charity based on this personal relation with the Other, a relationship which is divine in nature and cannot be destroyed. Of course, this relation happens in this world and in history, and a historical perspective in the Hegelian sense contributes to a better understanding of the concept hereby discussed, given the strong positive vs. negative influences on self-esteem, self-image and self-compassion through, for instance, imperialist-colonialist-cultural supremacist political propaganda, as evidenced in the studies by Beneduce, Carstairs, Cummings, Fanon, Fassin, Fried, Lin, Plotsky, Rechtman, Schechter, Seguin, Wittkower, Worthman as well by comparing other research contributions in parallel fields, for instance developmental and ecological psychology, for example, Roger Barker, Urie Bronfenbrenner, Howard Gardner, James Gibson, Edward Reed and Lawrence Shelton. On the other side, of course, a cultural, cross-cultural and trans-cultural Psychiatry which fully understands the social context of the relationship with the Other in time and history also needs to deal with the biological-matter part of the dichotomy nature-nurture. In this context, Hans Jürgen Eysenck stated in *Race, Intelligence and Education* (1971) that a variety of intellectual differences which we observe in our culture and much of the difference observed between certain racial groups depend to a great extent on genetic factors.

5.2.3 Cultural and Transcultural Neuroscience

Most of the conceptual elements discussed above can be applied to cultural and transcultural neuroscience, although in this case, the focus is obviously more on the neural underpinnings of culture and social interaction and—vice versa—to the cultural underpinnings of neural activity and behavior. Of course, given that the specificity of ἰατρεία is theoretically absent in this case, the attention covers both normal and abnormal functions. In fact, we could argue that such judgment on positive or negative aspects of functionality is absent from the scope of this type of research. However, given that it is quite difficult, 'to the point of missing the point', to avoid value-judgment when discussing culture, cultural and transcultural neuroscience target processes such as categorization, conceptualization, labeling and organization by linking them to social practices and neural factors. For instance, interesting research has been conducted on the visuo-spatial and object processing areas in the brain, to investigate whether the perception of details vs. broader views or direct vs. indirect gaze, especially between human subjects in a social environment could be linked to geo-cultural differences. The neuroscientific analysis thus covers, for instance, regions such as the bilateral middle temporal gyrus, the left superior parietal gyrus and the right superior temporal gyrus and compares their activities to processing and activation levels in limbic system. Of course, inferring that the above-mentioned areas are more or less activated in the execution of specific tasks such as watching people exchanging verbal conversations or simply gazing at a person's face directly implies the activation of more cognitive-interpretative areas and a deeper level of cognitive appraisal and judgment which is possibly the primary effector of the socio-cultural components (read: cognitive frameworks) at hand. An even stronger sociocultural component is found in the interaction between certain neurological functions and their connection with the experience of specific emotions and/or thought, particularly in relation to self-esteem, self-perception and lack thereof, as in some aspects of depressive states. Of course this represents the link between neuroscience and psychiatry, from a transcultural perspective.

5.2.4 Social Neuroscience

Similar views are shared by social neuroscience, a scientific discipline interested in the relation between biology and sociology, more specifically neuroscientific discoveries under the light of a sociologically based investigation of human behavior. In this case, the choice of methodology is truly a combination of quantitative and qualitative research at the crossroad of a neuro-bio-social science. Social neuroscience thus uses single studies and meta-analyses with the help of neuroimaging and other neurobiological technologies, for instance, including electrocardiograms, electromyograms, electroencephalography, endocrinology, facial electromyography, functional magnetic resonance imaging, galvanic skin response, magneto-encephalography, positron emission tomography and single-cell recording transcranial magnetic stimulation. Furthermore, social neuroscience obtains very useful information from therapeutic disciplines such as psychiatry and psychology, especially neuropsychology, particularly in the context of brain lesions and their outcomes to single-subject or multiple-subject behavior in a given society. Furthermore, social neuroscience draws theoretical assumptions and frameworks for debate from philosophy. As we have seen at the beginning of our analysis, human consciousness has been found on multiple levels of human existence and in different anatomical areas in the human body. Hegel, for instance, believed that nerves were communicating pattern of consciousness, allowing impulses to be delivered throughout the system, while the brain and spinal cord, which we in modern times defined as central nervous system, constitute "the immediate presence of self-consciousness, a presence self–contained, not an object and also not transient" (Hegel 1986). From the perspective of social neuroscience, the activity of this consciousness is to be found in the activation of this system in the social context, starting from the family, where the child develops the self in stages, including ethical stages, by leaving home and possibly marrying, In Hegel, marriage is where male and female, both antithetic, join in synthetic union and produce offspring to carry on the family estate. Where the female is the domain of Divine Law and the male is the domain of Human Law, the synthesis is a practical ethic which expands to include the whole community (Tomasi 2016). The connection between development, family, society and community is evident in the observed experience-dependent neuro-maturational processes which contribute to this maturation in childhood and adolescence. Moreover, the development of individual differences in processes underlying ethical and moral value judgments is connected to motivation, productivity, social engagement and social economy, where honesty, thrift and reliability are highly valued virtues already within the family. Consciousness, as well as self-consciousness, is in itself properly reason in an implicit form. However, when the object, the other-from-me, the character before consciousness is determined, consciousness possesses reason and that is why consciousness has reason in relation to the categorization of such character. Social neuroscience can therefore be very much helped by phenomenology in the Hegelian sense, as the scientific analysis used by this discipline is itself a phenomenon, linked via the correlation between *inner* and *outer* and the stages of thesis and antithesis to 'the life of the Spirit'. To be sure, the ultimate focus of this science is to show the *via* to absolute knowledge, which the spirit can attain by getting rid of its own alienation in the correlation/comparison between object and subject, through self-realization. However, for this Absolute Spirit to *be there*, this consciousness has to take different forms, and has to understand their forthcomings and remorph into a different shape in order to see itself, *als Ende und Anfang*, through the way of 'despair', the *Weg der Verzweiflung*, where the doubt is Zweifel and again represents 'two cases' (Zwei Fälle). These two cases are to be understood in the context of the principle of falsifiability and the scientific method at the basis of data analysis. A very useful presentation of the basic principles of data analysis and research methods in this sense is offered by the so-called Doctrine of Multilevel Analysis (Cacioppo and Berntson 1992), which is based on the following elements:

1. Multiple Antecedents (= preceding conditions): According to this principle, a target event on a structural level can have different triggers, either within this level or across multiple levels.
2. Non-additive determinism: The properties of the whole are not necessarily predictable from the properties of the constituents.
3. Reciprocal determinism: There is a reciprocal influence between biological and social factors with regard to the development of the target behavior.

While John Cacioppo and Gary Berntson are generally considered the founding pioneers of social neuroscience, Cacioppo also worked with Jean Decety to better define the best research approaches in this field, as evidenced in the 'golden triangle of human neurosciences' (Decety and Cacioppo 2010):

1. Behavioral data (i.e. response time, choice and judgment)
2. Physiological measurements (i.e. neuroimaging techniques providing correlative data of the human brain)
3. Animal-model and human experiments (including lesion studies, experimental pharmacology—for instance, testing oxytocin vs. placebo in social contexts—and transcranial magnetic stimulation): in this case the experimental investigations allow causal inference

The correlation between neural process and their localization in terms of neural mapping on one side and the creation and processing of social events is at the center of this triangle, more specifically because the combination of different approaches provides a much deeper and more accurate picture of what is actually happening in any of these processes. In other words, observed and verified correlations are in turn analyzed from a social perspective (including self-reports but also the Implicit Association

Test and others) and therefore require a scope-specific research design. The conceptual framework and main hypothesis underlying these perspectives is, of course, the view according to which human beings are 'social animals' and are therefore better understood in a social environment, rather on an individual-istic (isolated subjectivist) basis. Therefore, even from a mere reductionist utilitarian-behavioristic point of view, the development of organization and social structures beyond the level of individual subjectivity is an integral part of what it means of being human and is translated into multiple, and interconnected, even mutually influencing or merging into one another and often hierarchical entities. It is the case with dyads, families, small and big groups (from simple friendships and fellowships to very complex social groups such as the *Männerbünde*), all the way to ethnicities, societies, civilizations-cities (in the context we refer to the very concept of Civis) and cultures. The neurological underpinnings of these developments are found in the activity of multiple systems, especially the nervous and endocrine systems, but also involving metabolic and immune processes, which helped our species survive and reproduce.

5.2.5 *Neuroanthropology*

If social neuroscience investigated the relationship between society and the brain, the interaction between culture and the brain is the main focus of neuroanthropology. In this sense the two disciplines are closely connected to the point that the relative descriptors are often used as synonyms. Of course, in neuroanthropology the attention is even more focused on the historical processes, interpreted under the lens of general anthropology, that run parallel (some would argue 'cause' in an evolutionary biology-based outlook) to the evolution of brain functions mirroring social expectations. Thus, neuroanthropology applies the aforementioned research approach and integrates the investigative basis derived from social sciences with genetics and biology-based studies to promote ever changing theories such as biogenetic structuralism.

Once more, there is a parallel between neuronal location, activation and function, and the theoretical-philosophical analyses at the basis of our efforts to investigate ever improving therapeutic modalities. In the case of anthropology, the focus is on the location, activation and function of the subject in a social context. In the case of neuroanatomy, the focus becomes the true (biological) 'matter' of the locus of visual interpretation vs. (self) perception of the individual, that is, the human eye (Fig. 5.3). This has

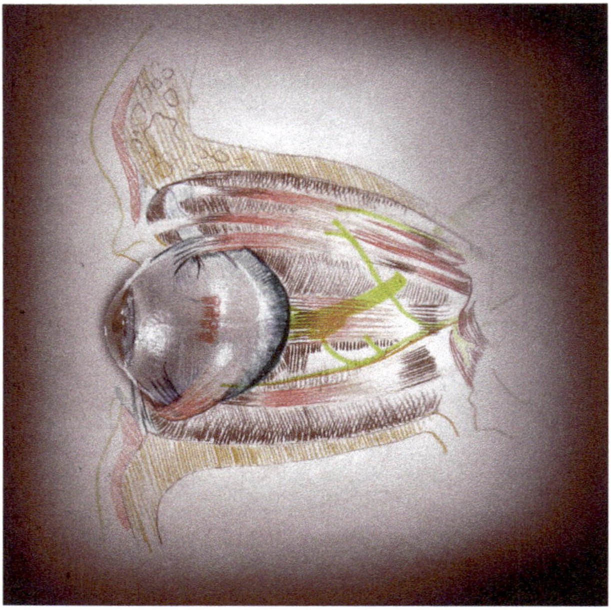

Fig. 5.3 Lateral view with the section of the eye including lateral rectus muscle, cranial nerves II, IV, V, VI and frontal, infraorbital, lacrimal, maxillary, nasociliary supraorbital, zygomatic nerves, as well as trigeminal ganglion

direct applications to therapy, for instance in the role that this type of visuospatial appreciation plays in training the human eye in the social context, with concept-associating/assigning and vertical vs./horizontal appreciation as examined by Doidge (2015).

5.2.6 *Sociobiology*

With social neuroscience and neuroanthropology, we investigated the relation between culture and society on one side and neural factors on the other. In sociobiology this general view holds true, in the sense that this discipline is the intersection between sociology and biology. However, the philosophical thought on the basis of sociobiology is completely different. This field originates in the 1970s as a follow-up, intended in negative terms as 'overruling and opposite' to the functionalism-dominated debate nature vs. nurture debate up to that point. Thus, in sociobiology (human) social behavior is a direct product of evolution, in the post-Darwinian sense, and as such very close to approaches found in evolutionary psychology (although psychology, by definition, is not limited to the study of behavior only, as in sociobiology) as well as Darwinian anthropology, and human behavioral ecology (the latter often used as a synonym for sociobiology). According to this framework, we can certainly see how the mind-brain problem could represent a false dichotomy, in the sense that the two terms would represent 'just' two different aspects (or perceptual views) of the same entity, that is, biological evolution on the level of neural activity producing social processes. Of course, as with the general views in evolutionary theories, the genetic evolution of such biological, especially neurological structures promoted advantageous social behavior. In this context, terms such as gene/meme (the latter interpreted as 'cultural replicator') and behavior, genotype and phenotype are understood as sense/direction-creating conceptual frameworks directly following natural processes, natural evolution and natural selection. Sociobiology contributed to the creation of a synthesis of the whole social behavior understood as the activity of a genotype directly influencing the phenotype on the basis of nature and in an adaptive sense. More specifically, given that, according to sociobiologists (even) social behavior is a phenotype, it is in its very nature linked to the interaction between environment (which creates a form of 'pressure', interpreted as selective social expectations on the individual or the group) and genotype and it therefore follows natural selection in the neo-Darwinian sense.

From the perspective of theoretical approach and research method, the common practice in modern sociobiology is to isolate a certain social behavior, and develop a related hypothesis on the basis of evolutionary conceptual framework, that is, a reliable strategy (as in the "interpretation of gene sequence/frequency") that would be juxtaposed to the observed behavior and perfectly match it (as in 'expressed in a given population'). Of course, the more moderate (in the philosophical sense of 'not fully deterministic') views within sociobiology identify the existence of certain behaviors as only partially inherited via natural selection and account for the major role that other forms of 'history' (i.e. not limited to natural history, where by natural we only refer to biologically related, matter based) and parallel and/or subordinate processes have in the development of (human) behavior. Although the most important founding fathers of sociobiology are possibly John Paul Scott and Edward Wilson, the development of this field in science and academia happened via precursors such as Richard Alexander, William Hamilton and Nikolaas Tinbergen, as well as through an animated debate between the thesis presented in "Sociobiology: The New Synthesis" (Wilson 1975), the criticism by scholars like Richard Lewontin and Stephen Jay Gould and others. Of note, some criticism is not necessarily related to the negative judgment of specific methodological flaws, but on the theoretical-philosophical assumptions used as premises for sociobiological analysis. Certainly some perspectives denote a truly sociological or political background (especially in regard to Marxism and Anarchy), and a response-debate on such perspectives comes from within the field of sociobiology, as in the case of Anne Campbell and Frans de Waal or through the support of specific sociobiological theses, whether in full or in part, by scholars from fields related but external to sociobiology, for instance, Steven Pinker and Noam Chomsky. In some cases, sociobiology has been criticized as close to reductionist, deterministic, or utilitarian-mechanistic perspectives and falling short of providing evidence, in a purely biological sense, for some of the claims made in regard to the direct connection, in terms of causal effect, between gene transmission and behavior or specific aspects of animal—especially human—nature, for instance, intelligence and creativity. Furthermore, some (including Gould) have argued that specific traits in human behavior could be literally passed on via cultural influence, and other perspectives include a broader (or deeper) definition of culture, that is, the effect of Tradition on (human) action and ideation. Sociobiologists reply that natural selection plays a much bigger role in the development of all these aspects, and that, even if some behaviors are only in part inherited through genetic transmission, behavioral traits (including complex descriptors such as altruism, but also in regard to concepts like the relation between 'ecological balance' and expansion/progression of combined temperamental traits) can be passed on to the next generation following the same rules found in general biology, such as reproduction and preservation of a specific gene in the population according to evolutionary fitness and adaptation—in this sense the perspective of psychobiology/biopsychology is very similar.

Of course, an important element of 'disruptive deconstruction', literally, is represented by a scientific process which is challenging to all we just discussed, in terms of nature vs. nurture. More specifically, what would happen if we could decide how to 'nurture nature'? We are obviously referring to the discoveries and application of the genome editing technology known as CRISPR/Cas9, first by the team of the Japanese molecular biologist Yoshizumi Ishino and very recently by the laboratory of Shoukhrat Mitalipov, a Kazakh biologist, currently director of the Center for Embryonic Cell and Gene Therapy at Oregon Health and Science University. At the center of these discoveries is a prokaryotic immune system whose main characteristic is the ability to provide acquired immunity via resistance to foreign genetic elements. This system is called Clustered Regularly Interspaced Short Palindromic Repeats or 'CRISPR' and represents repetitive palindromic patterns of prokaryotic DNA, more specifically cascade complex ('Cas', bound to single-stranded Deoxyribonucleic acid) genes followed by a leader sequence and finally by a repeat-spacer array, which allows these proteins to identify, isolate and literally 'cut off' exogenous DNA. In the latest version CRISPR/Cas9, scientists have finally reached the ability to 'genetically intervene' in the modification of cellular structures for the purpose of 'correcting' pathogenic gene mutations, for instance at the level of human embryos, or to 'repair' or 'edit' specific genetic sequences through the insertion of other DNA patterns. In the context of the nature vs. nurture debate, it is extremely important to understand that at this point of our technological development we need to make sure, once again that the totality of our scientific development, that is, *inclusive* of philosophical, especially ethical analysis, matches those discoveries that will literally allow us to 'play' (and we mean the most comprehensive acceptation of the term) with our "building (and *self*-building, with an added layer of complexity on multiple levels) blocks" in ways difficult to 'imagine physically possible' decades ago.

References and Further Readings[3]

Bibliography

Cacioppo, J.T., and G.G. Berntson. 1992. Social psychological contributions to the decade of the brain: Doctrine of multilevel analysis. *American Psychologist* 47: 1019–1028.

Decety, J., and T.T. Cacioppo. 2010. Frontiers in human neuroscience, the golden triangle, and beyond. *Perspectives on Psychological Science* 5: 767–771.

Dimitrova, M. 2011. *In Levinas' trace*. Newcastle upon Tyne: Cambridge Scholars Publishing.
Doidge, N. 2015. *The brain that changes itself*. New York: Penguin Books.
Dynes, W.E., ed. 1990. *Encyclopaedia of homosexuality*. New York: Garland.
Eagleman, D. 2015. *The brain: The story of you*. Edinburgh: Canongate.
Fone, B.R.S. 2000. *Homophobia: A history*. New York: Metropolitan Books.
Habermas, J. 1993. *Justification and application*. Trans. C.P. Cronin. Cambridge: MIT Press.
Hegel, G.W.F. 1986. *Phänomenologie des Geistes*. I ed. Suhrkamp: Verlag Frankfurt am Main.
Hill Collins, P. 1990. *Black feminist thought: Knowledge, consciousness, and the politics of empowerment*. Boston: Unwin Hyman.
Kearney, R., and M. Rainwater. 1996. *The continental philosophy reader*. London/New York: Routledge.
Kuefler, M. 2006. *The Boswell thesis: Essays on Christianity, social tolerance, and homosexuality*. Chicago: University of Chicago Press.
Levinas, E. 1947. *Time and the other*. Pittsburg: Duquesne University Press. English Trans: Richard A, Cohen, 1987.
Tomasi, D.L. 2016. *Medical philosophy. Philosophical analysis of patient self-perception in diagnostics and therapy*. New York: Ibidem Verlag/Columbia University Press.
Wilson, E.O. 1975. Dialogue. The response: Academic vigilantism and the political significance of sociobiology. *Bioscience* 26 (3): 183–190.

CHAPTER 6

Perception and Cognition

6.1 The Mind–Body Problem: A General Overview

Our analysis contains "a scientific re-examination of the mind-body problem". Let us begin this chapter with a disclaimer. We do not claim to be able to cover every possible research study, philosophical position or author, or psychological and neuroscientific method that has dealt with this very complex problem throughout history. For a thorough, albeit by no means complete, list of fundamental publications on this theme, we refer to the bibliography at the end of this chapter. In this chapter we will attempt to extrapolate the most important elements from each position and present them in a clear fashion, to set the stage for the explanation of our personal opinion and philosophical position in this matter. Furthermore, we also want to stress that our analysis is "philosophical in the scientific sense". What we mean by that is that, as we claimed in a previous work (Tomasi 2016), "Philosophy is the mother of all sciences" and thus already contains scientific examinations by definition. Our re-examination is scientific because it utilizes the commonly accepted classical form of scientific method, declined into the research studies presented in all the neuroscience-based fields we discussed, but it looks at these studies *from above* by way of applying a thoroughly philosophical method to understand its science.

As a starting point, we will quote here the work by John Eccles, the famous Australian neurophysiologist who thoroughly analyzed the mind–body problem, starting from considerations first presented by Karl Popper,

D. L. Tomasi, *Critical Neuroscience and Philosophy*,
https://doi.org/10.1007/978-3-030-35354-4_6

which we will encounter again below. Although Eccles does embrace many aspects of the tripartite worldview by Popper, his founding analysis of the relation between mind and body is generally considered to be truly dualistic, with a pre-eminence of the mind-self-soul which 'controls' or 'governs' the brain–body. In his case, we can certainly talk about a scientific re-examination of this connection (as well as on the problem of Free Will), as it proposes a new outlook on tri/dualist theories without sacrificing the scientific premise of energy conservation. To summarize his main theses, developed in collaboration with Friedrich Beck, we can identify the concept of a mind-triggered movement that acts upon the brain via the activation of cortical dendrons, more specifically dendrons in the six laminae of the cerebral cortex, which are in turn connected to *psychons*. The latter is a term coined to define a 'unitary conscious experience' that elicits an electrochemical (neural) response under the laws of quantum mechanics. This is truly connected with the concept of free will because this activation of dendrons *by* psychons happens in "willed actions and thoughts" increases the probability of action potentials in specific (thereby selected, i.e. by the mind) neurons through quantum tunneling effect in synaptic exocytosis. Of note, John Eccles also theorizes that the opposite of what we just stated, for example, a reverse process, is in place when we are dealing with perception, instead of willed cognition. Some similarities, not necessarily in the quantum mechanics-based explanation of the neural processes thereby involved, but on the primacy of mind, are found in the work by Vincent Torley (2013).

The mind–body problem can be better understood by realizing that Western medicine, in particular psychiatry and psychology is not only ethnocentric (as we previously discussed), but is also "cognicentric" or "pragmacentric" (Grof 1998), and it is focused on very different, albeit not entirely separated, layers of communicative aspects of delivery of such understanding. To provide a useful example in this context, Vygotsky (1962) offers a similar view when he writes that "Inner speech is speech for oneself; external speech is speech for others". The question is therefore on a possible unified theory of consciousness. The analysis by Vladimir Lossky in "The Mystical Theology" (1944), focuses on the relationship between gnosis and theosis, In this volume, N.O. Lossky writes about V. Lossky, pointing out that:

> [There is a] distinction between the ineffable divine essence and the inaccessible nature of the Holy Trinity, on the one hand, and the positive revelation

> of the Trinitarian energies, on the other. "When we speak of the Trinity in itself," said Lossky, "we are confessing, in our poor and always defective human language, the mode of existence of the Father, Son, and Holy Spirit, one sole God who cannot but be Trinity, because He is the living God of Revelation, Who, though unknowable, has made Himself known, through the incarnation of the Son, to all who have received the Holy Spirit, Who proceeds from the Father and is sent into the world in the name of the incarnate Son".[1]

This discussion moves far beyond rational-mechanistic approaches based on empirical experimentation, especially given that, in Lossky's view (and in classical Christian orthodoxy) theosis is above (human) knowledge. This perspective goes against the commonly accepted view in modern science, according to which research follows an artificially pre-constructed idea about a process, and then it tries to find out whether its existence is warranted, via experimental processes. The opposite process happens in both mysticism and postmodern approach, where "one understands something and then tries to get an idea about it" (Schaef 1992). But how can we relate to such deep spiritual and religious worldviews and still maintain the scientific mind when analyzing our nervous system? As we have previously seen, pyramidal neurons are involved in electrochemical activity, at a neocortical level, of synchronized action potentials which in turn are responsible for the memory-based foundations of consciousness. Thus, from this perspective, there are specific mechanisms involved in the creation, maintenance and development of consciousness, in the sense that we are conscious of explicit memories of external and internal phenomena. However, this relation between what is inside and what is outside is quite difficult to understand, especially because we are "both the inside observer and the outside observed", as well as all the combinations therein. If we took the Cartesian dualistic view, we would maintain that *both* mind and body (brain) are *separate*, yet united; diverse, but correlated; distinct types of (the same?) substance. To shed light on the essence of the substance and the matter-at-hand, we could also say that, according to Descartes, our brain is part of a spatially extended body, thus substance, while our minds are unextended. In any case, the relationship can only work if and only if there is (a) a clear separation between (these two) *things* and (b) a certain level of communication between them. Also, a series of problems

[1] As quoted in (Tomasi 2016).

might arise when we are attempting to identify the "hard problem of consciousness" on the level of subjective perceptions, feelings, emotions, thoughts, actions, reactions, and behaviors, and their neural correlates, objectively understood as mechanistically equivalent in every subject given the same external/internal conditions, starting from gross neuroanatomy, down to neurotransmitters and neuroreceptors, brain function and activity, including action potentials, and internal–external environment, upbringing, social circumstances and so on.

Furthermore, even when all these parts can be appropriately combined to give rise to a whole of objective/observable-turned-to-subjective apparatus, we still couldn't be 100% sure that neuronal correlates are actually causing-creating (our and subjective) consciousness. This is also because theoretically intended, we will have to identify product/agency-specific properties underlying the whole process and creating sequential follow-ups, at least on an algorithmic basis. At this stage the enormous complexity of the 'first-person experience' is truly difficult to replicate and understand from the outside, although some very convincing theoretical frameworks such as IIT appear to have provided very good starting points. In any case, if we took as a hypothetical assumption that the existence of the souls has warrant for belief and would possibly fit some parameters of proof according to the falsifiability model, we would have to justify the very movement/trajectory/target of this soul to/into/through the body and its effective-affective ability to "move from within" thus creating emotions, and also perform cognitive tasks, voluntary acts and understand and be understood by memories. We would therefore understand the substantial non-substance or para (as in 'beyond') substantiality of the soul in phenomenological terms first. This movement of the souls (Descartes talks about the pineal gland literally making the spirits move the limbs, Fig. 6.1) would thus be equated to a propelling propeller, possibly in turn propagated by another type of substance or essence, perhaps a higher-level impulse. This task would thus require another substance, matter or surface to act upon and by which and/or virtue of which to be 'felt' 'activated'.

This mental activation could however be of a non-'push/pull' type and be therefore not well explainable according to the mechanistic view, as causal forms could be very well happening according to other models, especially given the current approaches in physics (Robb and Heil 2009). That said, the matter-part of the mind–body problem still plays a fundamental role, as we have seen, in perspectives close to embodied cognition, especially in the thought and computation level analysis. In this sense,

Fig. 6.1 If the Cartesian analysis of the pineal gland did not survive modern neuroscientific research data, the pituitary gland is still referred to as the 'master gland' for its fundamental role in hormonal regulation at the intersection of endocrine and nervous systems. In this image, the pituitary gland is seen with anterior and posterior pituitary (in red and beige, respectively), optic chiasma (light orange), hypothalamic neurons in the paraventral nuclei (light blue), axons and neurons in the ventral hypothalamus and hypothalamic–hypophyseal tract through the infundibulum (aquamarine), hypothalamic neurons in the supraoptic nuclei (fuchsia) and venules (blue) for ACTH/FSH/GH/LH/PRL/TSH and Oxytocin/ADH in the anterior pituitary and posterior pituitary, respectively

simple algorithmic modeling or theoretical (even of the psychological type) abstraction no longer suffices for a deeper understanding of the role that the context, in terms of our bodiliness, plays in the development of our consciousness. This is not only an attempt to overcome the century-old dualistic view, but also to better integrate the underlying meaning of our thoughts, feelings and behaviors with the "world we live in".

More specifically, this world includes (and is included by, at level of perception) our body and thus we have to talk about 'integration' and not 'correlation', because those two sides of the equations cannot be separate(d). Furthermore, one difficulty in this area is represented by the multifaceted neural networks involved in any process, so that some artificially induced response, let us say via transcranial magnetic stimulation for instance, still relies on (by definition) observed-based or organic vs. artificial modeling, and the required manipulations therein are thus themselves subjected to further human-based processes, even when delivered via (human or self-programmed[2]) artificial intelligence-based technologies. Even when combined with genomic analysis or structural modifications, these efforts:

(a) Do not seem to solve the brain–body problem into a reliable monistic (in the materialistic–mechanistic reductionist–physicalist sense) model.
(b) Do not provide appropriate and reliable predictive framework, as necessary outcomes for this type of scientific inquiry.

We could certainly point out that, as with other areas of philosophical debate, consciousness cannot be fully understood without a preliminary, albeit not necessarily a priori, agreement on what the definition of the term entails. As there are possibly as many definitions of consciousness as there are perceptions of consciousness, we would start by delineating the main difference between the biological-mechanism/process related aspects as opposed to the content of consciousness. In the first case we would obviously analyze the neural activity in the nervous system, but also investigate its relations, in terms of mental state fluctuations and homeostatic processes, with other systems, most especially the endocrine, the immune and the cardiocirculatory systems. The second case is even more complex, as it involves both conscious states and consciousness *of* such conscious states, thereby opening up the discussion on the conceptualization of self, identity and personality in context. From the perspective of experimental observation however, we could claim that states of consciousness can still be investigated in a quantitative way, by associating levels, from high to low, of arousal and activation, perception vs.

[2] This "self" would of course account for a "second-level" human process as well, given the nature of the primary (chronologically and causally intended) programmer.

proprioception, wakefulness, vigilance and attention. We can also examine sleep patterns via wave analysis, REM states, circadian rhythm and external influences from pharmacological interventions—especially psychotropic medications—to psychotherapy and integrative/holistic approaches such as meditation, relaxation and mindfulness to parallel/alternate/impaired states such as general anesthesia and persistent vegetative state, multiple-level comatose states, locked-in-syndrome and multiple minimally conscious states especially given the neural activation of specific brain areas (Fig. 6.2) in the context of this type of therapeutic modalities.

As we previously mentioned, the quantifiable parameters here defined come as a total absence of consciousness, and persistent vegetative state as

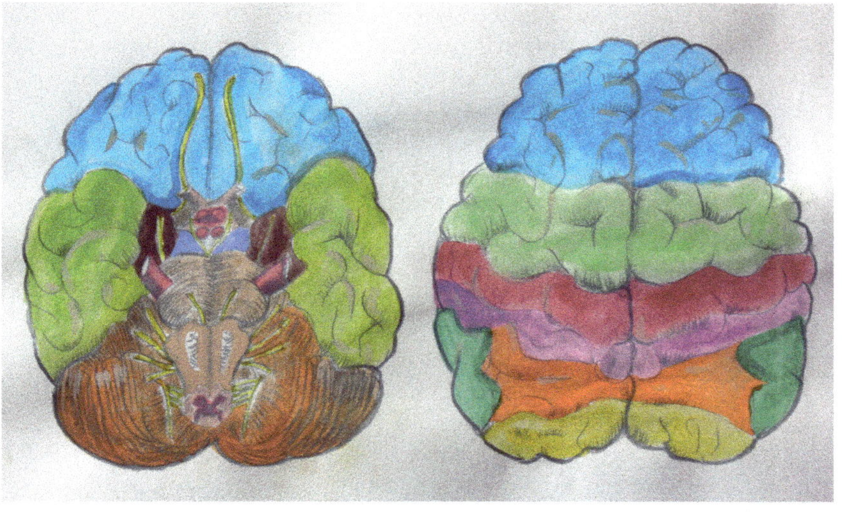

Fig. 6.2 Ventral and dorsal views of the human brain on the left and right side, respectively. The colors on the ventral view indicate the frontal lobe with higher mental functions (blue), the association area (in light green), the olfactory area (dark purple) and the motor functions in the cerebellum (orange), together with the brain stem (pink). The colors on the dorsal view indicate the frontal lobe with higher mental functions (blue), motor function areas, more specifically eye movement and orientation (light green) and initiation of voluntary muscles (dark red), followed by the sensory area (purple), somatosensory association area (orange), and Wernicke's area (dark green). Finally, in yellow, the visual area in the occipital lobe is indicated

a minimally conscious and swinging-fluctuating state and so on. Of course, really anything in the external/internal environment can have an effect on thereby-quantified consciousness (especially given its meaning of 'shared knowledge' as we have seen at the very beginning of our investigation), including the consciousness of others (or other beings-entities-things), and we can use this type of stimuli to elicit a response, even under a behavioral-conditioning scope, to study consciousness. We will return to this topic when discussing the views on human nature in relation to consciousness as evidenced by Giulio Tononi, Mario Beauregard, Francis Collins, and others. For now, we want to briefly examine the most important positions on the mind–body problem, starting from the analysis of monistic perspectives, in particular idealist(ic), physicalist(ic)-materialist(ic) and neutral, as opposed to dualistic perspectives. Generally speaking, we can see how the division between brain and mind, or between physical and mental substance(s) is opposed to the three main areas of monism, with the prevalence of the physical over the mental in physicalist–materialist positions, the exact opposite in the idealist position, and finally a third position, which is often used in spiritual/religious frameworks, that defines a third entity-essence-substance, bigger/greater than (thus influencing and activating/acting upon) both mind and body (brain). In this analysis, our neural activity will be actually mediating the impulses, stimuli and information coming from "*the other* substance and the substance *of the other*" (once again, also in Levinasian sense). As an example, we could think of the limbic system, and more specifically of the nuclei within the thalamic areas, but also extended to the pons and the midbrain as the neurofunctional structures that would allow us to feel anything at all, thus making us conscious of this 'thing'. In turn, neural activity at the cortical levels would literally encode, mediate (as in medium and literally '*social* media') amplify these conscious sensations. This is in fact, a very good argument used by many philosophers and neuroscientists, to challenge the purely physicalist–materialist position, comparing the brain to, for instance, a radio(transmitter) which does not create the transmitted messages, but encodes, amplifies and delivers these messages to the receptive ears (and consciousness) of the listener (Beauregard 2012). In other words, these messages come from something other/outer than the brain.

According to traditional Buddhist teachings, mind and body are separated but connected, interdependent and mutually supporting one another, metaphorically leaning against one another. More specifically, according to the Buddha, our mind is flowing like a river or a stream, constantly manifesting the so-called five aggregates, namely material form, feelings, perception, volition and sensory consciousness, which are themselves influenced by the "five causal laws", namely biological laws, psychological laws, physical laws, volitional laws and universal laws. This view of the mind–body problem, including the morphing properties of the universe, is at the basis of Buddhist meditation and mindfulness practice, very similarly to what happens in Catholic and Orthodox forms of prayer and *unio mystica* (conceptually a cognate of the nirvāna, and to some extent even to the *satori* vs. *kenshō* of the Zen tradition) although the perception of an ever-changing/morphing stream in which the present moment as *hic et nunc* also relates to much older spiritual traditions, often found under different labels in the *Nova Religio Romana*, *Родноверие*, *Ásatrú* all the way to the healing practices found in the *Traditionelle Europäische Heilkunde*. When we meditate, this parachronistic moment of eternal presence is achieved, the external shells of the ego or constructed self disappear and we can get rid of unnecessary illusions, which keep us trapped in this (material) world, thus a "Veil of Maya", preventing us from seeing clarity—seeing clearly, is lifted. A related form of shell or shadow is also discussed in Greek philosophy, especially by Plato, who thought that (primary) forms give rise to material objects, give them shape and meaning, identity and purpose, function and structure. In this sense, everything material we perceive with our physical senses is actually "just a shadow" of something deeper, more profound, a higher (form of) reality. In relation to the mind–body problem, his view was that the first could be linked to the soul, which is in itself party of the immortal "world of Forms/Ideas". This concept is at the center of the Greco-Roman (especially Hellenic) influence on Semitic culture in the conceptual changes on the concept of a soul from Judaism to Christianity, as a product of many processes, including the Galenian tripartite soul. Related perspective could also be found in many pagan or pre/para-monotheistic traditions, for instance in neoplatonic ideas, early Islamic philosophy and Sufi mysticism, ancient mysticism, Gnosticism, even Zoroastrianism.

Moreover, Plato saw the soul as both immortal and non-existing in time/space, as well as only temporarily united with the body, always communicating but eventually destined to leave it 'behind' in the last moment, the moment of our physical death. That is precisely why our soul can reach to the true/absolute reality of the ideas. Based on the information we can get from the *Phædo*, Socrates pointed out that (a) there is a separation between body and soul, (b) the soul is immortal and (c) there is an afterlife.

Socrates followed four main arguments to provide evidence in this sense: (a) the Cyclical (Opposites) Argument, (b) The Theory of Recollection, (c) The Affinity Argument and (d) The (Final) Argument from Form of Life. More specifically, the eternity and unchangeableness of Forms are explained in the first argument as basis for the two 'poles' body vs. soul. By 'Recollection' Socrates here refers to (as we have previously seen with anamnesis) a different form of knowledge we can obtain through the information the soul carries before (physical) birth. The Affinity Argument instead focuses on the external manifestations of body vs. soul. In the first case, we are talking about perceivable (through the physical senses, but especially sight, thus visible), mortal and corporeal/material elements. In the second, we define again the soul as the opposite of all those descriptors, and as such able to continue life after physical death. Finally, the Final Argument or "The Argument from Form of Life" defines the causal element of all the things we perceive. In other words, the Forms are the true reality, both incorporeal and static, in which all material things participate.

Aristotle's *ideas* in this area were somewhat different, in the sense that he viewed mind as a 'faculty' of the soul, thus mind and body are two aspects of the same thing, in an uncomplicated relation similar to the connection between the shape of a physical object and the object itself. Democritus and Leucippus instead are considered atomists in the sense that in their view everything we perceive in this universe is actually composed by being (made up by a-toms, indivisible particles the combination of which only makes the emergence of specific qualities possible) and void, a monistic point of view similar in this seen to Lucretius. This perspective could certainly be a challenge for the support of a soul existing (at least ontologically) separated from the body/brain, but it also carries important foundations for modern views such as embodied cognition, as well as clinically challenging ideas on the role of the mind in controlling the body in relation to apraxia (Goldenberg 2013). Cain Travis (2015) views the soul as an integral part of substance dualism, in the sense that a complete reduction of soul/mind-related phenomena cannot be entirely accounted for in matter, and is therefore opposed to both reductionism and epiphenomenalism. In regard to the latter, we will simply observe that in this view, mental phenomena and processes are caused by physical elements and mechanisms, especially at the level of neural transmission, and do not possess independent existence, but do possess (especially in the case of Thomas Henry Huxley) non-physical qualities (which would make this view part of property dualism). The opposite (a separate mind-controlling or at least acting upon the body/brain) is simply not true according to epiphenomenalism, but this view fails to account (i.e. explain in causal terms or at least in procedural mechanistic explanation) for self-perception or awareness of epiphenomenalism (although we should separate the concept of epiphenomenalism from epiphenomenalism itself), and therefore about the existence of (at least one) mind(ful-l) process. Criticism or further elaborations in this sense has been proposed by Daniel Dennet (with regard to metaphysics), and in part by Georg Northoff (for compatibility of mental causation with classical formal and final causality), Celia Green (interactionism), Walter J. Freeman (proposing circular causation over linear causation), David Ray Griffin (panexperientialism), John Searle (proposing, in analytical terms the false dichotomy of the mind–body problem) and of course A. N. Whitehead (panpsychism) and Gilbert Ryle (with specific criticism of the Cartesian-based *ghost in the machine*). Many discussions here are based on analytical stands on concepts such as properties, qualities and qualia, for which we necessarily need to refer not only to Dennet and the criticism by Gary Drescher, Marvin Minsky, Michael Tye, but also by David Hume and John Locke, but also to the philosophy of Immanuel Kant and Karl Popper. In fact, in the classical kantian view, the world of a priori forms exists beyond both mind and matter, in this sense bringing neutral monism perspectives into account. This world is actually responsible for the understanding of deep(er) truths, including the basic building blocks of (we would say in modern terms) neural perception, especially in the context of visuospatial apperception and aesthetics, but also cognition, and the most important "capacity to go inside/within" (our) intuition. Certainly, the so-called hard problem of consciousness as formulated by David Chalmers relates deeply to this issue, but in a non-explicability/non-basis function. More in detail, the (subjective vs. objective) experience represents its (own) complexity in ontological terms. This is because (a) by its nature, subjective experience is related to the subject perceiving it, and (b) because it is independent (in causal/explicatory terms) from the findings of specific neural underpinnings of (each of the) cognitive, perceptual, computational, behavioral (etc.) functions. Karl Popper targets the problem in three different ways, or better said, proposes three different 'worlds' *within* the mind–body problem. This perspective is again drawing some foundational elements in the context of neutral monism (as well as the previously analyzed metaphor of the radio receiver), albeit with very specific (and opposed) aspects. We thus have the world of matter, world of mind and world of *the creations of* the mind.

To briefly summarize the main positions related to what we have discussed so far, we could start from the fully reductionist-materialist perspective found in a monism fully focused on neural process, activities which are ***not underlying anything***, they are not neural underpinnings of psychological processes, because there is no distinction, whether ontological or even theoretical between psychological process and neural activities. Neural equals to psychological in this view, which—depending on the epistemological weight posited—can be fully defined as neurologically-based monism (it is the case of Hebb and Pinel, for instance) or "neurophilosophical eliminative materialism" (the '*Churchlands*'). Mario Bunge proposed a "systemic psychoneural monism" which, from a certain perspective is the common view in many contemporary neuroscientists, especially cognitive and computational, but on the other side fails, in our opinion, too.

(a) Add any major change to the classical positions on the mind–body problem.
(b) It is still very unclear on the combining–reacting–producing factors that link internal (even if fully neurological-matter-based) and external (even if fully justified in evolutionary-biological terms) environment. In this sense we can certainly talk of an 'explanatory gap' as in Levine (1983).

An "agnostic monistic view" is what is to be observed in neutral monism, where there isn't (again) any distinction between mind and matter, but there is also no knowledge, actual or possible (and achievable) of this "neither mater-based nor mind-based *entity*", a (form of) being which is perceived (by us and by our experimentation, thus appears) very material under the lenses of external-empirical, evidence-based science and yet has some 'ideal taste' as '*felt* from the inside of our minds, of our*selves*'. This view is, in practice, not very distant from the dualism found in Spinoza, where there are separate (thus independently existing) processes, mental (psychological) on one side, and cerebral (neural) on the other, but the function 'in parallel'. "Pushing dualism to the extreme" means to account for spiritual, mystical, mysteric, religious, transcendental, esoteric and so on experiences, which by definition (again, a very rooted form of experiential definition) are not against the proposals and verified proofs obtained via the scientific method, but they also go far beyond what reductionist views in the science (again, not in science *per sé e in sé*) claim. A more agnostic 'front' in this form of dualism (to some extent found also in Chomsky) views mind and matter as existent, yet unknowable, and in this sense it is also shared by certain mysteric perspectives with the aforementioned neutral monism, in some sense incorporating a 'yin-yang' outlook on the whole of reality. A view that more strongly suggests not only the existence of separate entities, but also their interaction in reality is found in interactionist dualism (at the center of the traditional forms of psychology and psychoanalysis, starting with Freud, and in philosophy with the Cartesian offspring). Within interactionism we should also account for naturalism, defining both sides of the spectrum as 'fully natural' in the sense that there is no such thing as supernatural, although this view would be rejected not only by some exponents of psychological-psychoanalytic theories such as Jung, but also by followers of Popper. In naturalism therefore, the view is that (depending on 'how strong' this naturalism is in the proponent):

(a) Yes, there is a mind-type of substance, but it is fully controlled by neural (i.e. materialistically intended natural) processes.
(b) —Which creates an added layer of uncertainty and logical problem—if mind is understood as separate from matter, and matter is fully equivalent with nature in the senses that natural processes are one and the same as neural processes (even considering the theoretical separation between them and psychological or sociological processes), then claiming that this view is not fully (a) monistic and (b) materialist/reductionist is somewhat of a misnomer.

Of course, the traditional, albeit not always common, way out of this impasse is represented by hylomorphism, either in the classical Aristotelian sense of connection between mind and form or in the developments of neoplatonic and scholastic views (especially Universal hylomorphism of Avicebron in the first case, Thomistic dualism of Aquinas in the second) as well as in the more recent developments such as the ones proposed by Jaworski (2016). As it follows, many of these positions present at least some contradictory aspects of comparison, whether on just one side, that is, specific characters related to matter, or the other, for example, defining elements of mind.

As we have seen, many of the challenges in examining the validity of each of these philosophical positions regarding the mind–body problem and the related hard problem of consciousness have to do with such

Table 6.1 Most common views in philosophy of mind, according to the model proposed by Baxter, more specifically dualism, epiphenomenalism, psychophysics parallelism and non-reductive physicalism

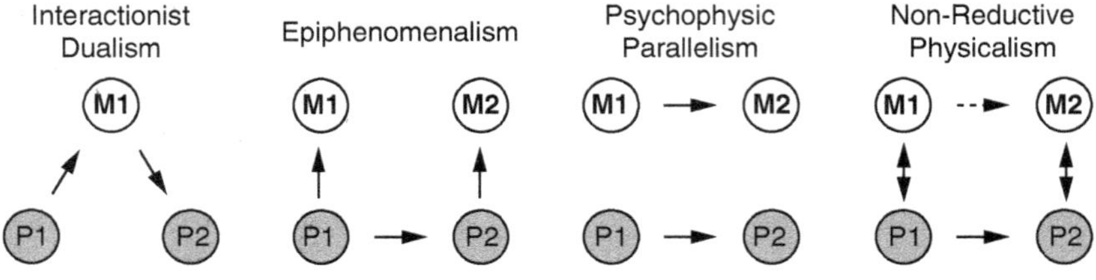

conditioning definitions (Table 6.1). Many theoreticians, including Stanislas Dehaene, Shiro Ishikawa, Leopold Kronecker, Colin McGinn, John Taylor and Sybil Wolfram, both classical and contemporary, have dealt with the connections between mind–body problem, consciousness, language and semiotics, philosophical logic and mathematics. It would be impossible to account for all the outcomes of each of the studies published in this context in this work, but we could try examining the most important elements of the philosophical discourse therein, by using very basic equation modeling as example.

For instance, let us hypothetically assign the coefficient *x* and *y* to mind and matter (we will not make any distinction with body and or brain, yet), respectively. We could start by a reductionist equation such as the one evidenced in the Hypothesis *a*1:

$$\mathrm{H}_{a1:}\ x \equiv y$$

In this case, we are simply stating that *x* and *y* are the same, that is, for every character(istic) of *x* we would find an exactly equal character(istic) of *y*. We would obviously call this type of equation 'reductionist' rather than 'materialist' because of the very nature of the statement "it is equal to", which does not allow for pre-eminence or prevalence of matter (for which we would use the term materialist) or mind (for which we could use the term mind-based or similar). In this view, mind and matter are simply identical. Of course, if there is absolutely no difference between *x* and *y*, then this equation would not make sense (i.e. it would not be needed) except from (abstract) philosophical or linguistic (especially semantical) considerations. These considerations are truly important as we will see, especially given some very interesting theoretical parallels between philosophical and neuroscientific research on consciousness and other areas of scientific investigation such as physics, in particular, quantum mechanics. In fact, using the term 'materialistic' in this context, thus following perspectives compatible with the Copenhagen interpretation, would refer to essential (ontological?) elements of matter. 'Matter' in this sense could be viewed—literally, given the observer's per*spective* at the center of quantum mechanics—as related to:

(a) An observable physical structure or process, thus empirically verifiable.
(b) The defining element needed for predictability purpose, as in the expected 'probability wave' or 'matter wave' inherent to an observed particle.[3]

Of course, in the context of critical neuroscience, that is, the application of quantum mechanics-based explanation of consciousness, we could say that, depending on the observer–observing instrument we would see behaviors we would expect from a particle as opposed to behaviors we would expect from a wave. In both cases, the crucial point is, in fact, this ex-*spect*-ation which truly involves "looking out for", 'waiting', 'anticipating', 'deferring action' and even 'hoping (that or for)' and 'believing (that or for)'. This is a fundamental point in that, claiming this identity between matter (in our case, the brain or the neurological processes) and mind (psychological process but also possible metaphysical or even spiritual, transcendental elements) truly depends on our perspective. Thus, not only philosophical perspectives are not useless because "they do not reach any ultimate (absolute) solution to the mind-body problem," but they are the determining factor for finding these solutions. We could even say that the solutions are to some extent created (in our *view*) by these perspectives, in that if we do not observe (for instance, in the case of quantum mechanics) an electron, it will behave like a 'matter wave', while if we look at it (i.e. we use experimental observation) it will behave like a particle. Of course, there are still doubts on the applicability of such theoretical assumptions from quantum mechanics to neuroscience, however, if assuming the complete equality between x and y in the equation above turns out to be

[3] This is an obvious reference to the "double slit" experiment, the Bohr's model, and the Schrödinger's equation.

wrong, we could incorporate such theories in morphing the equation into the Hypothesis *b* which simply states that the two variables are actually equal, not necessarily identical:

$$\mathrm{H}_{b:}\ x = y$$

Of course, generally speaking this could be view as a simple equation without any intrinsic value, as it simply compares the equivalent element in *x* and *y*. However, this does not truly solve the problem, for the basic reason that (a) "some equations are statements, some are meant to be solved, some are not," and (b) if we claim equivalency between *x* and *y*, we should still use Hypothesis *a*1. A possible (incomplete solution) would be "adding more to the matter", that is, relating the two variables in terms of material equivalence, as in Hypothesis *a*2:

$$\mathrm{H}_{a2:}\ x \leftrightarrow y \,/\, x \Leftrightarrow y$$

This material equivalence, however, presents two added aspects:

(a) It posits a biconditional logical connective (in the second case of $\Leftrightarrow$) between *x* and *y*, thus an "if and only if" carrying necessity and sufficiency.
(b) It involves (in propositional logic terms, or $\leftrightarrow$) further investigation on the concepts of 'valence' vs. 'value'.

For point (a) there is of course a theoretical assumption of time vs. space to be considered, which presents a practical application of Heideggerian consideration on essential *Being* of matter vs. mind. In other words, this "if and only if" presupposes at least two moments, either temporal-chronological or spatial between:

1. The observation of *x* and spatial between the observation of *y*.
2. Between the first/primary observation of *x* and *y* and the second/secondary observation of *x* and *y*.

Thus, observation is truly time-framed and space-framed. Therefore, trying to define this connection between mind and matter after (i.e. following) observation is, from this perspective, a fallacy, possibly even meaningless. Of course, in the examples above there is an intrinsic problem in even considering mind and matter as variables, both on a purely logical level (i.e. the primary investigation at this level is their interdependency, not their essence as defined by special-particular features, characters, or traits) and from the perspective of experimentability (i.e. we could possibly examine both of them only indirectly).

For point (b) we have to further examine valence and value in relation to psychological and neurological processes, and relate those processes to what we already discussed in terms of ethical and moral decision making. In this sense, our decision of looking at something, including processes themselves, is preconditioned by our decision of:

(a) Deciding to look in the first place (all meanings thereby included).
(b) Looking *for or toward* something and not something else.
(c) Looking "in a certain way" or "from a certain perspective".

Certainly, the same problems encountered on the free will debate apply to the mind–body problem as well, in that this decision is itself influenced and influencing—thus constrained—in these assumptions, not completely free from these parameters. In order to understand, at least from a purely logical point of view the levels of freedom of this decision, we have to go back to the level of freedom of the entity making this decision, that is, whether decisions are fully the product of matter (for which we will have to move back many steps following evolution in biological terms) and vs. or the product of mind, which (again) requires investigation on warrant for existence of two separate, independent variables, our x and y. Given the discussions above, we could at least argue that, for the sake of observable processes, x and y *appear* to behave as if they both existed independently and (dependently or independently) worked on different levels, using or creating different processes, and possibly following different (albeit not necessarily non-contradictory) laws. A possible approximation might be defined by the set:

$$\mathbb{U}1 : x \sim y, x \approx y$$

However, whether we chose a weak or strong approximation, the set would not make any sense at all, if (once more) this choice would not be justified in either temporal or spatial sense or given the context of possible applicability of the set in different circumstances. This would obviously represent a dualism in itself as we would (for instance) posit ***and*** apply a full identity-equality-equivalence of *x* and *y* when applying Newtonian physics to certain observations we make (it would be the case of neural correlates), and we would posit ***and*** apply a non-identity-equality-equivalence of *x* and *y* when applying Quantum mechanics to other observations. Following this dualism ***in and of*** theories, the same sate would need to be slightly modified:

$$\mathbb{U}2:(x\sim y)\cup(x\approx y);$$

$$\mathbb{U}3:(x\sim y)\cap(x\approx y);$$

$$\mathbb{U}4a:(x\sim y)\subseteq(x\approx y)\text{ vs. }\mathbb{U}4b:(x\sim y)\supseteq(x\approx y)$$

Let us analyze these sets for a moment. In $\mathbb{U}2$ we can see that a weak equivalence between *x*-mind and *y*-matter is united to a strong equivalence between these two variables, while in $\mathbb{U}3$ we observe an intersection of these two levels of strength. However, applying this significance in the context of consciousness would mean that the same order of magnitude found in the connection ~ would account for an approximation based on logarithmic calculation, while the isomorphism defined by ≈ would mean that mind and matter are *structurally* identical. To be more specific, if we are to say that *x~y* that the approximate measure of the size of (in this case) the mind (or the computational value vs. number associated to it), would be equal to the base 10-logarithm rounded to a whole number, and the same number would be what would *quantify* matter. Thus, we could see how mind and matter could be, at least mathematically, connected by a computational relation, or, in other word, via the analysis of computational valence providing the same number for both entities. Saying instead that mind and matter are *structurally* identical is actually going beyond their mathematical number, or that, in a hylomorphic sense, hyle and form present the same structure. It thus follows that in $\mathbb{U}2$ computational value and structure for mind and matter are united, in the sense that the connecting elements of mind and matter are found either in $(x{\sim}y)$ or $(x\approx y)$ or in both. The set $\mathbb{U}3$ instead clearly states that all those connecting elements are shared by $(x{\sim}y)$ and $(x\approx y)$. This is certainly no small difference given what we previously said about the quantum element of observation present in order to discern those connecting elements. Which set should we therefore accept as valid? If $\mathbb{U}1$ only presents these two type of connection, and $\mathbb{U}2$ 'only' provides a general (probabilistic?) function of presence of such elements, $\mathbb{U}3$ might ***appear*** too 'confident' in stating the existence of all connecting elements, whether on a computational or structural level in both sets. A possible redefinition of the significance of the set is to be found in $\mathbb{U}4a:(x \sim y) \subseteq (x \approx y)$ vs. $\mathbb{U}4b:(x \sim y) \supseteq (x \approx y)$, which simply states the comparison-contrast between the possibility that every element of $(x{\sim}y)$ is also an element of $(x \approx y)$ for $\mathbb{U}4a$, and the possibility that every element of $(x \approx y)$ is also an element of $(x{\sim}y)$. Of course these are all hypothesis, but the fact that they might not be verifiable 'objectively' might actually be connected to the "valence of decision and the decision of valence" in a philosophical sense, especially given (a) the multiple significance of each mathematical symbol used in this context and (b) the needed warrant for the application of mathematical constructs on subsets of neural vs. psychological constructs. Now, the assumption here is that we can theoretically create a separation between x and y to verify the existence of such separation, which truly appears to be a contradiction in terms. Furthermore, things are complicated by the fact that not only this separation might truly be an artificial one, but also that the artifice to obtain such result is created by the observer/decision-maker or observation/decision-making (cut-off) method. Thus, we would at least attempt to calculate the probability that each of these solutions might actually be "close to truth" in a statistical approximation sense. Thus, using the p-value to calculate the statistical probability for each of the hypotheses H_{a1}, H_{a2}, and H_b we could start from verifying:

(a) Pr (x), thus limiting the investigation on whether ether is (enough)[4] warrant for (independent) existence of mind.
(b) Pr (y), thus limiting the investigation on whether ether is (enough) warrant for (independent) existence of matter.
(c) Pr (Ha1), for identity (structural, computational, etc.).
(d) Pr (Ha2), for material equivalence (biconditional logical connective in terms of necessity and sufficiency for existence).
(e) Pr (Hb), for equivalence (as opposed to equality).

[4] Of course, the underlying question here is whether we can still use computational elements to even calculate not only the degrees of freedom in statistical sense, but also the levels of possible increase vs. decrease in evidence to support such warrant.

Certainly we could do the same at each intersection thus applying the above to all the sets $\mathbb{U}1$, $\mathbb{U}2$, $\mathbb{U}3$, and even $\mathbb{U}4a$ and $\mathbb{U}4b$, although we think that, besides this being an interesting (read: entertaining) opportunity for logico-mathematical analysis, it would still yield results to a test which we have previously chosen, thus an a priori decision in the full sense. More specifically, since we are talking about (a) a possible comparison between computation and structure (hyle vs. form), and (b) probability (p) value, it might be appropriate to remove, or at least account for fixed significance thresholds as opposed to incremental values. This would mean interpreting our results as grades/degrees, steps, or stages of the "strength of evidence of existence" against the null hypothesis. We should at the same time be very careful in not "assigning value to the value", given that our investigation is about the parameters which could help us define the existence of mind and matter as either completely separated entities, combined or interacting entities, or if we should only talk about one entity, in turn presenting with features appearing connected to a mind-based series of process or activities or a matter-based series of process or activities. Therefore, although starting from a test for probability value, is fundamental to understand that the p-value is not the probability that the null hypothesis is true, or the probability that the alternative hypothesis is false, which is especially important in the context of quantum mechanics, given that it represents the *prior probability* of an observed effect given that the null hypothesis is true, and not the *posterior probability* that the null hypothesis is true given the observed effect.[5] Again, "no observation after observation".

Going back to $\mathrm{H}_{a2:}\ x \leftrightarrow y\ /\ x \Leftrightarrow y$, we should really follow our previous discussion ad rewrite this hypothesis as a *set* of hypotheses:

$$\mathbb{U}\left[\mathbb{U}(Ha2)5a \text{ vs. } \mathbb{U}(Ha2)5b\right]: \mathbb{U}(Ha2)5a: \\ (x \leftrightarrow y) \subseteq (x \Leftrightarrow y) \text{ vs. } \mathbb{U}(Ha2)5b: (x \leftrightarrow y) \supseteq$$

To simplify, we would leave aside for now the differences between $\mathbb{U}(Ha2)5a$ and $\mathbb{U}(Ha2)5b$ in terms of subset vs. superset, and we would only focus on the connection in propositional logic terms between x and

[5] For further reference, see bibliography, especially Wasserstein R.L., and Lazar N. A. 2016. "The ASA's statement on p-values: context, process, and purpose". *The American Statistician*. *70* (2): 129–133.

y, that is, between mind and matter. In any case, we are interested in seeing whether the assumption of possible ontological separation between these two variables and the following (in purely logical terms) connection-mutual influence between them can be theorized in a simpler form. Certainly, this is an oversimplification of the very complex logical analysis of the mind–body problem, but we would like to be especially "mindful and parsimonious" in regard to the application of Occam's razor to critical neuroscience in the context of the hard problem of consciousness. Thus, let us go back to our original hypothesis by 'splitting' the sets into smaller components:

$$\mathbb{U}\left[\mathbb{U}(Ha2)5a \text{ vs.} \mathbb{U}(Ha2)5b\right]$$

$$\mathbb{U}(Ha2)5a:(x \leftrightarrow y) \subseteq (x \Leftrightarrow y) \text{ vs.} \mathbb{U}(Ha2)5b:(x \leftrightarrow y) \supseteq (x \Leftrightarrow y)$$

$$\mathrm{H}_{a2:}\ x \leftrightarrow y \,/\, x \Leftrightarrow y$$

$$\mathrm{H}_{c1:}\ x \leftrightarrow y$$

$$\mathrm{H}_{c1:}\ x \rightarrow y \text{ vs.} \mathrm{H}_{c2:}\ x \leftarrow y$$

H_{c1} can be defined as "mind over matter" and H_{c2} as "matter over mind" in the sense that the arrows represent an ***implication*** ('it implies'), a (logical, psychological, process-based and computational) direction, or function. In H_c these arrows are bidirectional, thus "implying mutual implication" between mind and matter. This view represents the core of dualism, although to better define which type of dualism we are talking about (for instance Substance dualism, Predicate dualism, Property dualism, but also Epiphenomenalism, Interactionism, non-reductive physicalism, Occasionalism, Parallelism, Universalism, etc.) we suggest we operate a further split, from:

$$\mathrm{H}_{c1:}\ x \leftrightarrow y$$

to:

$$\leftrightarrows$$

$$H_{c2:} x \leftarrow \quad \rightarrow y$$

With this hypothesis we actually suggest a third element to provide such explanation, in practice moving from different dualistic modalities to a position close to neutral monism, in the sense that:

(a) The variables *x* and *y* exist independently on an ontological level ***and yet*** they are influencing one another. This contradiction between independency and dependency is solved by (a^1) the dependency between *x* and *y* is not direct but indirect (the arrows do not touch), and (a^2) the paradox is represented by the 'third element' above (the arrows on top).
(b) This connection-mutual influence happens on two different planes/levels, intended *either* as horizontally vs. vertically *or* parallel vs. perpendicular.

Another way to represent this hypothesis could be this type of relation:

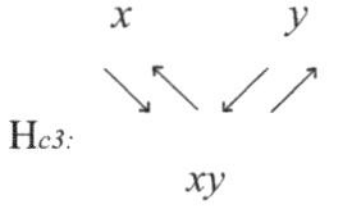

With this hypothesis[6] another type of neutral monism is presented in strong connection with interactionist perspectives. Thus, whether we are talking about a monistic (single) entity with two (or more, as we will see) interacting aspects (faces) or dualistic (two) entities interacting with one another, this interaction happen indirectly, thus (a) on a different/separate(d)/ulterior level or (b) ***neutrally***, with respect to mind vs. body. This perspective will obviously bring us back to other triads in the context of neuroscience, psychology, and sociology, most especially Social Cognitive Theory (SCT), as seen in Table 6.2:

Although the application of SCT is on the learning process, in terms of education and knowledge acquisition via observation, and this observation is directed toward (the) other/(the) external, thus through experiences, media/environmental influences and social interactions, the underlying idea of a 'third element' can provide further basis for a movement beyond dualism, not in the sense of a separation between two vari-

[6] A similar view and representation has been presented by Herald Atmanspacher, first in 2004 and then in 2015 in the Stanford Encyclopedia of Philosophy. For further reference, please see Atmanspacher, H. 2015. *Quantum Approaches to Consciousness*. Stanford, CA: Stanford Encyclopedia of Philosophy.

Table 6.2 Simplified description of Social Cognitive Theory (SCT) according to the model proposed by Bandura

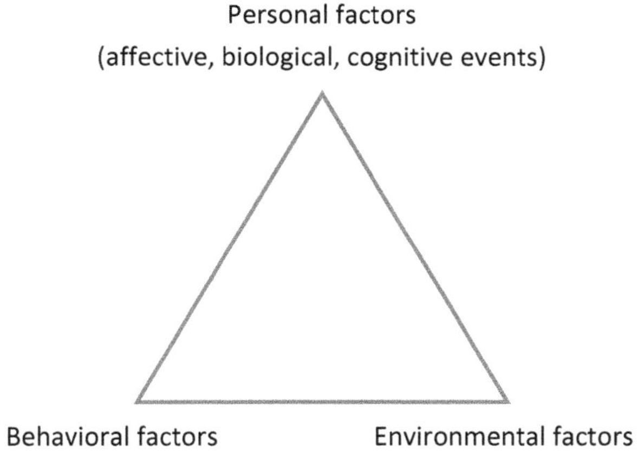

ables (in our case *x* and *y* for mind and matter), but a separation between planes, even between dimensions; in this context between two dimensions and three dimensions. Of course, considering a tridimensional conceptualization of the mind–body problem would immediately make us think of Trinitarian theories, especially in the metaphysical or theological sense, but also in relation to the physical dimension of structural features, for instance the double helix of the DNA. In this regard, we would like to briefly mention the work by Francis Collins, who focused on (a) human nature, (b) the divine origin of evolution, genetic (sub)structure of creation (including us humans) and (c) moral law (as perceived by us humans) and altruism through a philosophical analysis at the basis of his research on the Human Genome Project. The most important aspect of this perspective is the presumed scientific requirements that science demands atheism *from within* (although it would be interesting to see, also in reference to the related studies by Pinker and beyond evolutionary perspectives, if it would be correct to talk about 'faith' or 'hope' in the case of an atheist saying "Come on! Please! Score!" while watching a football game). In other words, science uses itself to claim that it is more scientific to expose a position in which there is no warrant for (a) god than the other way round. Collins presents numerous fallacious assumptions of the concept of a deistic vs. theistic and personal god, religion, religiosity and spirituality, including Mark Twain ("Faith is believing what you know ain't so"), Sigmund Freud ("the belief in God is just wishful thinking") and the one by Dawkins ([Faith is] "blind trust, in the absence of evidence, even in the teeth of evidence") (Tomasi 2016). Experiences in multiple areas of healing traditions, albeit possibly misguided under the aforementioned points of view, almost always involved a spiritual, mystical dimension based on Nonordinary states of consciousness, where the healer is the one who does indeed "attend and assist in the healing process" which is truly the origin and etymology of compounds of *iatry*. Modern science and modern medicine often forget these elements and splits the intervention on sides. More specifically, we have a biologically analyzed evidence-based data on one side, and a systematic, symbolic, introspective as well as transpersonal exploration of the human nature, starting from the psyche. However, since this side of the equation is not 'truly' observable with the means of evidence-based science, this represents (and presents, shows) a limitation defining our 'real', physical (material) nature. Grof writes:

> According to the Newtonian-Cartesian paradigm of traditional Western science, these restrictions and limitations are absolutely mandatory and definitive, since they result from the material nature of the world and are determined by physiological laws of perception. However, modern consciousness research has clearly demonstrated that in transpersonal experiences these limitations do not apply and can be transcended. This represents a critical challenge not only for psychiatry and psychology, but for the entire philosophy of Western science.[7]

In recent studies, these phenomena have been analyzed by many medical scientists, philosophers and physicists; good examples are found in the research by Beauregard on alternate states of mind, Moore, Jansen and van Lommel on Near-death experiences,[8] Penrose-Hameroff on Consciousness,[9] Stevenson on Reincarnation,[10] and many others. In fact,

[7] Ibid., p. 47.

[8] Jansen, K.L. R. 1998. *The Ketamine Model of the Near Death Experience: A Central Role for the NMDA Receptor*. Journal Article, The Maudsley Hospital, Denmark Hill, London.

[9] Hameroff, S.R. 2006. *The entwined mysteries of anesthesia and consciousness*. Anesthesiology 105 (2): 400–412. doi:https://doi.org/10.1097/00000542-200608000-00024. PMID 16871075. See also Penrose, R. 1989. *Shadows of the Mind: A Search for the Missing Science of Consciousness*. Oxford, UK: Oxford University Press.

[10] Stevenson, I. 1974 (2nd Ed.) *Twenty Cases Suggestive of Reincarnation*. Charlottesville, VA: University Press of Virginia.

as we will see, the main problem in addressing these issues is first and foremost philosophical in nature. Of course, empirical observation through evidence-based experiments can provide fundamental elements to support the existence of neural underpinnings to the aforementioned processes and events, but the analysis of the matter-at-hand still requires philosophical explanation. In this context we would like to quote the viewpoint offered by Jaworski (2016) in regard to (the concept of) structure, from a hylomorphic point of view:

> Structure matters: it operates as an irreducible ontological principle, one that accounts at least in part for what things essentially are.
>
> Structure makes a difference: it operates as an irreducible explanatory principle, one that accounts at least in part for what things can do, the powers they have.
>
> Structure counts: it explains the unity of composite things, including the persistence of one and the same living individual through the dynamic influx and efflux of matter and energy that characterize many of its interactions with the wider world.
>
> Structure minds: it provides us with resources for understanding the place of mental phenomena within the natural world.[11]

These observations are truly important, especially regarding what we have said about Pr (Ha2), in particular the necessity and sufficiency for existence of the equivalence between mind and matter. In the context of hylomorphism in the version proposed by Jaworski, the focus on necessities of formal structure determine that the constitution of matter is structured into form and thus (read: therefore or through this process) generates, or at least supports, mental activity. Of course whether this activity accounts for mindful or conscious activity is to be understood in psychological terms, nevertheless rooted in matter, according to naturalistic hylomorphism and its view of substance-attribute basis, in ontological terms, for its metaphysics. In other words and applied to critical neuroscience, our 'self' is such and perceived as such by us and others, when it is *supported* by (its) form. In this context we are (hylomorphically intended) "more than the sum of our parts." This metaphysics doesn't actually require anything particularly supernatural, in the sense that I would lose my self if all the components which constitute the (form of) my (for instance) neuroanatomical

[11] Jaworski, W. 2016. *Structure and the Metaphysics of Mind: How Hylomorphism Solves the Mind-Body Problem*. Oxford, UK: Oxford University Press, p. 97.

structures were to be 'crushed' so that each building block will still be there (thus, matter would be preserved) but the 'general organization' (i.e. the hyle defining-creating my self) would be lost. In any case, if the self is still intact, all those psychological processes at the basis of cognition, perception, memory, computation, behavior and other are also still there—albeit in multiple and diversified levels and with different characters or traits varying from individual to individual and (in a sociological sense) from group to group—and are at least in part and in the view of Atmanspacher (2015) *required* for an understanding of neuroscience, even in the context of ulterior theoretical models as in quantum mechanics:

> One rationale for the focus on psychological phenomena is that their detailed study is a necessary precondition for further questions as to their neural correlates. Therefore, the investigation of mental quantum features resists the temptation to reduce them (within scenario A) all-too quickly to neural activity. There are several kinds of psychological phenomena which have been addressed in the spirit of mental quantum features so far: (i) decision processes, (ii) order effects, (iii) bistable perception, (iv) learning, (v) agency, (vi) semantic networks, and (vii) super-quantum entanglement correlations.[12]

And it is exactly from observations of semantic nature in the context of quantum theory that proposals such as the one by Ishikawa (2017) for the final solution to the mind–body problem originate. More in detail, Ishikawa suggest that quantum language represents the only possible solution to the mind–body problem in that it provides a scientific method to (re)examine the problem under the lens and as a linguistic problem, rather than an epistemological one. We find this solution very appealing and presenting some very useful elements for a better understanding of the mind–body problem, but the proposed solution fail, in our opinion (a) under a Post hoc ergo propter hoc fallacy and (b) due to some tautological components. Ishikawa starts from consideration on the very nature of quantum language in relation to the Copenhagen and Born's probabilistic interpretations of quantum mechanics, the causality axiom, and posits that there is an equivalency between idealism (although it remains to be understood why the author puts a particular weight on dualistic idealism as "the mainstream of philosophy", p. 50) and metaphysics defined as

[12] Atmanspacher, H. 2015. *Quantum Approaches to Consciousness*. Stanford, CA: Stanford Encyclopedia of Philosophy.

non-verifiable experimentally. The philosophical analysis continues by presenting the ***measurement axiom*** [axiom 1][13]:

$$\in X \text{ obtained by } \mathrm{M}_N\, 0 := \left(X, \mathcal{F}, F\right), S_{[\rho]} \text{ belongs to a set } \Xi\left(\in \mathcal{F}\right) \text{ is given by } \rho\left[F\left(\Xi\right)\right]$$

In this axiom Ishikawa also uses quantum mechanics as a mathematical basis for the investigation of the mind–body problem. More specifically, in this case the X defines a 'metaphysical' (in parenthesis given the aforementioned definitions) space in relation to the *movement* vs. *presence* of a particle. In detail, the axiom proposes that the probability density of finding such particle at a given spatial-temporal (quantum-based) point is "proportional to the square of the magnitude of the particle's wavefunction at that point" which is the same (and Ishikawa acknowledges that) as Born's law for Schrödinger's equation, thus:

$$\hat{\mathrm{H}}\,|\,\psi\left(t\right)\rangle = i\hbar\frac{\partial}{\partial t}\,|\,\psi\left(t\right)\rangle$$

Of course, in the case above, Schrödinger suggested that, starting from the Hamiltonian operator $\hat{H}$ (defining the system's total energy/set-spectrum of possible outcomes when there is a measurement os the total energy of a system), there are quantum effects playing a significant role in the way a physical (most importantly, observ*able*) system changes over time. In fact, the equation above is time-dependent, as opposed to the time-independent version[14]:

$$\hat{H}\Psi = E\Psi$$

We should note that the equation focuses on $\hat{H}$ acting on Ψ (the wave function), where $\hat{H}$ represents a set/system/spectrum. In philosophical terms, these words represent three different levels of interpretation within the interpretation itself, respectively:

[13] Ishikawa, S. 2017. A Final Solution to the Mind-Body Problem by Quantum Language. *Journal of Quantum Information Science, 7*, 48–56, p. 50.

[14] I.e. "When the H-define Hamiltonian 'operator' is acting on the wave function Ψ, and the result is proportional to the same wave function Ψ, it follows that the wave function above is a stationary state, and the proportionality constant, E, is the energy of the state Ψ."

(a) $\rho[F(\Xi)]$
(b) σύν + ἵστημι [to stand(up)/(still), set (up) stop-stay, check, establish → *stō* + *bilis*], thus involving an operator-observer-actor *in* vs. *on* such system
(c) It is truly the center of a double hermeneutics in quantum mechanics, thus viewing, observing, reflecting and mirroring

The latter term mirroring is especially important in this context, as it relates to the applicability of the theoretical framework of quantum mechanics to the vast realm of neuroscience. Etymology aside (although we doubt that such operation could be fully justified in this context), in quantum terms, who/what observes also becomes integral part of who*m*/what is observed, in a mirroring process which makes a lot of sense on the neurological level as well, for instance in the case of the role mirror neurons play in terms of cognitive processes related to emotion appreciation, understanding, production-reaction, and response. Of course, we could argue that such (semantic vs. observable mechanism-based) 'jump' from theoretical frameworks to (level 1) neurological functioning and (level 2) further philosophical justification of mind-matter definitions of existence/essence might not provide warrant, or even logical sense, *tout court*. Now, Born's law states that the probability density $p(x,y,z)$ of finding the (vs. a!) particle at a given point is proportional to the square of the magnitude of the particle's wave function at that point. This means that the mirror/spectrum component of such probability (i.e. that a measurement on this system—defined in quantum terms/parameters—will yield a *given* result) is in itself part of such predicament, as it certainly (e.g. by definition) is:

$$\text{probable} = \text{able to be proven}; p = \mathrm{F_a}(p)$$

However, we could certainly go back to our first hypothesis $\mathrm{H}_{a1:}\ x \equiv y$ and tautologically infer that $p \equiv \mathrm{F_a}(p)$ since "just because the word says so, it must be true" which is again not true logical *stable* ground (upon which) to establish such paradigm, except if we are willing to consider metaphysical, even theological elements as in "ἐν ἀρχῇ ἦν ὁ *λόγος*." Of course, we cannot use this argument here, so we must rely on our human logic to attempt to entangle these problems. As it is, Born's rule states that—following what we said above—if an observable (i.e. a measurable dynamic variable) corresponding to $\hat{H}$ with discrete spectrum is measured in a system with normalized wave function Ψ, it follows that the measured result will be one of the *Eigenwerte* λ of $\hat{H}$, and that the probability of measuring a given *Eigenwert* λ will equal $\langle \Psi \mid P_i \mid \Psi \rangle$, P_i being the projection onto the *Eigenraum* of $\hat{H}$ corresponding to λ. Ishikawa starts from Born's rule to reach Axiom 1 and continue with the causality axiom (Axiom 2): "Let $t_1 \leq t_2$. The causality is represented by a Markov operator $\Phi\ t_1,\ t_2$: A_1."[15] What is truly interesting about this perspective is that Ishikawa talks about Axioms 1 and 2 as "kinds of spells"[16] in the sense that they do contain an important feature for practical use, that is, the use in itself, as opposed to a (true) understanding. To be sure, this does not mean (although Ishikawa does not state it openly) that they do not *mean* (i.e. do not have an intrinsic meaning *of* vs. *for* value), but simply that they cannot (this is what *does not make sense*) be verified experimentally. I we applied this interpretation, which Ishikawa defines as *linguistic*, our statement about probability would not hold anymore:

$$p \not\equiv \mathrm{F}_{\mathrm{a}}(p)$$

Thus, we will have to find another route. Very notably, Ishikawa also suggests that there is progression in philosophy, and this progression is scientific in the full meaning of the word, most specifically from Plato, through Descartes, Locke and Kant, to Quantum language, which is the only possible way out of the impasse, as Ishikawa suggests that the denial of substance dualism found in 'going' from Kant (K) to Husserl (H) is to be rejected, unless it moves on to a philosophy based on cognitive science, a scientifically based philosophy of mind (ϕ_M). This appears to be a contradiction in terms, as we just said that:

$$(\mathrm{K}) \rightarrow (\mathrm{H}), (\mathrm{K}) \leftarrow (\mathrm{H}) \Leftrightarrow (\mathrm{H}) \rightarrow (\phi_M)$$

Ishikawa suggests that there is actually compatibility—better, non-contradiction—between (the acceptance of) substance dualism and the denial of substance dualism, *because of* quantum language: "describe any

[15] Ishikawa, S. 2017. A Final Solution to the Mind-Body Problem by Quantum Language. *Journal of Quantum Information Science, 7*, 48–56, p. 50.

[16] Ibidem.

monistic phenomenon [...] by dualistic language (=quantum language)".[17] We could thus rephrase the above as:

$$(\mathrm{K}) \rightarrow (\mathrm{H}) \cup \left[(\mathrm{K}) \leftarrow (\mathrm{H}) \Leftrightarrow (\mathrm{H}) \rightarrow (\phi_M) \right]$$

Of course, this non-contradiction does not necessarily mean union (as in this example), as we could see how we might fall prey of tautology:

$$(\mathrm{A}_1 + \mathrm{A}_2) : (\mathrm{K}) \rightarrow (\mathrm{H}) \cup \left[(\mathrm{K}) \leftarrow (\mathrm{H}) \Leftrightarrow (\mathrm{H}) \rightarrow (\phi_M) \right] \leftarrow (\mathrm{A}_1 + \mathrm{A}_2)$$
$$(\mathrm{A}_1 \cup \mathrm{A}_2) : (\mathrm{K}) \rightarrow (\mathrm{H}) \cup \left[(\mathrm{K}) \leftarrow (\mathrm{H}) \Leftrightarrow (\mathrm{H}) \rightarrow (\phi_M) \right] \leftarrow (\mathrm{A}_1 \cup \mathrm{A}_2)$$
$$(\mathrm{A}_1 \cup \mathrm{A}_2) \leftarrow (\mathrm{A}_1 \cup \mathrm{A}_2)$$
$$\neg(\mathrm{A}_1 \cup \mathrm{A}_2) \vdash \bot$$

To be sure, Ishikawa appropriately points out that the definition of validity depends on the defining process itself, in the sense that language defines the limits of the perspective, because:

(a) Linguistic limits (limitations) ↔World(view) limits (limitations)
(b) Measurement ↔ Measurer (brain)

These considerations represent the main reason why in Ishikawa's view the mind–body problem should not be *viewed as* and epistemological problem, but a linguistic problem. Of course, Ishikawa is not the first scholar to attempt a solution to the mind–body problem by way of quantum theories. In fact, the term 'quantum mind' has nowadays become very popular, not always in a positive sense, given some truly pseudoscientific claims in regard to supposed existing processes justifying consciousness in this way. In fact, for some philosophers 'quantum mind' could also be synonym with a reductionist quantum theory, in the sense that this arbitrarily posited (without *contradictio in se*) quantum underpinnings of consciousness still do not provide warrant. A similar non-reductionist (non-reductibile) view is embraced by David Chalmers, and to some extent also by Walter Freeman, Victor Stenger, Giuseppe Vitiello and John Taylor, albeit with vast differences. More specifically, for Chalmers those proposed quantum

[17] Ibid., p. 54.

justifications of consciousness do not make much sense, not because they are intrinsically wrong, but because the *hard* problem of consciousness can be *hardly* solved by certain *hard* sciences, more specifically new physics. Freeman and Vitello instead provide a model to represent the mind–body problem from the perspective of a dialogue between the classical (with major differences, as in the rejection of psychoneural representations as basis for cognitive and behavioral processes) and the quantum parts (base on quantum field theory specific to their proposal) of the brain. Stenger's rejection of quantum theory for the solution of the mind-problem is as radical as Chalmers, but from a completely opposite point of view, as the very existence of a quantum mind is in his case viewed as complete pseudoscience. Taylor's views on the mind-body problem are to some extent similar, in the sense all the possible non-physical (immaterial) explanations are eliminated in favor of physical reductionism. Taylor's view moves from a thorough analysis of multiple views, including various degrees of dualistic and monistic perspectives in order to achieve a completely physical explanation *from* physical activity. In detail, Taylor identifies the Corollary Discharge of Attention Movement (CODAM) model as the best explanation for consciousness originating in physical processes. The main support for this view comes, in Taylor's opinion, from both physically based experimentation such as EEG/MEG as well as from philosophical analysis, most importantly from 'inner experience' in the conceptualization by Husserl. This analysis (Taylor 2012) presents CODAM first as a model for consciousness through sensory attention in terms of engineering control. Attention is thus interpreted as a function, or better as a series of functional modules, which in turn physically (procedurally) create the (subjective) experience of 'ownership' through efference copy/corollary discharge of the attention movement control signal. Of course, another whole set of definitions is open given these premises:

$$\left[\Xi C\left(\in A, O, F\right)\right] \cap \mathrm{S}_M$$

where Consciousness C is vs. belongs to a set identified as Attention A, Ownership O and Focus F (cognitive vs. computational) intersect with the concept of "Self as Monitor" (S_M). Of course, this is a general analysis of the connection between a common conception of consciousness and the model proposed within CODAM. Of course, for each of these elements, we could identify a (dualistic) dichotomy, one for Attention A, one for

Ownership *O*, and one for Focus *F*. Quite interestingly, Taylor starts his analysis of the mind–body problem exactly from this core-defining dualism (2012):

> The mind is composed of mental fragments – sensations, feelings, thoughts, imaginations, all flowing now in an ordered sequence, now in a chaotic fashion. There are also non-conscious components involved in early brain processing of stimuli (as in lower level processing in vision, such as in V1) or in emotions not yet in consciousness. On the other hand the body is constructed under the underlying laws of physics, and its components obey the well enumerated laws of physiology.[18]

The comparison here is obviously between the planes of 'effectors and interpreters as it is often found in other models of consciousness, such as:

- Coalitions of neurons
- Dynamic core & IIT
- Field models
- General Quantum (mind) models
- Global workspace models
- Higher-order thought (HOT)
- Information integration
- Multiple drafts theory
- Recurrence Model
- Sensorimotor theory
- Subcortical models
- Thalamocortical rhythms
- Internal simulation and self-modeling (Retinoid model, Self-model theory of subjectivity, World simulation metaphor and others)
- Cognitive/cognition (attention)-based models (Intermediate level theory, Cognitive and Affective—CogAff, Consciousness as Attention to Memories, Corollary Discharge of Attention Movement—CODAM, Supramodular interaction theory, Multilevel feedback, Radical plasticity thesis)

Let us examine them more in detail:

[18] Taylor, J. 2012. A Final Solution to the Mind-Body Problem. *Journal of Mind Theory. 1*(1)25–58, p. 25.

6.1.1 Coalitions of Neurons

The model by Crick and Koch (1990) uses the observation of a 40 Hz neural synchronization (oscillations vs. coalitions) across modules to define the creation of consciousness via feature binding (an attentive-attentional process) into object representation, thus content. In neuroanatomical/functional terms, unconscious processing in this model mainly consists of feed-forward cortical waves. Conscious processing instead is constituted of bidirectional signal flow standing waves. A big problem of this model (eventually abandoned by Crick and Koch, and reformulated by Lammer, especially in regard to the phenomenally conscious essence of reentrant cortical processing) is that it does provide enough evidence of causal factor for the emergence of the self and it overlaps with the same level of oscillation as observed in anesthetized animals.

6.1.2 Complexity of Activity (Including Dynamic Core and Integrated Information Theory: IIT)

This a series of models developed by Gerald Edelman and Giulio Tononi from the theoretical framework contained in TNGS or theory of neuronal group selection, itself considered by some as neural darwinistic model of consciousness. Among the main traits of this theory, we find the selectionist approach toward neural development, microarchitecture/anatomy and function. These aspects were first developed independently by Edelman (1993) and as part of the dynamic core hypothesis (Tononi and Edelman 1998, subsequently modified in Edelman and Tononi 2000), and finally in the IIT or information integration theory of consciousness (Tononi 2004, 2008). Core elements of all these models are the complexity and the computational vs. information-based essence of consciousness. More specifically, the theory suggests that for every conscious perception there is computational-selective discrimination in place, which follows and produces information. That is why these perceptions or 'scenes' are both integrated and differentiated through (ontologically) qualia or subjective conscious (emergent) experiences, as theoretically opposite of propositional attitudes. This complexity is translated in providing-promoting an ever increasing (as differentiating) series of information through the complexity of neural networks, especially in the central nervous system (CNS), which 'model' consciousness starting from a perceptual perspective. This means that a higher order of complexity is actually found (also by theoretical definition within the model) in higher cortical areas. More in detail, the dynamic core hypothesis sees these processes in the connection between the thalamus and the cortex, making this core constantly changing. This shift happens via an inside-out and outside-in movement (itself caused by external and internal environmental stimuli) by neural clusters, which create a series of interactions. What is interesting is that, despite the still unclear method—in terms, of quantitative vs. qualitative observational parameters—that could provide an accurate measurement of this complexity over time, is that these changes in state happen in a specific succession/order, but (in this theory) they are still able to maintain a unitary stability. Whether this stability means preservation of consciousness is however still a question, as specific aspects of the conceptualization of self on one side, and the "consciousness of this vs. 'self-'" activity (even in attentive terms, with possible exception of the 'remembered present') on the other, remain unclear. Some of these problems found solutions in the model of IIT, more specifically in claiming that there is a way to accurately measure consciousness, if the latter is understood as the (discriminative) information integration-capacity of a system. The 'quantity' (thus involving measurement of dynamics and capacity) of information thereby provides is what IIT defines as Φ, a direct measure of *causal* interactions within this system, and is thus interpreted also as 'quality' of information, because the differentiation-specification of conscious experiences are 'produced' by the intrinsic value of these variables. This means that there are different levels of Φ, corresponding to different levels of consciousness.

6.1.3 Field Models

In this view, consciousness is equated to the activity of fields, which can be in turn defined as having integrative electromagnetic neural properties. For example, Margolin (2001) sees these fields as resonating in a critical mass of neural areas (perceived in unity-integrity through thalamocortical loops) with a global negative entropy creating consciousness (better said allowing consciousness to emerge). Electromagnetic properties of the nervous system are supported (with some differences in terms of function and synchronicity aspects) also by (Kinsbourne 1988), McFadden (2002) and Pockett (2013).

6.1.4 General Quantum (Mind) Models

We already discussed the main components of the application of quantum theories (especially in the context of quantum physics and quantum mechanics) to neuroscience, particularly in regard to the mind–body problem, the hard problem of consciousness, and the related semantic-linguistic frameworks of causal justification. Beside Schrödinger, Bohr, Born, Planck, Einstein, Ishikawa, Heisenberg, Sommerfeld, Fermi, Dirac (although this would be included in field theories, more generally) and many, many others, we would briefly mention the position by Penrose (1989) in terms of comparison between (human) mind and computational (sequential) computers (Turing, specifically and offspring). This would, as we have said, fit into Utaum theory perfectly, using the same theoretical framework, modus operandi, rules and laws. The process (in a mechanistic-causal sense) of consciousness creation from (quantum) structures is however still missing (enough) detail.

6.1.5 Global Workspace Models

In the global workspace model developed by Baar (1998), consciousness is interpreted as cognition, more specifically as cognitive processes underlying attention, evaluation, language and memory with neural underpinnings which in turn can access (in a single stream of content) a centralized global workspace. The neural activity leads to 'universal report' in neuroanatomical cerebral areas, which in turn leads to consciousness. This model was further developed by Franklin and Graesser (1999) and Shanahan (2005), with a focus on software interaction in the first, and internal electrochemical stimulation in the latter. Duncan (2001), and Dehaene et al. (2003) also used this model to identify multiple sites for continued activity in higher neural center in the brain, a discriminating (in the attentional sense) neuroarchitecture to excite a pattern of neurons with long-range cortico-cortical axons, and a transitional giant omnicomprehensive neural component of previously disconnected subnetworks, respectively. These models, however, still do not provide enough causal links for (the emergence of) inner self vs. consciousness.

6.1.6 Higher-Order Thought: HOT

In the model proposed by Rosnow and Rosenthal (1997), the main hypothesis is that consciousness is created through a cognitive-based process, more specifically via a second-order thought activity about a first-order thought activity which becomes conscious (retrieved) in the process. Thus, sensory processes are considered to be lower than consciousness. However, the most complicated thing to explain is the causal effect from lower to upper order and it is also unclear how attention vs. introspection (only in the sense of inner self) can enter such process.

6.1.7 *Multiple Drafts Theory*

Daniel Dennett introduced the Multiple drafts theory in 1991 with the purpose of reassessing the view, most especially linked to Descartes, according to which perceptive stimuli are cognitively received by the subject. Dennter proposed that consciousness can be better understood in terms of calculating its impact on the system, as this impact depends on a neurocognitive content produced in parallel. Therefore, consciousness is better framed as 'conscious access' in a mechanistic, but also procedural, sense. More specifically, according to the "fame of the brain" parallel the author drew in later essays, time cannot be constricted in an aprioristic sense, since the fame discussed in this metaphor can only be assessed (perceived cognitively) in retrospect.

6.1.8 *Recurrence Model*

The model defined by Lamme (2003) focuses on the 'recurrence' of feedback as primary causal factor for holding of activity. This process in turn creates consciousness, as neurologically intended to be a correlate of attention and memory, especially short-term. However, many recurrent neural networks still do not appear to play a role in consciousness or the (creation of/a conscious) sense of self via recurring feedback, as in limbic system substructures, especially in the hippocampus.

6.1.9 *Sensorimotor Theory*

O'Regan, Kevin and Noë (2001) propose a theory in which sensorimotor skills are at the center of consciousness. In practice, consciousness would be created by the conscious exercise of such skills in the individual, thus discarding qualia as a purely subjective (experiential) phenomenon and suggesting that this conscious emergence is the same as a cognitive perception, even better 'access' to an element of activation, a control of such activation, ultimately a conscious, cognitive, perceptual understanding that the individual is him/herself mastering such skills. This model unfortunately does not fully clarify how this sensorimotor (also to be understood in the context of afferent and efferent neurons vs. pathways, as in Fig. 6.3a, b) contingency could account for every type of perception, and more importantly perceptions of perceptions by the self, also in relation to internal vs. external environmental stimulation, beyond the neural underpinnings vs. mechanisms in the sensorymotor cortex.

6.1.10 Subcortical Models

In a work on critical neuroscience, we cannot avoid a brief examination of subcortical models, especially in relation to all the philosophical considerations following neurosurgical methods and theoretical frameworks to understand the links between consciousness, integrated—united vs. unique/split sense of self, free will and so on and epilepsy. We mentioned the studies by Gazzaniga following corpus callosotomy multiple times in our analysis; here we would like to mention the experiments by Penfield and Jasper (1954) and the further elaborations by Merker (2007) and Ward and Moreton (2011). In subcortical models based on all these elements, the focus is on the brain, more specifically subcortical areas, in particular the midbrain but also the ascending value systems, ventral thalamus and the zona incerta as *locus* for consciousness. In the case of Merker consciousness emerges also due to the activity of the superior colliculus, while for Ward the Diencephalon, especially the thalamus, represents the primary aspect of consciousness. This view presents to some extent similar theories from psychology, especially in regard to cognitive process at the basis of attention and emotion. In the case of Ward, our conscious experience is supported by these elements:

(a) Neural synchronization (already discussed in connection to the dynamic core theory) is in this view the best neural correlate for consciousness, especially in perceptual terms.
(b) An individual first experiences the outcomes of cortical computations (in the thalamus), not the processes of cortical computations (on the cortex).
(c) The very function of the thalamus is connected to "a common brain locus of action among general anesthetics, and as a critical locus of damage leading to vegetative state" (Seth et al. 2006).

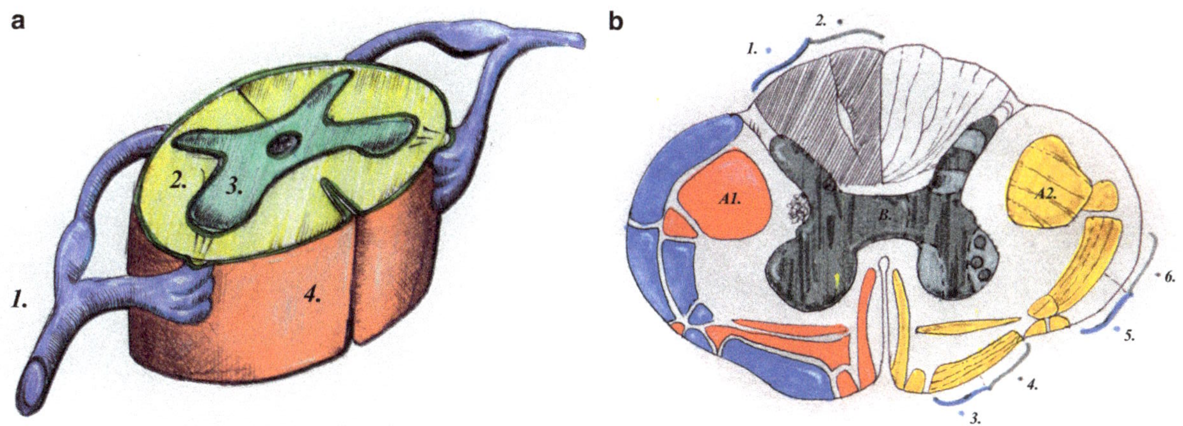

Fig. 6.3 (**a**, **b**) Sections of the spinal cord. In **a** (right) we observe the spinal nerve and sensory root (1, in blue) the white matter with corticospinal tracts (2, in yellow), the gray matter with sensory and motor horns (3, in green, ventral and dorsal, respectively), the external layer of the dura mater (4, in red), as well as the central canal (at the center of the image, in black). In **b** (left) we observe the fasciculus cuneatus (1) and gracilis (2), the vestibulo-spinal tract (3, 4), and the spino-cerebellar tract (5, 6), as well as the lateral corticospinal tracts (A1, A2), and the gray matter with the central canal (B)

(d) The anatomy and physiology of the thalamus appear to be "a perfect fit" for a centralized computer producing, amplifying, and controlling computations on cortical level.

In this context, we would like to stress not only functional, structural and procedural aspects, buts also order and sequence-based elements. While current neuroscientific research clearly indicates that the thalamus can be viewed as a computer, in the sense that it receives, interprets, amplifies and sends out to the cortex (for further interpretation vs. computation) all types of signals and stimuli, there is still uncertainty on some causal links between these neural components, especially in connection to the emergence of (a/the) self under the lens of perception/cognition/emotion. The latter in particular appears to be a crucial element in diverse and often conflicting theories, especially—as we have seen—in psychology. To provide a common example:

(a) Cannon–Bard Model:

$$\text{Event} \rightrightarrows \begin{matrix}\text{Arousal}\\ \text{Emotion}\end{matrix}$$

(b) James–Lange Model:

$$\text{Event} \rightarrow \text{Arousal} \rightarrow \text{Interpretation} \rightarrow \text{Emotion}$$

(c) Schachter–Singer (general/reasoning) Model:

$$\text{Event} \rightarrow \text{Arousal} \rightarrow \text{Reasoning} \rightarrow \text{Emotion}$$

(d) Schachter–Singer "Two-factor" Model:

$$\text{Event} \rightarrow \text{Arousal} \rightarrow \text{Cognitive labels} \rightarrow \text{Emotion}$$

6.1.11 *Thalamocortical Rhythms*

The concept of "thalamocortical resonance" is at the center of this model, especially in regard to the theory developed by Llinas, Ribary, Contreras and Pedroarena (1998) on the modulation of the thalamocortical rhythms by the brainstem, a type of gamma-band frequencies which are considered to be the origin of this resonance. These frequencies in turn are caused by two different systems of thalamocortical loops which interact to produce consciousness and disrupt consciousness when thalamocortical dysrhythmiae are present. Of note, in this model, one thalamocortical loop system has specific projections, while the other one has diffuse projections. In any case, the thalamus (Fig. 6.4) is the center of this analysis.

6.1.12 *Internal Simulation and Self-Modeling*

In 2002 Hesslow defined the internal self-modeling approach as a simulation-based process encompassing multiple functional activations, including the neural basis for cognition and sensorymotor processes at the core of (the perception of) an inner self vs. inner world. This perspective has also been, at least in part, embraced by Grush (2004) and Revonsuo (2005), particularly in regard to the so-called world simulation metaphor. Revonsuo views consciousness as interaction between the environment and the organism. This interaction as both internal and truly 'dreamlike' in the senses that it is best represented during neural activity in dreams and—in an adaptive sense—via mimicking situations. Another model is offered by Trehub, focusing on a perspective of the self (better analyzed as permitted by the very origin of such self through vs. in the 'self locus') which we could categorize as 'egocentric'. This perspective is functionally active/activated in the retinois structure that allows the conscious content to be perceived by the self. A close model can be found in the "self-model theory of subjectivity", by Metzinger (2004), in the sense that everything we perceive is itself a model—phenomenologically based, but not ontologically defined (−able) as 'true self' in the (external) world. This is, in fact, literally defined as 'phenomenal self-models' or PSMs.

6.1.13 *Cognitive/Cognition (Attention)-Based Models*

Attention, focus and cognitive process are some of the most common underlying processes in many models for consciousness based on neural

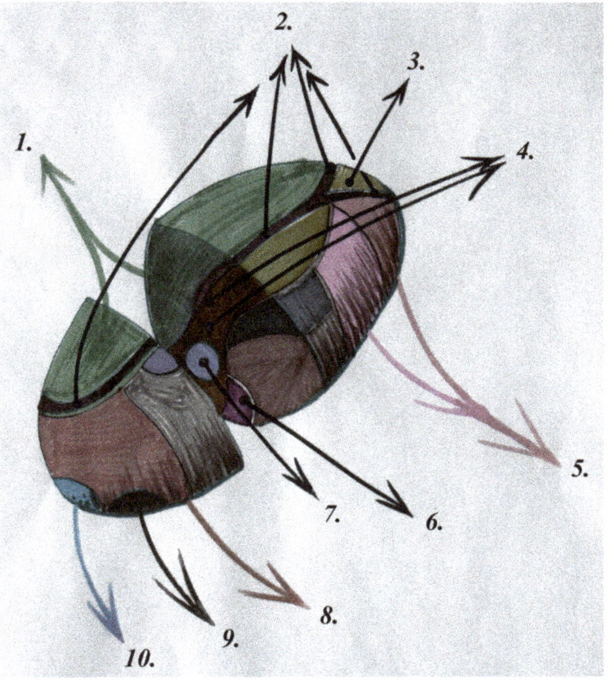

Fig. 6.4 The thalamus with the connection to the prefrontal cortex (1), the internal medullary lamina (2), the anterior nuclei (3), the intralaminar nuclei (4), the connection to the motor cortices (5), the VPM (6), the centrum medium (7), the connection to the posterior association areas (8), the lateral geniculate (9. to V1) and the medial geniculate (10. to A1)

underpinnings although, as we have previously said, they do not represent neural underpinnings per se. In this context we will discuss again the cognitive and affective computational schema or CogAff proposed by Sloman and Chrisley (2003), and the Corollary Discharge of Attention Movement or CODAM developed by Taylor (2012). However, these are only two of the cognitive models still in use in contemporary consciousness research; we can think for instance about the multilayered representation feedback of Haikonen (2003), the differentiation between consciously penetrable processes and consciously impenetrable processes in the Supramodular Interaction Theory (SIT), and the polychronous groups by Izhikevich

(2007). In CogAff consciousness is seen as been differentiated in many types and computationally quantified on the basis of deliberative, meta-management and reactive processes. In CODAM, we find a position quite similar to the Global Workspace model by Baar (1988), in the sense that the inner experience is added to the coding of content. However, CODAM is also similar to intermediate level theory in the sense that consciousness is mediated by a system which is aware of its own world representation in a connective sense. Furthermore, this model of sensory attention views neural processes as coordinated into a set of functional modules and the inner self as originating from a "corollary discharge of the attention movement signal". This view is very close, in supportive terms, to certain philosophical positions in phenomenology, particularly in Husserl, given the concept of 'temporal flow'.

All these models present very interesting arguments for the possible connection between biological, especially neurobiological, processes and other matter-related (i.e. 'based') mechanisms with the 'emergence', 'product(ion)', 'generation' or 'creation' of consciousness. Of course, given the focus and content of this work, we did not discuss in depth models of consciousness from very different perspectives, most especially the mystical, spiritual or religious models of consciousness, which are truly as rich and as complex (in many cases far more complex) than any of the models presented above.[19] We decided to cover the most important models of consciousness from the perspective of critical neuroscience, for example, from a *critical* standpoint *on* neuroscience. Our analysis takes into high consideration the totality of these models to try a better understanding not only of the possible connections between consciousness and all those neuroanatomical structures, functions, processes and properties, but also to understand (a) the very need for such explanation-modeling both in the sense of the need to justify in neurological terms and the psychological (or other/further) processes involved in the developments of such needs, and (b) to be aware of the most recent, cutting-edge techniques, technologies, and methods to 'dig deeper' into the neural matter before, through and after philosophical investigation. However, those questions we posited at the beginning of our work are still unanswered; better: unanswered via physicalist-reductionist-mechanistic explanation:

[19] For further investigation of these models, Please see references at the end of this chapter.

> If we *give it some thought*, we *realize* it is highly unlikely that a pathological process afflicting the brain could, in and of itself, *generate* the incredibly rich experiential spectrum of the states currently diagnosed as psychotic. How could abnormal processes in the brain generate such experiences as culturally specific sequences of psychospiritual death and rebirth, *convincing identification* with Christ on the cross or with the dancing Shiva, an episode involving death on the barricades in Paris during the French revolution, or complex scenes of alien abduction?[20]

Of course, this would represent an opinion, albeit coming from a psychiatrist and researcher with vast experience in this field, but it would nevertheless remain a well-informed opinion. Thus, if specific processes were found on a purely biological-mechanistic level that would explain *away* all these factors, we would have to make sure to store this opinion as interesting intellectual effort and yet no longer valid except as a form of literature, crushed by the weight of scientific evidence. However, such evidence is simply not her, or—if we still wanted to defend the most reductionist view possible—not yet. Moreover, even if we took—as we always *should* take—into account all the studies published on the connection between consciousness, the self, the mind–body problem and neurobiological factors, our interpretation here is that there is still no enough (and we mean in both qualitative and quantitative terms) evidence of such links in reductionist-causal terms, that is, that all these experience were only produced by—rather than channeled through, or mediated, amplified by—the brain. A similar view is embraced by Beauregard, who writes, in regard to the research by Wilder Penfield at the Montreal Neurological Institute:

> At the end of his scientific career, Penfield concluded that higher mental functions –such as consciousness, reasoning, imagination, and will- are not produced by the brain: mind is a nonphysical phenomenon interacting with the brain.[21]

Beauregard covers a wide range of experimental evidence to support this positions, from the studies on hypnotic trance by Fabienne Roelants, to the research by William Braud on psi phenomena, from Ganzfeld telepa-

[20] Grof, S. 2012. Healing Our Deepest Wounds: The Holotropic Paradigm Shift. Newcastle, WA: Stream of Experience Production, p. 125. Italics added.

[21] Beauregard, M. 2012. Brain Wars: The Scientific Battle Over the Existence of the Mind and the Proof That Will Change the Way We Live Our Lives. New York, NY: HarperCollins, p. 11.

thy experiments to the research by Charles Tart, and many, many others. Therefore, a complete discussion on all these elements of para-mechanistic explanations needs to be very well understood and contextualized in a thorough research on neuroscience, consciousness and the mind-body problem. A good example in this area is that the research on Nonordinary states of consciousness (NOSC) within the field of medicine, especially in the subfields of neurology/neuroscience and anesthesiology has to begin with philosophy. In fact, there are ways to induce these states artificially, in the laboratory with synthesized forms of chemical elements, or by using psychedelic substances (isolated or wholly extracted) from plants, on the other. Can altered perceptual states controlled, or at least partially understood? These states can certainly occur spontaneously, but we still do not have a clear idea if this concept of "spontaneity" makes sense at all in this context. We don't fully understand, with the modern scientific method, whether these phenomena could be in turn elicited by a higher or deeper level of consciousness, whether spiritual or divine. Moreover, the most recent scientific techniques there are many means of altering consciousness through brainwave biofeedback, sensory isolation and, in a psychiatric sense, techniques such as Transcranial Magnetic Stimulation (TMS), Transcranial Direct Current Stimulation (tDCS) as well as Electroconvulsive therapy (ECT). According to Wade et al. (2014) "the brain can no longer hide from researchers behind the fortress of the skull" but the very concept of "hiding" pre-assumes the existence of a purposeful action aimed at hindering the abilities of the observe to investigate what lies beyond the surface, and idea that certainly appeals to quantum physics-based investigations. As it is the case, we are far from being able to claim that the brain *generates* these mental experiences. In this regard Satel and Lilienfeld (2013) warn us against "mindless neuroscience: the oversimplification, interpretive license, and premature application of brain science in the legal, commercial, clinical, and philosophical domains."

Elsewhere (Tomasi 2016) we addressed the notion of faith in relation to "jumping into the unknown" and the "God of the gaps" in the context of neuroscientific analysis. Here we extract the perspective on neuroscience by Satel and Lilienfeld (2013), especially the observation that "[it is not] it logical to regard behavior as beyond an individual's control simply because the associated neural mechanisms can be shown to be in the brain."[22] Through this association-correlation, we have for decades now

[22] Satel, S., Lilienfeld, S.O. 2013. *Brainwashed. The seductive appeal of Mindless Neuroscience.* New York, NY: Basic Books, pp. 149–150.

assumed the existence of an 'illuminated brain'. Illuminated in three different ways: (a) because of the functional magnetic resonance imaging of the areas illuminated through this technique and others, (b) because of the light it shed on this very association, undermining the value of anything non-neurally (matter) based and (c) because it is "enlightenment through enlightenment". In other words, if we appropriately use the scientific method, which is this sense is the mere translation of neuroimaging, we are indeed 'enlightened' in our scientific procedures, and we obtain more 'light' shed on the brain (read: 'us'), which in turn becomes enlightened and fostering more enlightenment. Previously analyzed references to a spiritual-type of enlightenment (and we could pretty much apply most of the assumptions herein, in points (a), (b) and (c) of the discussion), as well as historical and social interpretation of the term 'illuminated' cannot but makes us more aware of the complexity of the issue. The other problem stems from putting human brains, computers and immaterial souls in the same category is a quite reductionist and imprecise summary. Given these premises, we could infer that in this process of 'illuminating' the brain, we are attempting to shed more light on the very connection between the entire series of concepts mind-brain-body-soul-spirit, in the sense that, from a purely logical point of view, as well from a correct theoretical analysis of our scientific method in this area of medicine, we cannot simply avoid dimensions beyond the physically evident. In the views of Stanislav Grof, this is where our modern 'psychospiritual crises' originates: people who *experience* these states, *suffer* these states according to modern medicine; these manifestations (a term that suggests a vast array of metaphysical and theological interpretations), these symptoms don't need to be fully experienced, perceived, embraced. On the contrary, they need to be suppressed or possibly annihilated. Modern medicine and modern science "emphasize experience rather than talk",[23] a process recently questioned by disciplines such as Narrative Medicine.

Confuting the position of Sam Harris "The more we understand ourselves at the level of the brain, the more we will see that there are right and wrong answers to questions of human values," Satel and Lilienfeld state that neuroscience is very useful in addressing and answering (from the

[23] Grof, S. 1998. *Rethinking Basic Assumptions about Psychology and Psychiatry: The Role of Spirituality and Nonordinary States of Consciousness.* In: Bassman, L. 1998. *The Whole Mind: The Definitive Guide to Complementary Treatment for Mind, Mood, and Emotion.* Novato, CA: New World Library, p. 45.

scientific point of view) questions of neural processes in moral decision making, but it does not focus on the ethical aspects of these decisions: "[…] it is not at all evident how such discoverable facts could ever constitute a prescription for how things should be" (Satel and Lilienfeld 2013). Note the language here: first of all we talk about 'evidence', a term that, once again, brings us back to the basic concepts of Medical Philosophy and of science in general, as well as (human) perception. Second, we follow one of the first mandatory requirements for scientific research: facts have to be 'discoverable'—the principle of falsifiability follows quite naturally—together with the (understood by some, misunderstood or completely rejected by others, as we have seen) scientific 'faith' in the possibility of finding a result (Tomasi 2016). All these discussions on the scope of neuroscience in the context of perception of self, of one's role and purpose in this world, and of morality brings us back to Jürgen Habermas (1993), who defines intersubjective dialogue as indispensable to understand morality or "communicative ethics of need interpretation" and presents many topics, including value and importance as *ulterior* to social phenomena, but also social norms at behavioral baseline and peaceful alternative to further conflict, in a truly metaphysical sense. But moral rules can certainly change over time, and we need to understand the social and historical activation of this change to understand where our values and principles come from. Of course, the assumption of the syntactical as well as cognitive conception of plural selves vs. single-individual self leads to the definition and awareness of society as a plurality, a combination of more personalities, or selves. We think about ourselves and about our selves, and we wonder whether this cognitive process can indeed change the "individualization of identity". As we have seen, neuroplasticity is the ability of the brain to change its structure and its functionality in terms of pathways and synapses, expressed in degrees and in different ways throughout the nervous system, depending on the activity of their neurons; for instance a neuronal activity related to stimuli received from the external environment, or in reaction to traumatic injury or pathological changes. We have also seen how neurological underpinnings to such outcomes are visible through Magnetic resonance imaging technologies (*MRIs* and *fMRIs*), showing an increase in the size of certain brain regions following their repeated use. More in detail, we have seen that neuronal cells have increased activity and consequently formed more synapses between them in enriched environments, especially during learning, and in case of cerebral reorganization. This is very important for the definition of a 'self' which takes in consideration the possibility that *today's self* could be different from *yesterday's self*, and could represent an internal-external riff and split at a social level due to the fact that "the lack of any coherent identity for the 'public' creates futile rituals of expert advice and disagreement(s) on what is morally justified (and justifiable)," as expressed by Langdon Winner (Winkler et al. 2017).

In this chapter we are discussing the mind–body problem and its philosophical implications for perception, identity and neural underpinnings of perception and conscience/consciousness. In regard to this combination, we mentioned the differentiation of terms between different (stages or levels of) consciousness exactly in relation to conscience and awareness. These differentiations happen due to qualitative and quantitative parameters used in medicine as well as in clinical ethics and in the legal system to assist, sometimes with positive outcomes, sometimes with negative outcomes, the patient and the patient's family in very hard decision-making processes. In this regard, we would like to focus on a very controversial and contemporary issue related not only to legal and clinical aspects, but also on theoretical examination as well as philosophical/ethical debate where "the exact science of the hard matter" is truly difficult to understand. That is why "examining, controlling, and verifying" all the implications, in clinical, philosophical-ethical, social, political and legal terms of our understanding of 'being conscious' is so important. To provide some examples in this area, we refer to the very difficult cases such as the ones of Michael Farley (Veterans Administration Medical Center in Manchester, NH, USA) as a result of wrong medication administered causing a 'locked-in syndrome' following 2 strokes, Terri Schindler-Schiavo (Humana Northside Hospital, St. Petersburg, FL, USA) following a cardiac arrest, Jahi McMath (Benioff Children's Hospital in Oakland, CA, USA) after complications from surgery to cure her sleep apnea, and similar.[24] Some of the cases mentioned above were nationally (in the United States of America) and internationally publicized, and need therefore an extended analysis of possible double-hermeneutics effect in the efficacy and effectiveness of the therapeutic intervention provided. The patient involved in the first case is Michael Farley, a 60-year-old US Navy veteran, who sought treatment at the Veterans Administration Medical Center in Manchester,

[24] In this context, we decided to simply mention some of the most famous cases related to this discussion. For further information on Terri Schindler-Schiavo and Jahi McMath, we refer to the Bibliography at the end of this chapter.

NH following a painful headache and losing his peripheral vision. Clinical tests were performed at the facility, and the VA physicians determined that the patient had suffered a stroke. The Veterans Administration Medical Center's physicians prescribed the wrong medications to the patient. This medical malpractice ultimately led to a Locked-in Syndrome. Furthermore, the physicians ordered an ambulance transfer to a VA facility in West Roxbury, MA. Once there, the patient should have been admitted to a hospital and evaluated by a neurologist. In fact, following this decision, the ambulance was ordered to turn back. The physician did not provide any reasons for this order. The allegations include the failure to use proper monitors, perform appropriate clinical testing, and refer the patient to qualified medical specialists. The medical malpractice lawsuit contends that Locked-in Syndrome the patient is suffering from could have been prevented if the VA physicians in Manchester, NH had properly diagnosed and treated the patient after his first stroke. US District Court Judge Landa McCafferty ruled that:

> It is a basic principle of medicine that a patient who has suffered a stroke is generally at an elevated risk of suffering a second stroke. Therefore, doctors who are treating stroke patients must be cognizant of this risk, and they must take steps to prevent a second stroke from occurring. Unfortunately, Mr. Farley's doctors at the Manchester VA did not adhere to this standard of care. They failed to provide him with an adequate diagnostic evaluation, and as a result, they carelessly prescribed him the wrong medication. (Schmidt and Clark 2011)

The combination of current, up-to-date, peer-reviewed medical studies and expert testimony clearly indicate that the physician who evaluated Mr. Farley should have prescribed the anti-coagulant drug Coumadin (warfarin) after the first stroke occurred, as the Medical Guidelines recommend this drug for ischemic stroke patients at risk of cardioembolic blood clots (Goldstein et al. 2011). Per official medical records and court documents, the VA physicians sent the patient home after his first stroke. In this context, he had been prescribed two baby aspirins QD. However, after six weeks, the patient was found unresponsive and taken to Elliot Hospital. Here physicians examined the patient again. This time, a massive stroke exactly in the same cerebral area where the first stroke occurred was (re) discovered. In defense of the physicians caring for Mr. Farley, Drs. Lamphere and Lombardi, per Court Rule the four expert witnesses testifying on behalf of the government offered Inconsistent and Non-Credible Theories on the likely cause of Mr. Farley's first stroke. More specifically, Dr. Kim indicated that his first stroke was likely caused by a cardioembolic blood clot; Dr. Greer stressed the importance of the result of a dissection on the brain artery as primary cause; Dr. Caplan disagreed on the basis that the patient's second stroke occurred in the exact same brain area of the first stroke, therefore limiting the likelihood that a dissection was the possible primary cause (Mujkanović 2016). Furthermore, the government sought defense by providing evidence of the patient's use of narcotics in combination with a history of noncompliance, and asserted a defense of comparative negligence to mitigate or eliminate damages awards. In the latter, the government claimed that the plaintiff's own negligence contributed to the patient's injury. Following the Standard of care applicable to the treatment of ischemic stroke, the physicians caring for Mr. Farley did not use the "ordinary skill and care that all medical practitioners must use" (Fremgen 2012, p. 61). Mr. Farley should have been admitted to the hospital and consulted with a neurologist and a cardiologist and undergo an echocardiogram to verify the high risk of a cardioembolic stroke. Thus, by failing to prescribe Coumadin, Drs. Lamphere and Lombardi violated the standard of care. Michael Farley won a lawsuit against the Veterans Administration Medical Center in Manchester, NH (which has been ruled as both prescribing the wound medication and leaving him "medically abandoned") and has thus been awarded a $21 million malpractice verdict. The amount was further subdivided in $12 million (possible medical care required in the future), $8.1 million (non-economic damages), and $1.3 million (past and current medical expenses). The medical mistakes should have been avoided by providing the appropriate standard of care. However, based on the court documents, it is important to note that specific aspects of patient's behavior such as noncompliance and use of narcotics, and the follow-up activities, both medical and legal might have contributed to the condition. These aspects might have run in favor of the VA's defense. In any case, the very complex presentation of a 'Locked-In' Syndrome following the stroke require a further investigation on the very relation, both neurological in terms of electrochemical response and neurogenesis, and psychologically–philosophically intended, between mind and body and quality of life vs. quantity of response, interpreted as patient's ability and capability to interact appropriate with others. The latter is especially important due to the impossibility, on the patient's side, to appropriately communicate with the healthcare providers important pieces of information. This is why, even if some medical circumstances could have been different, more specifically in relation to some of the parallel or alternative causal scenarios offered by Drs. Kim, Greer, and Caplan, the clinical judgment and actions by Drs. Lamphere and Lombardi would still constitute medical malpractice, and we would agree with the ruling that the patient was 'medically abandoned'.

6.2 Perceiving, Seeing, Evidencing, Understanding

6.2.1 Cognitive Neuroscience

Cognition, especially human cognition, is where the *attention* of cognitive neuroscience lies. As in the previous chapters, we are again faced with a multidisciplinary approach, involving fields such as general and computational neuroscience as well as cognitive, neuro-, and physiological psychology, to better investigate the neural connections/underpinnings of cognitive activity. The same can be said about the techniques and technologies behind the experimental methods used in this field at the crossroads with the scientific fields using functional neuroimaging and electrophysiology but also deriving observational elements from genetics and physics. Furthermore, as with the histories of the disciplines discussed in Chaps. 1, 2, 3, and 4, the history of cognitive neuroscience overlaps with the history of its predecessors, from physiognomonics-physiognomy, phrenometrics and phrenology to the precursors of modern neuroscience. The evolution of these non-verifiable (from the perspective of the principle of falsifiability and hypothesis-testing in the evidence-based sense) disciplines led to diverse theoretical approaches, among which the localizationist view (for instance with John Hughlings Jackson) and the aggregate field view (especially via the studies by Pierre Flourens). The first attempted to connect cognitive abilities to specific areas in the brain, first in the cortical and subcortical areas, then in the limbic system. The second view promoted a more cooperative-integrated activity in the brain as primary cause for specific mental functions, especially in the case of cerebral cortex, cerebellum and brainstem. A complete or almost complete mapping of the brain started with the work by Pierre Paul Broca, Carl Wernicke, Angelo Mosso, Eduard Hitzig, Gustav Fritsch and Korbinian Brodmann who were instrumental in the rapid evolution of neuropsychology-based investigations which would support the further developments such as the competing perspectives of behaviorism and functionalism on the side of theoretical framework and conceptualizations such as the 'Neuron Doctrine' on the other. Given these premises, psychology, and especially experimental (neuro)psychology served as strong foundation for the emerging field of cognitive neuroscience, which truly developed at a much faster rate with the new brain imaging and brain mapping technologies, especially MRI, fMRI, SPECT and PET. However, this discipline was not recognized as a separate field until the 1960s, particularly in the United States and especially due to the new perspectives offered by Michael Gazzaniga, George Miller, David Marr and Ulric Neisser. The classic "three levels" of (neuroscientific-psychological) analysis offered by Marr, that is, the computational, the algorithmic/representational and the physical still represent a very popular framework among cognitive neuroscientists.

6.2.2 Computational Neuroscience

This is possibly one of the most interesting subfields of neuroscience, particularly for a philosophical audience. The main reason is the theoretical investigation of quantitative information-learning processes of the neural structures which, aside from providing the rational for the synonym 'theoretical neuroscience', determines the hypothesis that neural–mental activity is a question of quantity, and, therefore, of computation. To be sure, this field also overlaps with disciplines at the intersection of general neuroscience, psychology, engineering, physics, mathematics, computer science and artificial intelligence, but is definitely separate from both field-related theories of learning and from psychological connectionism. In fact, cognitive neuroscience focuses primarily on the biology, function and activation-activity-processes of neurons and neural systems. It utilizes real and model-based computational systems to create theoretical hypotheses which can in turn be tested using laboratory experimentation with the help of disciplines such as biology or psychology. For instance, frontal and parietal lobes are analyzed as brain integrators for the data coming via multiple sensory processes and are investigated with the help of animal models, especially single-unit recording in primates. Animal models are truly important to test disorders at the intersection of psychiatry and neurology—as in clinical computational neuroscience—or to debate the possible application of computational approaches to consciousness, for instance in the work by Crick and Koch on neural correlates of consciousness or NCC. Modeling and simulations are at the center of integrative neuroscience-based investigations of neural processes and visuospatial recognition and memorization processes and the main goal of understanding how to replicate the complexity of the neural computation found in biological structures into 'intelligent machines' and theoretical models for the analysis of decision-making processes such as Neural Modeling or the Computational Representational Understanding of Mind or CRUM. Thus, computational neuroscience also comprises mathematical descriptors like the biological neuron model (spiking neuron model, subdivided in electrical input–output membrane voltage models and natural or pharmacological input neuron models) used to investigate the properties of specific cells in the nervous system and the generation of action potentials across the membrane. From the biological neuron models to the outcomes in terms if modulatory-dynamic aspects, to processes such as adaptation, sensitivity, and shunting, this field serves as cornerstone to the now popular topic of synaptic plasticity. This is of fundamental importance, due to the application of information gathered in the analysis of such topic to areas such as clinical psychology and preventive medicine. The mechanisms at the center of synaptic plasticity demonstrate that synapses play a far deeper and more important role than 'just' delivering the messages from multiple stimuli and connect neurons and other cells and structures, which truly represents a groundbreaking perspective for neuroplasticity, especially in its application for pain management, as we have previously suggested (Doidge 2015). First of all, these processes represent the neural underpinnings of attention, learning and (short-, mid- and long-term) memory—as in the models by Hebb and Hopfield—and as such they constantly morph, strengthen and rearrange according to the two basic types of synaptic plasticity, that is, intrinsic/homosynaptic and extrinsic/heterosynaptic. From the viewpoint of computational neuroscience, multiple time scales functionality is a very interesting aspect of this type of plasticity, especially in the context of neurophysiological memory analysis. This approach in fact provides new insight on the quantitative impact of the aforementioned changes (cascades) in synaptic transmission elicited by external stimuli. However, there are still many unanswered questions such as the ones pertaining the lack of detailed information on low-connectivity areas, in comparison to the side-specific neural networks found throughout the brain and appearing very complex and yet well organized, at least at the level of biological neurons, less specific than their artificial counterparts. The analysis of these neural structures relies on theoretical model presenting the complexity of neural interactions to simplified-basic-constituting sub-models which can lead to pairwise interactions and/or parallel functions, for instance following algorithms (themselves reducing the complexity of such systems into single or coupled neuron-based theoretical structures, without biological accuracy but allowing for cost reduction of experimental technology) artificial models like Genesis or Neurons (software platforms) or the Ising and (vs.) McCulloch-Pitts models, the latter especially useful for low neural network activity. This activity of course depends on specific computational functions and geometric structures vs. properties at the level of neuronal cells, axons, and dendrites (possibly the most important area for the execution of computational activities) but also including—it is the case of site-specific and interactive models—synaptic clefts, spines, as well as receptor sites. In synthesis, computational neuroscience contributes to the analysis of the nervous system by focusing on quantitative, structural and procedural aspects of neurons, their interaction and network. Common foci of research are therefore the production, development and transfer of information through neurotransmission, itself a combination of electric impulses and chemical reactions. From a theoretical perspective, computational neuroscience utilizes several frameworks, spanning from structural and process-describing models to mathematical modeling through wiring hypotheses and functional mechanisms, for instance following Bayesian schemata.

6.2.3 Sensory Neuroscience and (Theory of) Perception

Perception is a fundamental concept in philosophy and in science. Sensory neuroscience targets the physiological–anatomical aspects of perception—especially the audio-visual and olfactory systems—while theory of perception investigates its theoretical underpinnings. Furthermore, philosophy of perception concerns ontological aspects of experience, existence and essence, more specifically the nature and status of perceptual information and data. These are thereby analyzed in lieu of the valence and influence of broader philosophical viewpoint (confirmation-perceptual) filter and bias, and/or system of belief. Following Kant's theory of Pure Understanding, Hegel introduces the concept of 'Understanding Consciousness' as mind's own image of itself, thus seeing itself as a unifying principle, an omnicomprehensive Greater Self in which the separation of subjects–objects becomes one, embraced by its very singleness. In comparison to the impersonal element of Kant's theory, so perceived by Hegel because of its description of the Self, which has no personality, the author of the Phenomenology of Mind introduces the Self Consciousness (Tomasi 2016). In the context of sensory perception, the philosophical framework evidences a combination of four narratives containing hidden dialectics. In the case of Hegel in particular, we observe Master and Slave (Lordship and Bondage), Stoicism and Skepticism. Furthermore, we encounter the concept of Unhappy Consciousness, which interestingly related to the concepts of Melancholia and Depression via the perception, sensorial, psychological, and philosophical of the identity and structure of an individual Self-persona, also in relation to quantifiable physical effects on this form of consciousness as it related to psychiatric disorders. Furthermore, if the theory of Kant is guilty of missing the mark of realism (fundamental in the context of sensory neuroscience), this type of consciousness is linked to a 'real' person, evolved in sub-moments, thus drawing ontological elements from philosophical tradition predating kantian speculation.

From the perspective of biology, perception happens via the neural activity allowing the brain to react-respond to internal and external stimuli by firing multiple action potentials following a 'neural code' at the center of the analysis of sensory neuroscience. There are certainly differences, however, between internal and external stimulation, although the combination of both provides the 'bigger image' of our thought, emotions, feelings and perceptions, including neural mechanisms of attention (Fig. 6.5). With internal perception or proprioception, we identify the response to and information gathered from or bodies, in terms of location, movement-motion, posture and activity-ability, as well as basic physiological needs. External perception (also called exteroception or sensory perception) provides us with information from the external 'world' through our senses. Furthermore, there are two main directions and two main process underlying perception. We have afferent and efferent directions and thus have afferent and efferent neurons, the first responsible for the delivery of nerve impulses from sensory stimuli toward the CNS, and the latter responsible for carrying impulses from the CNS to the muscles (they are, in fact, motor neurons). From the perspective of neuroanatomy, afferent neurons enter the spinal cord through the dorsal root, while efferent neurons exit the spinal root from the ventral root. The two aforementioned processes instead are the processing sensory input, responsible for transforming low-level to high(er)-level information (at the subconscious or at least pre-conscious level of awareness), and the cognitive-attentive processes connected to representation, expectation, and selection. Of course, this translates into localization processes

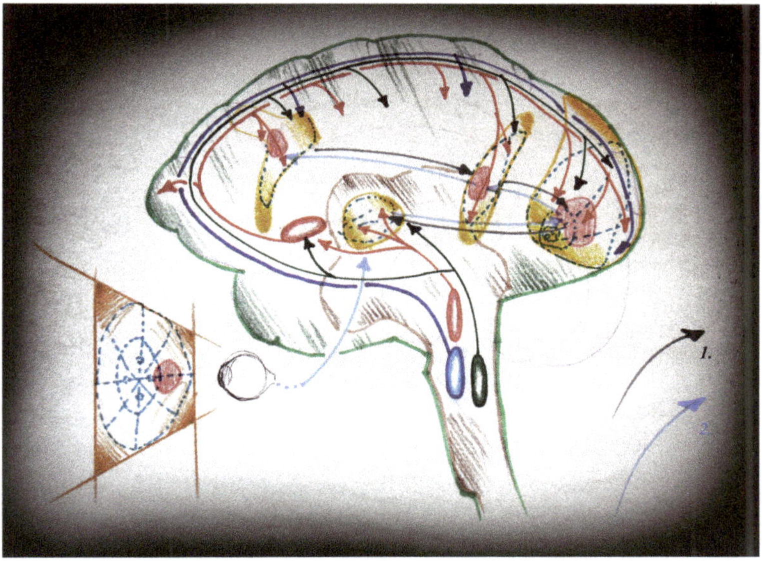

Fig. 6.5 The neural mechanisms of attention according to the model by Mangun and Hillyard (2012). In detail, the pathways originating in the brain stem cholinergic nuclei are here represented in red, the pathways originating in the locus coeruleus in green, and the pathways originating in the Raphe nucleus in blue. The feedback is indicated by the black arrow (1) and the feed forward by the light blue arrow (2), with the eye perspective and the focus of attention in the side plane-view (red)

in the central nervous system, especially in the brain, which require further analysis with the help of computational and socio-cognitive approaches. In fact, the earliest studies in neuroscience the general view was that sensory processes contained a somewhat passive component, in the sense that receptors were receiving information 'fully' and only depended on biologically based activity of our senses and were, therefore, "at the mercy" of environmental stimuli. In more recent times the open debate on "perception as active process of hypothesis testing" appears to provide more insight to the issue-at-hand. In fact, the same point of view is to be found in theoretical medicine, with special regard to confirmation bias and placebo/nocebo effects. In this case, possible bias at the conscious or subconscious level cannot only be fully embraced (as a theoretical assumption) but it could lead us to a much better understanding of the very role that subjectivity might play in sensory processes and perception. In this sense, the very connection that a subject has with his/her own environment, the link between man and nature, needs to be investigated to capture the extent to which we can perceive the world around us as "stable, but not still" by way of ever-changing, modular, combinatory (as in the case of mutually influencing senses, for instance smell and taste)[25] and sensory map-based activity in our nervous system. Of course, within the nature vs. nurture framework, we can also hypothesize an evolutionary base for the development of sensory systems. As such (human) perceptive processes could have been developed based on the development, itself a product of evolution, of external factors and stimuli. This explains the adoption of natural stimuli with mathematical modeling in neuroscience. Of course this can be very challenging due to the complexity of natural stimuli (and the intrinsic difficulty of using them in practice, in natural environment and/or controlled settings)—as opposed to artificial ones, a complexity which theory of perception has to take into high consideration, given the transferability, or lack thereof of simplified equations. In any case, experimentation in sensory neuroscience usually happens by monitoring the subjects' neural activity—including the analysis of spike shapes, frequency (for instance, with post stimulus time histogram), time (especially latency between stimulus and the effect) and patterns—in response to a single stimulus or a set of stimuli, an approach, at least theoretically similar to behaviorist investigations, especially in conditioning, albeit with the help of techniques such as electroencephalography, electrophysiology or the now common functional magnetic resonance imaging. For instance, in the case of an observation of a single neuron, the experimenter will isolate a neuron on the base of the stimulus it reacts to in the form of an action potential. It is thus important to understand that spiking patterns might actually react-respond to more than one stimulus, and thus the experiment will have to be conducted over time and "following the patterns around" to precisely monitor neural variability. This approach, together with linear regression, will allow a very accurate image of the neurons' spatio-temporal receptive (perceptive/perceptual) field and thereby account for excitation and depression processes in the neurons.

[25] Although, as we have seen, the olfactory system plays a very special role given that it almost completely bypasses the thalamus, in order to reach cortical areas.

6.2.4 *Systems Neuroscience*

Following the analysis of perception in "Sensory Neuroscience and (Theory of) Perception", we arrive at the field of systems neuroscience, responsible for the study of neural circuits, networks and systems, including—the associative circuit, the limbic circuit and the motor circuit (Fig. 6.6). As with many other disciplines, systems neuroscience combines multiple theories, techniques and technologies from different fields, especially general neuroscience and systems biology. More specifically, the focus here is on the connection between internal and external environment, between the subject and his/her world, and between molecular and cellular approaches as biological underpinnings for higher functions including cognitive and rational(ization) processes, language production, understanding, and communication, emotion, memory and awareness-consciousness.

6.2.5 *Neuroinformatics*

Continuing with the computational analysis in neuroscience we land on the field of neuroinformatics, a discipline focused on the application of such analyses to areas like development of theoretical frameworks and databases for the collection, investigation, storage and sharing of information on the nervous system, amelioration and application of computational modeling features and tools for neural networks and processes. More specifically, the information thereby collected involves different typologies of inputs, including cognitive, cellular and molecular data, as well as developmental, disease/disorder-related and neurobiological information, and artificial-neuromodeling/neuroengineering data. In the case of information gathered via neurons and neural networks, the main focus covers cortical and subcortical areas, especially sensory-related areas linked to visual, auditory, taste, hearing and olfactory systems, organized in neuroinformatics maps and atlases to foster precise classification and averaging-standardization (in terms of modeling-based simulations). However, neuroinformatics can also draw information from the analysis of neurotransmission, as in action potentials and channels, as well as in neural and protein pathways, to better understand physical and emotional pain but also behavioral/lifestyle or genetic-based disorders within general medical health and mental health. Neuroinformatics thus investigates aspects of neural cytology and neuronal (and motor) differentiation, axon

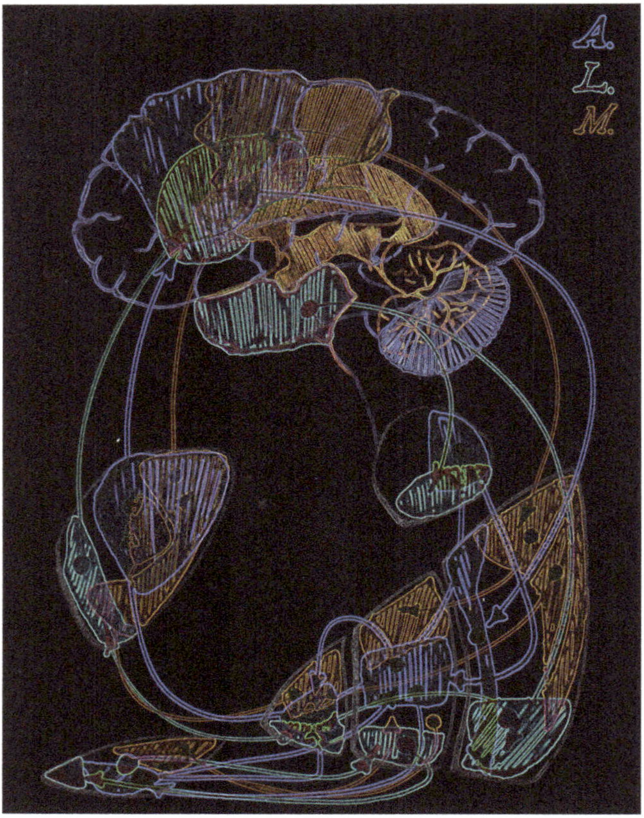

Fig. 6.6 A comparison of the associative circuit (A., in light blue) involving the dorsolateral prefrontal and lateral orbitofrontal cortex, the limbic circuit (L., aquamarine) involving the limbic and paralimbic cortex, as well as the amygdala and the hippocampus, and the motor circuit (M., in orange) involving the sensorimotor and premotor cortex. This model has been investigated in detail by McBride in *Trends in Neuroscience* (2014)

guidance, synaptic formation and growth factors in general, including clinically relevant aspects such as healing processes after injury and cellular/tissue regeneration. Of course, a problematic aspect here is the applicability and integration of mathematical models to the biological base of the systems investigated. Therefore, neuroinformatics strongly relies on studies in biology, chemistry and genomics on one side, and to information science and information technology (IT), and software/neuroimaging research (brain-computer interface) on the other side. The theoretical framework is moreover investigated using inputs from experimental psychology and medical science. In this sense, digital models accompany biological observations to better describe brain anatomy, structure, and function, also with the use of morphological analysis techniques such as neuron tracing and reconstruction. Technologies include a multitude of engineered schemata, atlases (including descriptors-related atlases with socio-epidemiological values), and combinatory database. For instance, we can think of the Neocortical Microcircuit Database or NMDB collecting and editing cellular and microstructural data, the web-based digital atlas BrainMaps, with a very high resolution and virtual microscope, the standard XML metaformat BrainML system, the Budapest Reference Connectome, especially useful to trace neural connections with a three dimensional view, based on the Human Connectome Project, but also the multidisciplinary modeling approach of SenseLab, also a part of the Human Brain Project, which provides a multilayered series of neuroscience tools for analysis and information database modeled on the olfactory pathway, especially targeting information mediating properties in neurons. Finally, we find the GeneWays system to investigate molecular substances and actions, monitor their interaction and integrating it via observation molecular pathways and improving the research atlas, and the virtual community Biomedical Informatics Research Network or BIRN. All these technologies truly help neuroinformaticians come together to develop new ways to look at the nervous systems, and thereby to better understand ourselves. However, as an independent field, neuroinformatics began it history only in the 1990s, with the first research grants awarded in this field in Europe (European Union and the Office for Economic Co-operation and Development or OECD) and the United States (National Institutes of Health [NIH]). Of notable importance in this context is the "Human Brain Project" or HBP led by the National Institute of Mental Health (NIMH), the United States European Commission Committee on Neuroinformatics, the OECD Mega Science Forum (MSF) first and the—Global Science Forum (GSF) after, and the US/European Commission Biotechnology Task force. The International Neuroinformatics Coordinating Facility at Karolinksa Institutet was created as a result of these shared efforts, following the exact recommendations presented to the GSF (INCF 2005):

1. National neuroinformatics programs should be continued or initiated in each country should have a national node to both provide research resources nationally and to serve as the contact for national and international coordination.
2. An International Neuroinformatics Coordinating Facility (INCF) should be established. The INCF will coordinate the implementation of a global neuroinformatics network through integration of national neuroinformatics nodes.
3. A new international funding scheme should be established. This scheme should eliminate national and disciplinary barriers and provide the most efficient approach to global collaborative research and data sharing. In this new scheme, each country will be expected to fund the participating researchers from their country.

The work by INCF represents a cornerstone in neuroinformatics, especially given the close and strong collaboration at a truly international level, involving countries like Australia, Canada, China, the Czech Republic, Denmark, Finland, France, Germany, India, Italy, Japan, the Netherlands, Norway, Sweden, Switzerland, the United Kingdom and the United States. Furthermore, given the direct support and supervision by the EU Commission, INCF was and is still able to promote academic resources and programs running in parallel with the scientific research area, specifically the production, development, and maintenance of computational data and neuroscientific applications in the laboratory and, in a translational sense, to the community. A good example in this sense is the Program in International Neuroinformatics or PIN. Elsewhere, we find initiative such as the Neuroinformatics Portal Pilot at the Humboldt-University Berlin, which focuses on computation elements, more specifically brain-computer-interface, distributed systems, eye-tracking experiments, machine vision, robot-force-control, artificial vs. natural signal processing capabilities and virtual reality. In Australia, the work of Howard Florey Institute at the University of Melbourne has been fundamental in analyzing cognitive processes using brain imaging technologies, especially at the level of organized networks, and in Japan, the Visiome Platform is the primary center for mathematical models, experimental and data analysis in neuroinformatics. The Laboratory of Computational Embodied Neuroscience (LOCEN) in Italy serves as the experimental section of the Institute of Cognitive Sciences and Technologies at the Italian National Research Council (ISTC-CNR) and analyzes neural underpinnings learning and (higher-level) cognitive processes and expression of sensory-motor behavior, with an embodied-cognition/embodied-computation framework. The focus of LOCEN is also translational, in the way it strives for the developments of artificial intelligence. Close to Italy, we find the Institute of Neuroinformatics (INI) at the University of Zürich and the Swiss Blue Brain Project, which uses the IBM Blue Gene/L supercomputer for the digital processing of three-dimensional versions of neural structure-based models and also provide statistical analysis, predictive simulations and experiments of cortical and neocortical circuitries to account for normal and abnormal functions. In the United States, the Society for Neuroscience Brain Information Group developed the Neuroscience Database Gateway (NDG), followed by the Neuroscience Information Framework, which allows for a more accessible database for web-based research. Other important initiatives in this area are the Cognitive Atlas project, the GeneNetwork, NeuroLex, and the Neuronal Time Series Analysis. The Genes to Cognition Project A is a UK-based neuroscience research program investigating cellular and molecular structure and function, as well as neurogenesis, and synaptic plasticity. The United Kingdom is also home to EBI Computational Neurobiology, EMBL-EBI group for neuroinformatics modeling, and the university-based interactive CARMEN project.

In Denmark, The Center of Functionally Integrative Neuroscience (CFIN) at Aarhus University hosts the MINDLab labs and research groups, and the Technical University of Denmark hosts the THOR Center for Neuroinformatics at its Department of Mathematical Modelling, and studies biomedical signal processing, functional neuroimaging, and multimedia signal processing, again comparing the biological model of neurotransmission with axons (Fig. 6.7) with artificial-theoretical models. Finally—but this list is by no means complete or definitive—a strong emphasis in neuroinformatics is found in Pakistan, more specifically at the NUST-SEECS Neuroinformatics Research Lab, at the interface between neuroscience, medicine, policy making, society and technology. The broad range of aspects related to the concept of informatics allows for an in-depth analysis of its definition, understanding and applicability within the scope of healthcare strategies, in particular clinical practice. In this sense we should differentiate between *applied informatics* and *theoretical informatics*. The latter could in fact be considered the philosophical-scientific base of the development of applied informatics and computer technology

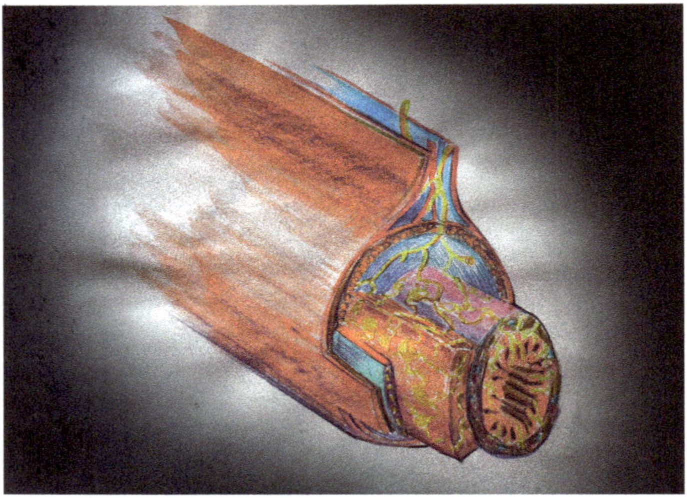

Fig. 6.7 Cross-sectional analysis of peripheral nerve with neural axon, surrounded by endoneurium, perineurium, fascicle, blood vessels and epineurium

in a wide array of fields. Thus, theoretical informatics intersects with the field commonly known as information technology, itself subdivided in:

- Graph theory, including analytical/analyzed data-based structures and (re)search algorithms—analysis, classical and genetic algorithms, and data structures.
- Mathematical logic, including logical/sequential modeling systems—especially common, comparable and distributing systems, including networking and parallel computing.
- Type theory, from base data analysis to epistemologically based inferences in programming—including translating/translational compiler modalities and programming language.
- Theory of computation, itself subdivided into classical theory of computation—including, as Steven Pinker points out (2010), philosophical perspectives such as critical neuroscience and cognitive neuroscience. Furthermore, the definition also contains the perspectives of automata theory, and the theory of computational complexity (a further re-elaboration of the latter, especially focused in digital time/space saving).

Finally, Bioinformatics, also referred to scientific/medical informatics, moves from all the theoretical considerations above to the perspectives of applied informatics, and therefore to the demands of clinical care. Thus, Bioinformatics allows for a precise collection, categorization, analysis, and storage strategies for biological/medical data. Cleary, the very definition and field separation of informatics, medical/nursing informatics and information technology are rooted in the base and scope of the terms, in particular the understanding, specific (field) applicability and technological development (Mitchell 2011). To be sure, in this subdivision we cannot notice some theoretical aspects which are themselves related to the very history of philosophy, especially in the epistemological sense. For instance, the separation in terms between informatics, medical/nursing informatics and information technology brings us back to the Aristotelian separation between *episteme*, *techné* and *phronesis*, and the conceptual framework of (respectively):

- Knowing: especially in truly scientific, rational and analytic terms, thus carrying universal, context-independent and invariable characteristics.
- Technology: a term wider than the modern common use, thus including evidence-based method and principles, art and craftsmanship.
- Practical applicability: the definition of which moves from ethical consideration to purely pragmatic, action-oriented and context-dependent intervention.

This analysis of terms brings us to the core of our question, focused on the understanding of the specific aspects of each term and their possible connections and/or differences. We answer this question in positive terms, namely by evidencing that they are deeply interconnected and focused on different aspects (in a field-related way) of the scope of intervention. The definition of Nursing Informatics provided by the International Medical Informatics Association (2009) supports this point: "science and practice (that) integrates nursing, its information and knowledge, with management of information and communication technologies to promote the health of people, families, and communities worldwide." Therefore, we could state that Informatics is the base and ground framework, in a theoretical and applied sense, of Nursing Informatics, which in turn utilizes Information Technology to:

- Collect, store and manage data and general information in direct (clinical) patient care.
- Analyze, monitor and evaluate the quality of patient care and clinical outcomes.
- Apply the information collected in the broader field of translational medicine and healthcare policy-making (including planning and budgeting).
- Foster the scientific and clinical expertise in the nursing profession by enhancing nursing research and education, including certification and continuing education modalities.

6.2.6 Neuroimaging and Neurophysics

The final disciplines discussed in this chapter are neuroimaging and neurophysics. The latter is a subfield of physics, more specifically biomedical physics (also called simply medical physics, as well as medical biophysics or applied physics in medicine), focused on the application of concepts, theories, techniques, and technologies, and methods and models on the nervous system, to investigate the physical underpinnings—at the cellular, molecular and system networks level—of its activity. Neuroimaging instead is one of the most important centers of attention in our analysis, as it provides with some of the most relevant techniques used to investigate the neural underpinnings of processes involving cognition, memory, emotion, (re)action, computation, expression, communication, conscience and consciousness, and many others. It involves not only the observation and analysis of anatomical structures, but also the investigation and experimental monitoring of function and pharmacology, with special regard to the central nervous system, especially in the brain. As such, neuroimaging deals with normal and abnormal structures for clinical and diagnostic purposes of gross intracranial disease and injury (it is the case of structural imaging), as well as with fine diseases and lesions or for more research-oriented efforts, especially in the areas of cognitive science and psychology, artificial intelligence, neurology, and neuroinformatics (functional imaging). As we have seen, this type of analysis is somewhat the product of the "decade of the brain", that is, the research years comprised between 1990 and 2000, and the following "Decade of the Mind" (2000–2010), in the sense that the claimed groundbreaking discoveries in the structural-functional links to human behavior and cognition we certainly fundamental to understand the neural underpinnings of such process, but ultimately did not 'substitute' other types of analysis, such as the psychological, philosophical and other investigative strategies, in part because of the low fidelity-resolution of such technologies (which has vastly improved to these days) or due to the relatively small sample size of many studies, but much more because they were based on non-verifiable (thus truly non-scientific in the modern acceptation of the term) assumptions on neural connectivity vs. functional subdivision and circuitry and the overimposed causal effects beyond the only observable factor, that is, correlation. In any case, the history of neuroimaging almost completely overlaps with the history of neuroscience in general, exactly because of the aforementioned emphasis on brain imaging techniques. From the 'human circulation balance' of Mosso, through the ventriculography (VG) by Dandy and the subsequent pneumoencephalography (PEG), from Moniz and cerebral angiography (CAG) to computerized axial tomography (CAT or CT scan) developed by Allan McLeod Cormack and Godfrey Newbold Hounsfield and the further developments of single photon emission computed tomography (SPECT), positron emission tomography (PET), all the way to the always improving technologies at the base of magnetic resonance imaging (formerly Nuclear Magnetic Resonance spectroscopy or NMR, now MRI or MR scanning, including MRI and fMRI) developed by Mansfield, Lauterbur and many, many others, neuroimaging has been instrumental in the development of new research studies and clinical-diagnostic tools used in the vast area of mind-brain sciences, for instance in the investigation of the neural processes in mechanisms of addiction (Fig. 6.8). For the purpose of our analysis, we will briefly examine each of the most important neuroimaging techniques still in use in current times: (a) Computed axial tomography, (b) Cranial ultrasound, (c) Diffuse optical imaging (DOI), (d) Event-related optical signal (EROS), (e) Magnetic resonance imaging (MRI), (f) Functional magnetic resonance imaging (fMRI), (g) Positron emission tomography (PET), (h) Single-photon emission computed tomography (SPECT), (i) Magnetoencephalography.

Of course, this list is by no means omnicomprehensive, but it provides a solid overview on the current state-of-affairs of neuroimaging:

(a) *Computed axial tomography* is a scanning x-ray technique, widely utilized for the analysis of brain injuries, which provides a series of head pictures taken from many different directions using the inverse Radon transform, a numerical integral calculation performed by a computer for an estimate of the levels of x-ray absorption in the brain.

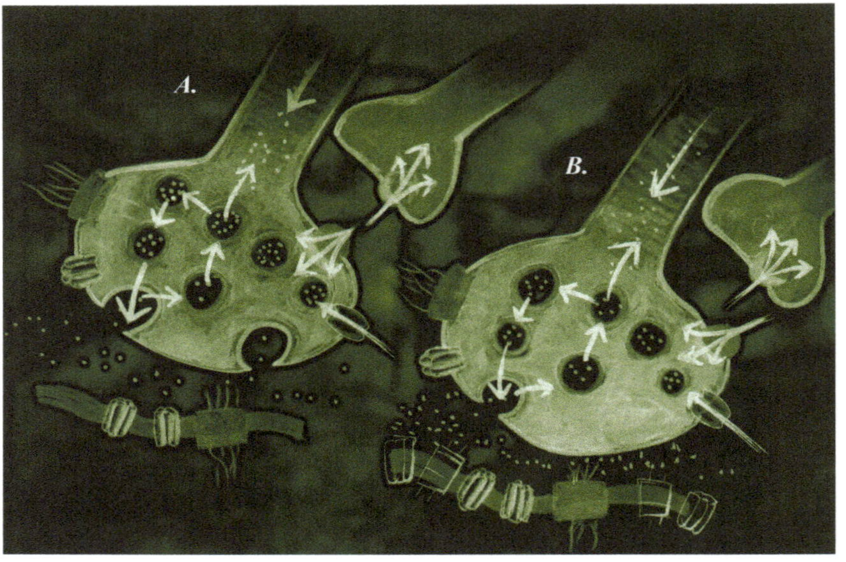

Fig. 6.8 Analysis of the process of neurotransmission in developmental ethanol exposure according to the model of Valenzuela, Puglia, and Zucca (2014), with views of axon terminals, dendrites, glial cells, and Ca^2 channels in phase A and B

(b) *Cranial ultrasound* is a diagnostic sonography technique generally used to create ultrasound imaging of the brain in the open fontanelles of babies. It is a safe technique, as it does not involve ionizing radiation, although it provides a lower resolution in comparison to Magnetic resonance imaging.

(c) *Diffuse optical imaging (DOI)* or *diffuse optical tomography (DOT)*, including High-density diffuse optical tomography (HD-DOT), uses near-infrared light as imaging technique to investigate the optical absorption of hemoglobin. DOI relies on the absorption spectrum of hemoglobin varying with the levels of oxygenation.

(d) *Event-related optical signal (EROS)* focuses on the scattering properties of the neurons via infrared light through optical fibers. It is used to measure spatial and temporal variations in optical properties of cortical areas which are active, especially at the cellular level. It is non-invasive, although it only provides information on a relatively superficial (cortical-subcortical) level.

(e) *Magnetic resonance imaging (MRI)* is possibly the best-known and used neuroimaging technique together with its sister fMRI. Magnetic fields and radio waves are used in this technique to develop high-quality 2D or 3D images of brain structures. As MRI does not use x-rays, it is generally considered a safe and yet very precise alternative to other brain imaging techniques.

(f) *Functional magnetic resonance imaging (fMRI)* shares many aspects with MRI, but allows for a much more precise monitoring of the activation, function, and performance of neural areas, also in comparison with baseline (resting) state, utilized both in diagnostics and in research Together with arterial spin labeling (ASL), fMRI relies on the so-called oxygenation hypothesis, according to which variations in oxygen levels (usage) in cerebral blood flow in specific brain areas thereby analyzed, is correlated (via hemodynamic response to neural activity) to cognition, emotion, behavior, memory, and other processes.

(g) *Positron emission tomography (PET)* focuses on brain metabolism in diagnostic settings via a 30-second 2D and 3D computer processed and measured emissions (radioisotopes) from radioactively labeled metabolically active chemicals (radiotracer) injected into the bloodstream of the subject. PET is a fundamental technique for the analysis of oxygen and glucose metabolism in neural tissues, at the base of multiple diseases such as dementia and Alzheimer's.

(h) *Single-photon emission computed tomography (SPECT)* is considered to be an offspring of PET, more specifically via injection of radioactive tracer, with a nearly 100% complete uptake from the brain within 30–60 seconds. Therefore, SPECT precisely monitors cerebral blood flow (CBF) at the almost exact time of injection of the SPECT agent, and it is considered to be even better than Neuro-PET, as the latter uses tracers with half-lives of at most 110 minutes, although its resolution is far inferior to MRI.

(i) *Magnetoencephalography (MEG)* utilizes superconducting quantum interference devices (SQUIDs) to measure the magnetic fields produced by electrical activity in the brain. Therefore, it can be used in diagnostic terms to analyze the exact location of pathologies, but also to support studies at the intersection of clinical and research efforts, for instance to study Neurofeedback. This technique is superior to common electroencephalography (EEG) because it does not encounter the same level of image distortions

> due to surrounding tissue, and it also provides a very high temporal resolution. Unfortunately, MEG is inferior to fMRI in terms of a much lower spatial resolution.

As we have seen, neuroimaging gave us an incredible amount of information of the structure and the function of the brain, and each of the aforementioned techniques allowed and still allows us to better understand how specific processes at a neural level determine who we are and how we perceive the world around and inside us. In this chapter we focused on *Perception and Cognition*, which are truly two fundamental aspects of our being in the world. In regard to this element, we can certainly relate to the *Gewesenheit* as the *Schon-sein-in-der-Welt*, in the definition of Heidegger. This 'being-in-the-world' depends, on a phenomenological level, on *Vergangenheit*, defining "something that passes" (Tomasi 2016). The concept of *Gegenwart* (present) instead is intended as *Sein-bei*, better described by the concept of "*dem momentan zu Besorgendem*", in Heidegger. In the definition of the Duden Dictionary of German Language, the term is the "Bezeichnung für einen nicht genau bestimmten Zeitraum zwischen vergangener Zeit (Vergangenheit) und kommender, künftiger Zeit (Zukunft)," and *Zukunft* (Future) as *Sich-vorweg-sein*, 'im Entwurf'. Regarding this *künftiger Zeit*, Saint Augustine of Hippo presented the view according to which time has been created by God together with the universe, this "togetherness" to be interpreted from conceptual, historical/ temporal and chronological point of view. In his *Confessiones*, time is an "extension of the soul", which is also a dis-tention linked to the perception of the individual. Thus, man has perception (conscience, awareness) of the past only thanks to memory and of the future because of expectancy. Man lives in the present moment, with attention/conscience (Tomasi 2016). In antiquity, Zeno of Elea thought the metaphysical essence (use, inner structure, ontological anchorage) of reality as eternal (beyond time in the classical Greek etymology). Thus, time was itself an opinion, a position (an imprecise, incomplete, truth) of the *doxa*. In the Hegelian conception, the *Dasein* appears as an element in between birth and death, able to choose among its possibilities (and also forced to, as it is thrown into the world, which is also its world). Linking the activation of specific neural areas to the personal element of human existence is necessarily linked to its historicity. Dasein's potential and very essence has to do with the concept of *Sorge*, the care, the meaning of concern, whose temporality is defined by the above mentioned three historical moments of past, present and future.

References and Further Readings[26]

Bibliography

Atmanspacher, H. 2015. *Quantum approaches to consciousness*. Stanford: Stanford Encyclopedia of Philosophy.

Baars, B. 1998. *A cognitive theory of consciousness*. New York: Cambridge University Press.

Beauregard, M. 2012. *Brain wars: The scientific battle over the existence of the mind and the proof that will change the way we live our lives*. New York: HarperCollins.

Crick, F., and C. Koch. 1990. Towards a neurobiological theory of consciousness. *Seminars in the Neurosciences* 2: 263–275.

Dehaene, S., et al. 2003. Imaging unconscious semantic priming. *Nature* 395: 597–600.

Doidge, N. 2015. *The brain that changes itself*. New York: Penguin Books.

Duncan, J. 2001. An adaptive coding model of neural function in prefrontal cortex. *Nature Reviews. Neuroscience* 2: 820–829. https://doi.org/10.1038/35097575.

Edelman, G.M. 1993. *Topobiology: An introduction to molecular embryology*. New York: Basic Books.

Franklin, S., and Graesser. A., 1999. A Software Agent Model of Consciousness. *Consciousness and Cognition* 8: 285–305.

Fremgen, B. 2012. *Medical law and ethics*. 4th ed. Boston: Pearson.

Goldenberg, G. 2013. Apraxia: The cognitive side of motor control. *Aphasiology* 28 (3): 377–379.

Goldstein, L.B., C.D. Bushnell, R.J. Adams, L.J. Appel, L.T. Braun, S. Chaturvedi, et al. 2011. *Guidelines for the prevention of stroke in patients with stroke or transient ischemic attack*. Guidelines for healthcare professionals from the American Heart Association/American Stroke Association. *Stroke* 42: 227–276.

Grof, S. 1998. Rethinking basic assumptions about psychology and psychiatry: The role of spirituality and nonordinary states of consciousness. In *The whole mind: The definitive guide to complementary treatment for mind, mood, and emotion*, ed. L. Bassman, 45. Novato: New World Library.

Grush, R. 2004. The emulation theory of representation: Motor control, imagery, and perception. *Behavioral and Brain Sciences* 27 (3): 377–396. https://doi.org/10.1017/S0140525X04000093.

Habermas, J. 1993. *Justification and application*. Trans. C.P. Cronin. Cambridge: MIT Press.

Haikonen, P. 2003. *The cognitive approach to conscious machines.* Exeter: Imprint Academic.
IMIA Special Interest Group on Nursing Informatics. 2009. *International Medical Informatics Association.* Retrieved from: https://www.amia.org/programs/working-groups/nursing-informatics
Ishikawa, S. 2017. A final solution to the mind-body problem by quantum language. *Journal of Quantum Information Science* 7: 48–56.
Izhikevich, E.M. 2007. Solving the distal reward problem through linkage of STDP and dopamine signaling. *Cerebral Cortex* 17: 2443–2452.
Jaworski, W. 2016. *Structure and the metaphysics of mind: How hylomorphism solves the mind-body problem.* Oxford: Oxford University Press.
Kinsbourne, M. 1988. Integrated field theory of consciousness. In *Consciousness in contemporary science*, ed. A.J. Marcel and E. Bisiach, 239–256. Oxford: Clarendon Press/Oxford University Press.
Lamme, V.A.F. 2003. Why visual attention and awareness are different. Trends in Cognitive Sciences 7: 12–1810.
Levine, J. 1983. Materialism and qualia: The explanatory gap. *Pacific Philosophy Quarterly* 64: 354–361.
Llinás, R., U. Ribary, D. Contreras, and C. Pedroarena. 1998. The neuronal basis for consciousness. *Philosophical Transactions of the Royal Society of London. Series B, Biological Sciences* 353 (1377): 1841–1849. https://doi.org/10.1098/rstb.1998.0336.
Lossky, V. 1944. *Essai sur la theologie mystique de l'Eglise d'Orient* as discussed in: N.O. Lossky. 1951. *History of Russian philosophy* (Original: *История российской Философии*). London: Allen & Unwin, London/New York: International Universities Press.
Mangun, George R., and Steven A. Hillyard. 2012. Modulations of sensory-evoked brain potentials indicate changes in perceptual processing during visual-spatial priming. *Journal of Experimental Psychology: Human Perception and Performance* 17 (4): 1057–1074.
Margolin G., Gordis E.B., John R.S. 2001. Coparenting: A link between marital conflict and parenting in two-parent families. *Journal of Family Psychology* 15: 3–21.
McFadden. 2002. Does the efferent system aid with selective attention? *The Journal of the Acoustical Society of America* 135: 2384.
Metzinger, T. 2004. *Being no-one: The self-model theory of subjectivity.* Cambridge, MA: MIT Press.
Mitchell, Jennifer K, MSN, A.N.P.-B.C., G.N.P.-B.C. 2011. Nursing informatics 101. *ONS Connect* 26 (4): 8–12.
Mujkanović, L. 2016. *Nursing research studies.* Ethics, community, and healthcare. Hooksett: SNHU.
O'Regan, J. Kevin, and Alva Noë. 2001. A sensorimotor account of vision and visual consciousness. *Behavioral and Brain Sciences* 24 (5): 883–917.

Penfield, W., and H. Jasper. 1954. *Epilepsy and the functional anatomy of the human brain*. Boston: Little, Brown & Co.

Penrose, R. 1989. *Shadows of the mind: A search for the missing science of consciousness*. Oxford: Oxford University Press.

Pinker, S. 2010. *Ethics and the ethical brain*. In *The cognitive neuroscience of mind. A tribute to Michael S. Gazzaniga*, ed. P.A. Reuter-Lorenz, K. Baynes, G.R. Mangun, and A.E. Phelps. Boston: Massachusetts Institute of Technology.

Pockett S., Purdy S. C., Brennan B. J., Holmes M. D. 2013. Auditory click stimuli evoke event-related potentials in the visual cortex. *Neuroreport* 24: 837–840.

Revonsuo, A. 2005. *Inner presence: Consciousness as a biological phenomenon*. Cambridge, MA: MIT Press.

Robb, D., and J. Heil. 2009. Mental Causation. In *Stanford encyclopedia of philosophy*, ed. E.N. Zalta. Palo Alto: Stanford University Press.

Rosnow, R.L., and R. Rosenthal. 1997. *People studying people: Artifacts and ethics in behavioral research*. New York: Freeman.

Satel, S., and S.O. Lilienfeld. 2013. *Brainwashed. The seductive appeal of mindless neuroscience*. New York: Basic Books.

Schaef, A.W. 1992. *Beyond therapy, beyond science*. New York: HarperCollins.

Schmidt & Clark. 2011. *Manchester VA to pay $21 million for malpractice that left vet with 'locked-in-syndrome'*. Washington, DC: Schmidt & Clark, LLP. Retrieved from: https://www.schmidtandclark.com/manchester-veterans-affairs-malpractice-lawsuit

Seth, A.K., E. Izhikevich, G.N. Reeke, and G.M. Edelman. 2006. Theories and measures of consciousness: An extended framework. *PNAS*. https://doi.org/10.1073/pnas.0604347103.

Shanahan, M., and Baars, B. 2005. Applying global workspace theory to the frame problem. *Cognition* 98 (2): 157–176.

Sloman, A., and R.L. Chrisley. 2003. Virtual machines and consciousness. *Journal of Consciousness Studies* 10 (4–5): 133–172.

Taylor, J. 2012. A final solution to the mind-body problem. *Journal of Mind Theory* 1 (1): 25–58.

Tomasi, D.L. 2016. *Medical philosophy. Philosophical analysis of patient self-perception in diagnostics and therapy*. New York: Ibidem Verlag/Columbia University Press.

Tononi, G. 2004. An information integration theory of consciousness. *BMC Neuroscience* 5 (42): https://doi.org/10.1186/1471-2202-5-42.

Tononi, G. 2008. Consciousness as Integrated Information: a Provisional Manifesto. *The Biological Bulletin* 215 (3): 216–242.

Valenzuela, C.F., Puglia, M.P., and Zucca, S. 2014. Focus on: Neurotransmitter Systems. *Alcohol Research Health* 34 (1): 106–20.

Vygotsky, L.S. 1962. *Thought and language*. Cambridge, MA: MIT Press.

Ward, G., and B.J. Moreton. 2011. Time scale similarity and long-term memory for autobiographical events. *Psychonomic Bulletin & Review* 17 (4): 510–515.

Wade, C., C. Tavris, and M. Garry. 2014. *Invitation to psychology*. 6th ed. New York: Person.

Winkler, C., J. Reis, N. Hoffmann, A.K. Gellner, C. Münkel, M.R. Curado, and B. Fritsch. 2017. Anodal transcranial direct current stimulation enhances survival and integration of dopaminergic cell transplants in a rat Parkinson model. *eNeuro* 4 (5). https://doi.org/10.1523/ENEURO.0063-17.2017.

Further Reading

Cain Travis, M. 2015. *The existence of the soul: Philosophy, not neuroscience*. Apologetics, Worldview, and a Pebble in the Shoe. Available at: https://hcchristian.wordpress.com/2015/04/25/the-existence-of-the-soul-philosophy-not-neuroscience/

INCF. 2005. *About us: INCF History*. The international neuroinformatics coordinating facility. Stockholm: Karolinksa Institutet. Available at: https://www.incf.org/about-us/history

Torley, V. 2013. *Do split-brain cases disprove the existence of an immaterial soul?* Available at: https://uncommondescent.com/intelligent-design/do-split-brain-cases-disprove-the-existence-of-an-immaterial-soul-part-two/

CHAPTER 7

Conclusion: Philosophy as Basic Approach Toward Neuroscience

7.1 Preliminary Discussion

At the beginning of this volume, we discussed the possibility of extended approaches to the scientific inquiry at the center of our investigation. We stressed on multiple occasions how it is very important to follow the scientific method when dealing with scientific investigations, that is, investigations of matter that can and should be observed and interpreted by science. In this context, Matthew Rampley (2017) advocates "for [a] wider recognition of the human motivations that drive inquiry of all types, [arguing] that our engagement with art can never be encapsulated in a single notion of scientific knowledge." We tend to agree with this view and we also embrace some elements from the well-known theory of multiple intelligences by Howard Gardner (2006). Thus we could propose a simplified scheme of the different levels/layers of "Philosophy as basic approach toward Neuroscience" and science in general. For instance, we could list the following models-methods:

1. Scientific (evidence-based, observational-empirical, double-blind, case-control, the principle of falsifiability and hierarchy of evidence-based) model and method.
2. Philosophical model and method (each of which depends on vast supragroups and positions, as we evidenced several times in this analysis, including thought experiments in experimental philosophy).

D. L. Tomasi, *Critical Neuroscience and Philosophy*,
https://doi.org/10.1007/978-3-030-35354-4_7

3. Artistic model and method (from theories of perception to art movements and philosophies, to the "sheer enjoyment through the senses", and to the social, political, sense/meaning-making, affirming-affirmative and activating-activist component of performance art).
4. Religious/spiritual/meditative model and method (with special reference to mysticism and mystical experiences, but also related to NDEs, OOBEs, alternate states of consciousness, neurotheology, etc.).

Following this thought, we can relate to John Gray's view on contemporary science (also in relation to Kurzweil's Singularity), which in modern times resembles what magic was for ancient civilizations, that is, it provided a general sense of hope toward the achievement of eternal life. According to Hegel, through religion, the absolute knowledge presents itself, though not in a conceptual form, proper of the Begriff, but in a pictorial form, as described in and by the *Vorstellung*. It is interesting to consider the poetic-imaginative-creative German (almost idiomatic) expression *stell' dir vor*. This pictorial form is essential to comprehend many perspectives and theoretical basis of a quantity, probably the vast majority, of alternative therapies. In order to understand what absolute knowledge really is, a fundamental concept in "Philosophy as basic approach toward Neuroscience" we should focus on the dual meaning of the term *Aufhebung*, indicating cancellation/annihilation as well as preservation, which represents the focus of Traditional perspectives. To be sure, absolute knowledge is not the combination of *An sich sein* and *Für sich sein*, but is actually the absolute distance that cannot be reduced between the two terms. Thus, the appropriate philosophical methodology should follow a formal dialectical process, but should engage in an "introspective rememorization, recollection" or internal absolute, which brings 'to the surface' this memory; a method called *Erinnerung* (Tomasi 2016) but truly connected with the main concept of 'speculative philosophy' in the sense given by Verene, through the analysis (which is ultimately and truly deeper perception and poetic creation, from this perspective) of the work by Grassi, Vico, Tagliacozzo (Verene 1997), and the very comparison of what the philosopher defines as "Vico's tetrad" (i.e. Plato, Bacon, Tacitus and Grotius) and his own tetrad (Hegel, Vico, Cassirer and Joyce).

Criticism of traditional (albeit not necessarily traditionalist) religions often identified the perceived delusional-illusional aspects of belief and practice. In this view, revealed religion is an illusion, for its truth is always something to be represented, not conceptualized, not conceptually understood, because the content of religion is correct, but the form is wrong, and that is the perspective of Verene in his analysis of Hegel. This certainly defines the "problematic relation between illusion, representation, perception, conceptualization and revelation of metaphysical, spiritual, and religious, especially messianic and/or prophetic type such as premonitions, apocalyptic visions, clairvoyance etc., is at the center of the debate on the ontology and phenomenology of psychiatric events" (Tomasi 2016). Philosophical, psychological and psychiatric definitions of 'manifestation' and 'epiphany' are particularly relevant in analyzing the cognitive-rational perspective as the preferred method of investigation of such issues. For example, a rationalization/cognitivization of (the figure and/or *imago* of) Christ or the Spirit would not make sense according to Verene because of their nature-substance. In the attempt to unify the two elements and moments, the Father and the Son, problems appear on both strictly cognitive levels (because cognition necessarily has an object) as well as from the theological perspective, for instance in the *filioque* doctrine at the center of the Great Schism. The Holy/Whole Spirit vs. the Ghost (*Geist*) appears in time as substance, and unsuccessfully tries to unite the elements, which is the same as saying it tries to annihilate (reduce) time. Under this framework, *Erinnerung* is a denial of time but recollection. That is precisely why cognition is 'less than absolute knowledge', and our understanding can only move beyond the cognitive stage, through the previously analyzed 'beautiful soul' which is both the very intuition of the Divine, and the Divine's intuition of itself, the self's own act in contraposition to what was 'content' in religion. This is the reason beyond the divine nature of language (learned at the level of *das Meinen*, since the real cannot be said), of the poetry of the very poetic art of intuition, which ultimately leads to philosophy, to a communion with itself (Tomasi 2016). Now, all these discussions on communication, creativity and intuition are to be understood, in this context, as areas of investigation for neuroscience. More specifically, as this conclusion involves 'Philosophy as basic approach toward Neuroscience', we want to spend some time justifying why we should care at all about these ideas. With the ever increasing specialization and subdivision of science into multiple subspecialties, we can observe how very often this process happens not for a genuine interest in the best possible practices or form of scientific inquiry, but simply due to an ever increasing demand for added 'labeling' at the scientific level in academia.

In other words, modern science, ever since post-enlightenment developments (as we have seen multiple times in this study), has increasingly become distant, even 'foreign' or 'alien' to disciplines such as natural philosophy. While it is understandable and to some extent justifiable, that, once a solid scientific paradigm, in terms of method and technology, has been found to be effective and accurate in the efforts of new discoveries, science will focus more and more on areas in which results can be found, these efforts are themselves part of (could read: 'prisoners of') the general goal of such paradigm. This has a direct impact on funding research studies, especially in the context of grants. If the amount of money is limited, and 'everyone wants the money' one of the best ways to access these funds is exactly this further subdivision of fields, a subdivision which promotes the idea that is actually better to know a lot more about a very limited area of investigation (a view which can certainly be defended given the increasing complexity of science in all areas) and 'not so much' about the 'general connections' between fields. While we do agree that is *humanly* very difficult, almost impossible, to be a specialist in one field and at the same time know 'a lot about everything' without losing focus on and expertise in that very field, we also see an increasing negative bias toward researchers who are actually trying to bridge these two perspectives. To give a quick example in the form of mere anecdotal evidence (the reader will not find any bibliographical reference in this case), we have observed, personally and professionally, that an expert in diverse fields of science can sometimes be viewed as someone like a 'popular science author' but (at times) with not enough knowledge of the complexity and multitude of details in any field. A similar bias we observed in the field of medicine, where terms/titles such as 'general practitioner', 'family physician' and 'generalist doctor' are at times 'looked down upon' by the (just as an example) "neurosurgeon-researcher who only specializes in the substantia nigra selectivity, with special reference to the pars cumpacta in relation to sleeping patterns observable in animal modeling". Let us restate that we completely understand the incredible difficulty of mastering a lot of knowledge in such a vast area like modern science without sounding (and often truly being) 'superficial' or 'general', but we want to promote the idea of "Philosophy as basic approach toward Neuroscience" to at least try to be flexible in moving the focus back and forth from the smallest detail to the bigger picture. Furthermore, as we also mentioned many times, an "expert is someone who experiences", someone who has or makes an experience, certainly with the methods of science but also via relations with other scientists, and human beings in general, all the way to those experiences that need to be quantified/qualified as personal (at times even subjective), artistic, spiritual and even mystical. Once more, this is the very nature of consciousness as 'shared knowledge'. If we do not 'keep this mindset in mind' the risk is that the aforementioned subdivision of fields will not be motivated by true scientific (and human) curiosity (broadly indeed to include genuine interest for fellow human beings, and the world in general), but by money. There can be in fact very dangerous consequences when science is used to foster not clinical research, but the monetary gain for pharmaceutical companies, not green energy solutions, but the oil industry, not universal wellbeing, but class-divided health coverage. To provide an example in this sense, when analyzing socioeconomic status, healthcare and medicine, education and literacy, science and technology, and equal distribution of wealth in the United States, for instance, this country does not score very high globally (Mujkanović 2016). However, although the general quality and affordability of the US healthcare and medicine in general is still far away from the standards of other countries (especially other American countries like Canada and European countries like Austria, Denmark, Italy, the Netherlands and Norway), the United States were certainly on the right path, and the implementation of the Patient Protection and Affordable Care Act (PPACA) and the Health Care and Education Reconciliation Act have been successful in achieving the goal of quality, affordable healthcare and reduction of uninsured rate for all citizens of or residents in the United States, until the change of powers in policy making. In this regard, we can certainly state that not everything was perfect with 'Obamacare'. First of all, many opponents felt that the centralization of powers in the hands of the Washington federal government might take away some of the liberties and abilities of single states to provide good (health) care to their citizens. Of note, the same fear is still found throughout Europe when discussing the centralizing power of the European Union in terms of financial structuration of society, which many perceive as globalizing in the sense of increasing the privileges of the banks, of corporations over people and of economy over morality. In any case, the vast majority of European countries, EU-members or not, were able to keep a relatively free healthcare system, with paid maternity leave, and very inexpensive higher education tuition rates, especially in the fields of medicine and medical science, again relatively to the US-based higher education system.

7.2 The Triple-S Model: Self, Soul, Spirit

7.2.1 Critical Neuroscience

In this final chapter we finally came back to the very definition of critical neuroscience. As we mentioned at the beginning of this volume, all the previously discussed fields in neuroscience, psychology, psychiatry and so on have made major contributions to our understanding of human nature. Each contribution provides a fundamental piece which is needed to solve the mysterious puzzle of our essence and existence as human beings in this world. Of course, critical neuroscience is also limited in its understanding of the whole, but it serves the fundamental purpose of bridging the thereby achieved knowledge and the perspectives of the last four areas of investigation we will discuss, namely neurolinguistics, neuroheuristics (or neuristics), neuroeconomics and artificial intelligence. The last discussion we will encounter is on 'Sense, Purpose, (and) Meaning'. As we have seen, the main reason behind this choice has to do with the very essence of our analysis, namely the twofold interpretation of 'critical' and the related discussion spanning the scientific analysis of the matter-at-hand and the following suggestion on new perspectives and viewpoints on our lives. Critical neuroscience is therefore a direct heir of epistemology and heuristics, although we want to stress here the relevance of all three elements of scientific knowledge, that is, the Aristotelian episteme, techne and phronesis, the latter especially representing a fundamental cornerstone of our investigation. In fact, we want our analysis to be 'mindful and aware' of neuroscience's potential, informative abilities and also limits. Limits that are created not in order to necessarily narrow the scope of observation found in this discipline, but that are naturally created by the existence of other elements such as the social, psychological, physical and metaphysical, spiritual and transcendent(al) parts of our self, the self of all the people and other beings around us, and the self of the world(s) we live in. Thus, critical neuroscience has the very important duty to investigate the fallacies such as deriving "[a] prescription from [a] pure description" (Moore 1903), also in connection to—we could argue—computational-mathematical perspectives such as Gödelian information not reducible to bytes, and provide elements to foster this level of understanding of the human mind-brain. In the research by Sperry and Gazzaniga, we have referred to on multiple occasions in this examination, one of the most important aspects is exactly this link between pure description and:

(a) What our mind and soul (both of which, at this level do not make sense at all, anymore), really are (reality=observation=description, following the same schemata).
(b) What (in the absence of a self, intended as connected to mind and soul) our brain *should* do.

In this regard, Vincent Torley thoroughly analyzed the significance of corpus callosotomy on the possible existence vs. non-existence of an immaterial soul, also quoting John Eccles and presenting the substance dualist, the thought control dualist and the formal-final dualist perspectives. To be sure, critical neuroscience has to be a subgroup of 'critical science', thus containing all the philosophical considerations (especially from the philosophy of science and natural philosophy) focused on (at the very least) epistemological, ontological and phenomenological (in our opinion) elements of investigative methods. However, given that our primary focus has been on issues such as the mind–body problem, the existence of free will and possible neural underpinnings for cognition, computation, action, interaction, communication, attention, memory, behavior, perception, awareness and consciousness, our study has targeted neuroscience first and foremost. This premise has also been the center of our previous discussions on ethics in the context of medical research and practice. This is due to the existential (here intended in a broader sense) components of the application of the insight gained via neuroscientific experimentation on clinical decision making. Therefore, we have to spend some time discussing what these components represent in our *Triple-S Model: Self, Soul, Spirit*, starting from the clinical context. As we have seen, there are multiple models for the application and significance of neuroscience in the investigation of these terms. Leaving aside etymological, semantic and historical considerations, this tripartite model does not represent any new idea in philosophy or neuroscience, but simply refers back to the tripartite component of our existential experience as:

(a) Individual and/or subjectively ***understood***
(b) Transcendental and/or internally *felt*
(c) Metaphysical or/or divinely ***inspired***

As we previously mentioned, we do not embrace any specific philosophical or spiritual position in this context, as our primary goal is to provide a common framework for the interpretation of the problem of consciousness in the context of neuroscientific research. However, this framework is critical *by definition*, in that it examines the first level of data collection from experimental studies in neuroscience, the second level of scientific (mechanical/process-based, statistical, epidemiological, etc.) and philosophical interpretation, and the third level of direct implications for the 'life lived'. Thus, the main point here is that similarly to what happens with the awareness of placebo/nocebo effects, the 'mind model' and 'model of mind' we use in our everyday life and in the laboratory truly influence our rationale for a concept such as 'empathic cruelty'.[1] It is exactly this ethical component that connects a scientific re-examination of the mind–body problem in the context of critical neuroscience with social, political and legal aspects of healthcare. Stephan Schleim (2014) accurately describes the possible outcomes of a moral discourse originating in a very specific (and we might add, narrow vs. narrow-*minded*) "philosophical assumption deprived of philosophical understanding" in modern neuroscientific research:

> Generations of anthropologists and moral psychologists before had gathered evidence on the development, cognitive-emotional mechanisms, and cultural diversity of morality, but suddenly in 2001 with the publication of the first neuroimaging experiments the situation seemed to have changed. It seems fair to say that of the seven different psychological-neuroscientific theoretical accounts of morality distinguished by Jorge Moll et al., all the evidence gathered hitherto does not unequivocally favor any particular one (Moll et al., 2005). While the science communication accompanying the original study by Greene et al. suggested the philosophical relevance of the research, even putting forward the idea that the new findings could make moral philosophers superfluous (Helmuth, 2001), so far the opposite has been the case: theoreticians of all kinds responded to the prescriptive/normative claims and emphasized how these neuroscientific reports rely on theoretical presumptions and individual interpretation. While theoretical in its scope, moral neuroscience is used to provide the ultimate answers of human right and wrong that Sperry and Gazzaniga called for. More applied/technical implications are promised by the complementary research that

[1] Young, A. 2012. "Empathic Cruelty and the Origins of the Social Brain." in: Choudhury, S., and Slaby, J. 2012. Critical Neuroscience. A Handbook of the Social and Cultural Contexts of Neuroscience. Chichester, West Sussex, UK: Wiley Blackwell.

might be coined "immoral neuroscience": the investigation of what makes us behave immorally or criminally.[2]

Moral neuroscience is a subfield and a sub-theoretical framework which can be applied to many other fields in neuroscience and to science in general. To provide further explanation of the direct implication of a 'moralizing neuroscience' in those areas where medicine intersects and interacts with ethics, morality and end-of-life decisions, we would like to briefly discuss the application of such ethical aspects as found in acute care settings. Medical Science, including those areas covered by neuroscience and neurology, is structured on a vast array of clinical, theoretical and ethical standpoints, upon which the combination of medical, patient-centered theory and practice of care is based. In this context, the analysis of advance directives is a fundamental cornerstone of critical neuroscience, because—in research, academia and healthcare—we need to understand those parameters which can guarantee patient's informed consent from both the quantitative perspective (patient's capacity for autonomy), and the qualitative perspective (the patient's ability for autonomy and voluntariness). These aspects are always present in clinical efforts at every stage of care, from the initial treatment to palliative care. More specifically, end-of-life decisions should be made once patient autonomy is assessed, thus determining a series of considerations including the broader conceptual umbrella of good and evil categories, particularly good life versus good death. In this regard, Schicktanz and Schweda (2009) helps us understand the planning of advance directives in juxtaposition and comparison by the way they are interpreted by a third party. Critical neuroscience faces bioethical demands by addressing questions from a multilayered, multifaceted and multicultural perspective. In this context, the focus on diversity and universality is key to determine a structured process which fosters a better understanding of the patient in his/her individuality. In fact, advance directives indicate a document addressing the specific needs and preferences of care in the partial or total absence of ability and/or capability in health-related decision making, including (self) care, due to illness or incapacity. To be sure, these decision-making abilities are influenced by an interplay between personal

[2] Schleim, S. 2014. Critical neuroscience – or critical science? A perspective on the perceived normative significance of neuroscience. *Front. Hum. Neurosci.* 8:336. doi: https://doi.org/10.3389/fnhum.2014.00336. For specific bibliographical references, Please see references at the end of this chapter.

and cultural identity (Schicktanz 2009). Certainly, the underpinning ethical considerations are part of the broader theoretical debate in philosophy, most particularly in the philosophy of medicine. Johnson (2009) refers to *universalism* versus *casuistry* in addressing the problems arising by considering ethical questions under the lens of the individual case/patient and the context/case-situation/environment, including universalization and generalization (as in the perspectives of Immanuel Kant vs. situation ethics). Furthermore, this analysis helps reframe the problem of the *continuity of self* also in terms of our connection with the sensory apparatus (Fig. 7.1), which is central to our efforts, as researchers and healthcare providers in general, to better address the needs of each patient, especially when a full

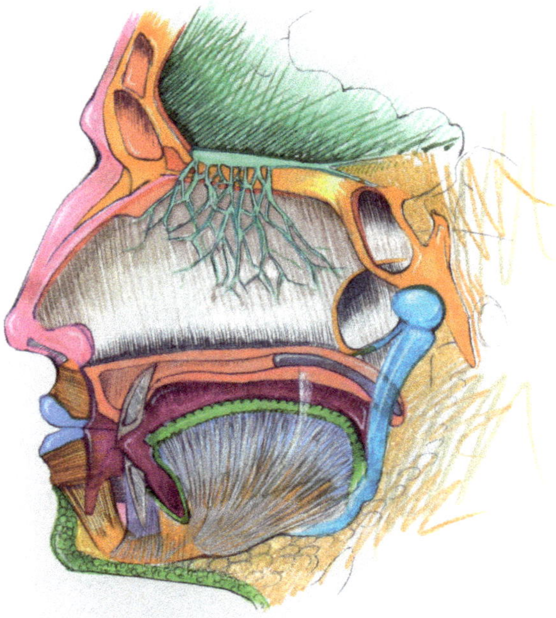

Fig. 7.1 The concept of continuity of self and the interaction of the individual-subject with the external world of stimuli via sensory perception. As it is well known, the olfactory system has the particular feature of almost completely bypassing the thalamic analysis (decoding, amplifying and transmitting) in its paths toward cortical areas

understanding, diachronically intended, is compromised due to issues related to the ever changing medical situation. It is certainly the case of degenerative disorders, including comorbidity of causes and effectors, such as neurological disorders and mental health disorders in general, for instance in patients with Alzheimer's disease or specific psychiatric diagnosis (schizophrenia, schizoaffective disorder, etc.). The conceptual issues regarding autonomy, consent and personhood are addressed by Coleman (2013) when discussing yet another philosophical concept, the *harm principle* by John Stuart Mill. In fact, utilitarian principles are in place in analyzing physicians' attitudes toward clinical decisions within advance directives. This perspective is certainly a core issue in the patient-provider, especially patient–physician relationship. Scientists and clinicians can and should be able to address this issue by fostering open, positive and competent communication with and between patient and physician.

This is another essential aspect of this study, that is, the promotion of appropriate ideas—thus, appropriate for the target population, for example, our patients—for the amelioration of care. Let us restate this principle once more. To provide better care we need to (a) provide the patient with better tools to foster self-care in the healing process, (b) identify those tools as parts of the mind–body connection, in terms of strategies the patient can *think of* and *use* for his/her wellbeing and (c) define the terms 'mind' and 'body', via the thorough analysis of the mind–body problem and the justification-proof of the existence of these two terms, their interaction, and *its* modalities.

As healthcare members provide direct clinical care to patients, the most effective strategies in this area include both the patient and the patient's family as active participants in health care decisions. According to Kroning (2014), there are several aspects underlying these strategies. In particular, physicians, clinicians and nurses on hospital units participate in the coordination and integration of care and are in need of educational modalities, specialized training, certifications or re-certifications to guarantee quality and safety within the framework of a shared decision making on the unit. This in turn originates from an organizational culture of safety, continuing education, positive approach and (mutual) support between staff members (Hinderer and Lee 2014). To be sure, both evidence-based practice and well-grounded clinical roles are required by healthcare institutions to adequately assist patients with advance directives. In particular, as team members become patients' advocates throughout the care they provide, there are several arguments that could be directed against the implementation of advance directives. In this regard, the delicate balance between *autonomy* and *self-determination* (not to mention the broader concept of *free will*, especially when certain neurological issues are in place) is at the center of a possible lack of support by some medical professionals, as noted by Mitchell (2011). According to this view, advance directives should not be considered a valuable source of information in the formation of treatment plans. To address these problems and suggest possible solutions and related practical implementation of the above, the American Bar Association (ABA) has analyzed the legal aspects both from the perspective of medical ethics, as well as from the epidemiological-statistical analysis, monitoring advance directives completion rates, and discussing in-depth state recognitions and liability. This analysis includes the instructions on advance directives by the American Medical Association (AMA) as well as the Patient Self-Determination Act (PSDA). Nurses especially, as healthcare members working in direct patient care need to follow basic human rights (in fundamental ethical term), as evidenced by Aiken (cited in Kearney-Nunnery 2012):

- Beneficence—What is in the best interest for the patient while focusing on safety in healthcare environments.
- Justice and Fidelity—Fair access to quality healthcare with a focus on trust and loyalty to the patient.
- Self-determination and Autonomy—Ethical obligation to patients; What measures to take on the patient's behalf.
- Full disclosure and Veracity—Full disclosure of truthful information allowing the patient to make an informed decision.
- Informed consent—A voluntary choice to accept or refuse treatment.
- Privacy and Confidentiality—Exercising trust in health care professionals to disclose private information.

Aside from the legal component of the utilization of advance directives in healthcare, medical professionals need to be aware of the psychological and sociological aspects of their clinical decision-making abilities to provide better care for their patients. More specifically, medical professionals can draw upon social contract and individual rights to understand the point of view, needs and abilities/capabilities of the patient throughout the lifespan, with a special understanding of the different developmental stages in young and older patients. This can be achieved for instance by analyzing Kohlberg's theory of moral development, especially the III-level model for morality, from heteronomous to universal, as discussed by Ma (2013), as well as through the implementation of principles of justice and human rights into the required coursework or training for healthcare professionals. A similar approach would foster a Universal Ethical System as an ethical, practical and clinical basis or the role and profession of researchers, within healthcare groups and systems. These aspects will all be fundamental parts of the legal decision-making process determining the hierarchy of health professions by reducing conflicts and dysfunctional patterns of communication, and clarify the role of clinicians, degrees and licensures. To suggest an alternative conceptual framework to Kohlberg's theory, parallels can be drawn between the psychologies in place in the patient-provider relationship, focusing in particular on the therapeutic self-care demands and needs as well as the specific evaluation of the professional role of the nurse in this context. Helpful in this context can be the hierarchy of human needs proposed by Abraham Maslow and discussed in Thielke, Harniss, Thompson and Patel (2012). In detail, being an expert and competent medical professional means:

- To be able to meet her/his full potential (self-actualization).
- To feel competent, with strong self-esteem, self-perception and self-worth (esteem).
- To feel psychologically and professionally supported in the system (love and belongingness).

The practical implications of understanding and implementing advance directives in acute care settings are linked to a general improvement in therapeutic outcomes and ethical considerations. As we have seen, from the patient's perspective, a fully informed and autonomous conceptual framework is necessary to guarantee the patient's informed consent from both the quantitative perspective and the qualitative perspective. This is possible if medical/nursing education will include an apt analysis of teamwork, collaboration and communication between healthcare members, to guarantee a strong and universal ethical basis and promote therapeutic effectiveness. From the perspectives of critical neuroscience applied to medicine, the psychological analysis of medical/nursing theories helps structure effective methods of clinical intervention following specific stage of action. Based on the evidence-based literature review and the theoretical analysis of the issues discussed above, we can posit that healthcare team members educated in advance directives will:

- Better understand their own needs, values, role and identity as they apply in a healthcare system, and better relate to the patient's own needs by clinically addressing them.
- Be more prepared and ready to check all the clinical tasks required as part of their clinical interventions, and verify their self-actualization process while performing those tasks.
- Be clinically and ethically more competent in addressing issues such as clinical decision-making and advocating for patients' actions in the case of illness or impaired capacity and autonomy.
- Be legally responsible for the application of specific ethical guidelines in advance care decision making (e.g. end-of-life and palliative care).
- Be supported by healthcare and higher educational system in which the role and scope of nursing will be better understood and respected by other healthcare providers and team members.

7.2.2 Neurolinguistics

In linguistics, the main area of interest is the analysis, in the form of scientific enquiry, of language, more specifically its structure, form, mutations and changes as well as of the context-at-large, including the history, development and environment as 'sense' or 'meaning'-creating effectors on the form and the use of language. Of course, the assumption here is based on the general (Western, evidence-based, empirical-experimental, etc.) scientific assumption that all these processes have a natural (in terms of matter-essential *res*) origin, and are not thus the product(ion) of a—for instance—a supernatural/paranatural entity and/or being which (who) gave the 'gift' of letters, alphabets, language and communication to mankind as a form of 'divine providence'. The same assumption is kept in neurolinguistics, although the focus here is on one specific side or part of nature, that is, the neural underpinnings beyond the production, comprehension, understanding, acquisition, recollection, categorization and communication of language, including speech-language pathology. Of course, there is a vast array of intersections and influences from other fields, especially from psycholinguistics, biolinguistics, sociolinguistics and many other subfields in applied linguistics. There are also enormous differences between the science-based field of neurolinguistics and some 'not-as-qualified' offspring. Monitoring the validity, strength and causational inference levels between what we can observe in the biological matter and mental-psychological processes in regard to the aforementioned linguistic elements is part of the primary questions posed to neurolinguistics and even more to neurolinguistics-linked (whether approved/accredited or not) approaches such as Neuro-linguistic programming (NLP), whose scientific validity (especially in terms of goal-oriented strategies applied to focus in attentional modeling) remains to be verified. In fact, the history of neurolinguistics starts from the neurobiologically-based scientific discoveries by Broca and Wernicke in the area of aphasiology first and dyslexia after, to the rise of an entirely new field with Henri Hecaen, Alexandr Luria and Edith Crowell Trager. From a philosophical point of view, neurolinguistics is a fundamental approach to provide a deeper analysis of important issues such as the debate on determinism vs. indeterminism on free will. For instance, the study of the very development of language from and evolutionary biology point of view contains some theoretical inference on the predictability of specific linguistic-morphological structures, and thereby a contextualized environment in which the options to choose from might be contained-constrained or conceptually limited into a therein produced set. Whether the analysis of complex linguistic forms such as sentences as indicators of underlying psychological mechanisms vs. processes such as priming vs. meaning, or the observation of neural activity as correlational activator of semantic sequences and algorithms, we are here faced with one of the defining component of science, namely the predictability aspect, which originates in the translation of neurological processes—especially cortical response to sets of stimulations, but also more generalized neuroimaging techniques in the context of linguistic exposure—into psychological theories and models explaining language, as well as their relation to auditory stimuli (Fig. 7.2). Good examples in this area are such as the 'unification model' by Gerard Kempen and Theo Vosse, or the 'serial model' by Janet Fodor and Lyn Frazier.

The philosophical discussion at the center of the aforementioned considerations also provides the framework for the semantic analysis of meaning vs. value judgment, especially in relation to the development of morality, as well as the theoretical basis for mismatch design, subtraction paradigm and violation-based studies. Charles Leslie Stevenson examined the cognitive use of language, thereby presenting the perspective of human self-knowledge through communication with other human beings in a specific context. The first pattern analysis focuses on the two parts of ethical statements, the speaker's declaration, that is, the declaration of the speaker's attitude and an imperative to follow it in a specular way: to mimic, copy and mirror it. The intersection of philosophy and

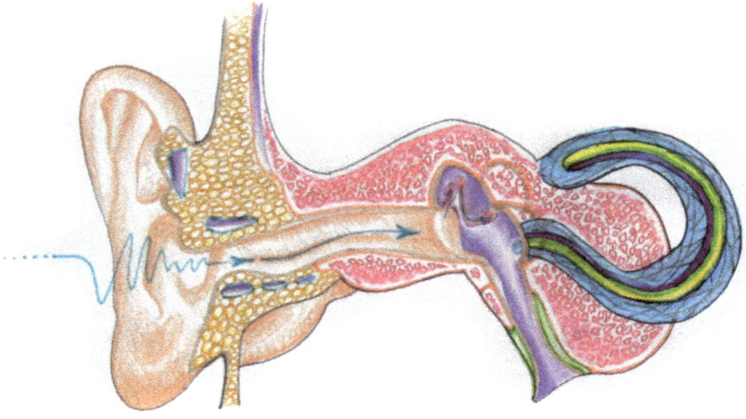

Fig. 7.2 Anatomical view of the ear in relation to auditory transmission (signaling in blue). Highlighted we can observe the ear canal and the eardrum or tympanic membrane (pink), the labyrinth (orange), the middle ear space (purple) and the cochlea (blue)

neurolinguistics thus helps us frame the way human beings shape their meaning and significance, whether to or through self-reflection or external factors. In the case of Stevenson, the translation of an ethical sentence remains a non-cognitive one, but it raises existential(ist) questions on personal responsibility and action. Since imperatives cannot be proven, they can be supported, and the purpose of this process is to make the listener understand the consequences of the action they are being commanded to do.

Furthermore, cognitive-attentive/attentional task monitoring studies originate from this semantic analysis design, including the probe verification in which a series of sentences are analyzed by presenting the subjects with a 'probe word' following each statement. In this type of experimental study, the subject then has to identify the presence of such words in the previously administered sentence. From this perspective, this experimentation is akin to acceptability judgment task-based research studies, and other types of research areas—for instance lexical decision tasks used in priming studies, or research on grammatical acceptability or semantic acceptability—shared by linguistics in general, including lexicology and morphology (the study of the relationships between related words, their independent and codependent structure, their formation and storage/accumulation and access/accessibility, thus related to recall and memory-based processes and LTP), semantics (concerned with encoded meaning, value and significance), syntax (concerned with combinatory patterns), phonetics and phonology (the study of speech sounds and—in neurolinguistics and psycholinguistics—their neural and psychological underpinnings, including the ability of separating language-based sounds from background noise), pragmatics (analyzing the context, and its role and conceptual weight in the interpretation of meaning, also in relation to sociocultural conventions at the base, for instance, of orthography).

Of course, in each of these areas of neurolinguistics, the 'neuro' element has to be experimentally understood with the help of evidence-based studies based on a series of tasks. The responses, reactions and/or outcomes following the tasks can then be analyzed both theoretically and with the support of technologies in neuroimaging and hemodynamic techniques such as functional near-infrared spectroscopy (fNIRS), or diffusion tensor imaging (DTI), especially useful to track neural connections between brain areas while the task is performed, and others. This type of research in neurolinguistics also monitors quantitative data in relation to (reaction/decision/acquisition) time and ordering—processing of linguistic patterns by the subject. More in detail, if we observe (via electrophysiological techniques such as EEG) a certain pattern of neural activity following a specific task, we can identify (again, we are still within the same theoretical framework) discrete values of brain response, on a computational level. In this context, we refer to the common neural responses reflecting semantic processing (event-related potentials or ERP) ELAN, N400 and P600. A less passive-observational and more active-stimulatory approach comes from other techniques such as transcranial magnetic stimulation and direct cortical stimulation. Furthermore, in neurolinguistics, we also find the usage of outliers to the standards for this type of investigation. More in detail, observing grammar-syntax anomalies in the order, disposition and conceptual structure of words within sentences, providence evidential relations for, as an example, the N400 effect or the P600 response. Studies by Embick, Hillyard, Kutas and Osterhou were able to shed new light on the neural processes and their location in neuroanatomical and functional terms behind these language-use anomalies. These studies also provide important information on the processes behind language acquisitions, more specifically around the various stages of linguistic development in connection to neural development (including 'babbling' stages), in relation to the acquisition of multiple languages and the interactions between them, also in terms of crossing-violation. The use of Neuroimaging techniques in this context allows for a deeper comparison between language processing and specific area neural activation at baseline (for instance, in experimental observation of subjects reading complex vs. basic sentences), to monitor not only area-specific developmental stages, but also to incorporate a broader analysis on neuroplasticity and increase in gray and white matter. The so-called subtraction paradigm focuses exactly on such comparison. Electrophysiological techniques and electrocorticography have also been used in the study of language processing, but they are generally limited, due to the nature of these techniques, to the analysis of the mechanisms at the basis of these processes and their time-sequence (often implemented with studies on event-related potentials, which provide detailed information on amplitude, latency and broad scalp/cortical topography) rather than on their exact location.

Certainly, when combining observation-type and experimental technology-based research with quantitative vs. qualitative theoretical, statistical/epidemiological analysis, we need to avoid, as much as possible, any type of subjective bias, or attempt to isolate this component to monitor possible contributors or distracters to the processes involving a final result. In fact, some researchers will use an artificial, experimental 'distractor' to (a) better investigate working memory in language processing and (b) avoid bias originating from the subjects (over) focusing on (in other words, orienting more attention to) the experiment itself and its stimuli. The distraction in this context might come from multiple and non-related stimuli to elicit a mismatch negativity response or MMN, or by asking the subject to engage in multiple tasks at once, as it is the case of the double-task experiment.

7.2.3 Neuroheuristics or Neuristics

A strong philosophical approach is presented in the heuristic examination of neural underpinnings. Such an approach is embraced by the field of neuroheuristics, also called neuristics. Therefore, research in this area analyzes the scientific information on neural activity from within, adopting a problem-solving framework including complexity, non-reducibility, deduction-induction-intuition-based debate and (abstract vs. extract) philosophical speculation, especially in relation to the cognitive examination of decision-making procedures. Neuroheuristics combines many transdisciplinary approaches in both 'hard' and 'soft' sciences; however, the philosophical baseline of such investigation is non-binary in an experimental sense. More specifically, neuroheuristics utilizes the fundamental data and information pieces gathered by neurosciences, neurobiology, in particular, to follow a 'bottom-up' process to have a deeper understanding of the structure and function of neural areas, especially in the CNS. However, a challenge to this approach is represented by the difficult task of monitoring (and avoiding bias at the same time vs. simultaneously) multiple neuroanatomical areas and activities and relate them to internal/external variables. To compensate this experimental difficulty, neuroheuristics also relies on theoretical frameworks such as double heuristics, quantum physics debate and black box theory, the latter in particular regard to the understanding of essence vs. existence in terms of (just vs. only) functionality. In fact, the term 'black box' is somewhat akin to the concept of 'camera obscura' although in this context the narrowing focal process is not mirroring, (re)representing, (re)coding or portraying an image, but is aware of the absence of knowledge of its internal functions, processes and workings; it is 'aware of being unaware'. This black box could thus represent the computational or at least quantitative aspect of the (human) brain as a calculator, better understood via algorithms and equations.

The neuroheuristic input comes from the realization of the importance of the observer in observation, and the attempt to quantify and qualify such impact in the experimental studies, with different degrees of evidence, of neural activity and psychological/mental processes. Neuroheuristics therefore attempts to bridge the gap between nature and nurture and monism (especially of the reductionist type) vs. holism by shedding light on the thereby interpreted lack of permanence of value in the binary analysis within the principle of falsifiability, because positive vs. negative or 1 and 0 results are already constructed in one-sided evaluation of the process-at-hand and are influenced by observation. It follows that the prereflective understanding of Hermeneutics is also a *para*reflective understanding. This understanding is at the center of the debates on the interpretation of medical disorders, and it therefore a funding component not only of medicine but also of psychology and bridging fields such as medical humanities or narrative medicine, as evidenced by Tricia Greenhalgh and Brian Hurwitz (1999) or Maria Giulia Marini (2012). Thus, Neuroheuristics help us understand illnesses, diseases and disorders by determining the neural basis for patient-provider, provider-medical science/knowledge, interaction, confrontation and communication as well as *quid/qualia* effects. Within these parameters, we can certainly remember Plato, as his definition of illnesses as originating in metaphysical ignorance, with a special connection to the soul. This view, expressed in the *Timaeus* where we learn about the role of *anoia* and *novs*, the contemporary concepts of placebo vs. nocebo effect and the underlying mind-brain connection, can thus be interpreted by taking into account *Mania* and *Amathia* as derivative from anoia and generate an imbalance in the soul. From the perspective of neuroscience, this imbalance deprives the mind-soul of its virtues connected to the absolute truth, being and the divine essence. Philosophically understood, the human soul-mind is in constant motion toward becoming more rational. Rationality obviously involves awareness and its related concept of consciousness. The latter is also the *unknown variable* for psychoanalytic interpretations, but can be perceived via the action of other layers of essence, as in the description of Paracelus, obviously modeled on archaic forms of traditional medicine:

Ens Astrorum or *Ens Astrale*, representing the influx and influence of the stars
Ens Veneni, through the poison absorbed, inhaled by the body
Ens Naturale, or the natural predisposition and constitution
Ens Spirituale, representing the influx and influence of the spirit
Ens Dei, defined as influx and influence of God

Whether we can identify some of this analysis as still valid in the context of modern neuroscientific research, or we completely reject all the teachings by Paracelsus as pseudoscience, it is still interesting to relate to the multi-layered conceptualization of essence in our attempt to understand the effects of each of the element on our psycho-physical and possibly spiritual health, whereby spirit we are willing to accept, theoretically assuming, the *ratio* it carries, as it helps us the reach higher levels of wholeness and perfection, also in reference to the *narratio* described by Verene (1997). According to this view, there is a constant battle of the soul against the influence of evil which ultimately causes physical and psychical disorders, a battle that can be won through education. This education is actually self-education, not in the sense of self-referring intellectual efforts (although certainly self/internally-directed, at least in part) but 'education of the self' as in the complex multilayered meanings of culture, as we observed in Transcultural and social neuroscience and psychology. Christopher,

Wendt, Marecek and Goodman (2014) described this aspect very well when they argue that every form of psychology is indeed folk psychology or in the analysis by Drew Leder on clinical interpretation (and) the hermeneutics of medicine (1990). Leder argues in fact that there is a need for hermeneutics in medical sciences, in order to 'interpret the interpretation of the interpretation' and thus provide insight on the role and relation of the physician-clinician/scientist-researcher in the diagnostic apparatus.

7.2.4 *Neuroeconomics*

What we have discussed so far in relation to cognitive and computational decision-making analysis is used by neuroeconomics to further investigate the relationship between neural activity and behavior involved in the creation-production, distribution-sharing-selling and use-usage-consumption of (primarily material) goods and services. More specifically, behavior is interpreted from the perspective of single vs. multiple decisions, or a single (less-least or more-most) better/worse choice among many options. Given this premise, neuroeconomics is fully part of a much broader philosophical debate, especially in regard to (hard/soft) determinism vs. indeterminism and compatibilism vs. incompatibilism in the free-will debate. However, in mainstream modern neuroeconomics, the main assumptions follow the ones of contemporary economics, especially in regard to expected utility, utility maximization (as in Bernoulli) logical base of informed and rational agent-based decisions, and standardized/single-currency/system models on overall utility value. A heuristics-based criticism of mainstream economics applies even more strongly to neuroeconomics, given the assumption of the validity of some animal modeling-based investigation of decision-making processes at a neural level. A similar criticism is found in the definition of risk and avoidable or unwanted outcomes, and on whether firing rates of individual neurons can be understood under the lens of 'better choice' or 'decision to avoid', as in the studies by Padoa-Schioppa and Assad on the orbitofrontal cortex of monkeys. As animals, especially humans, make the decision in a social environment, elements of social neuroscience and psychology or sociology are used in this field to account for the number of social effectors in each decision or series of choices. Strong moral-ethical and even theological-eschatological elements of discussion are present, as universality of value and judgment, cooperation, prize/praise, retribution, punishment and altruism are the direct outcomes of such analysis. To provide an example, in the so-called prisoner's dilemma by Flood and Dresher (1951) the concept of trust plays a fundamental role, in that it determines the level of cooperation between/among individuals. On the level of neuroeconomics, the increased outcome in terms of spread/shared benefit within social cooperation is compared with individual/single gain and it is modulated by the presence of the hormone oxytocin and the activation of the reward pathway in the CNS, most specifically the ventral striatum (as well as the tegmental area) in the brain. Therefore, the theoretical assumptions in neuroeconomics are tested using multiple Neuroimaging technologies. These studies rely on the analysis of blood-oxygenation levels as well as on the presence, absence or increase/decrease levels of specific, task-related chemicals during activation-action-function and at baseline, often by comparing any such activity with a control activity or comparing average subjects with subjects affected by neurocognitive damage, especially in the case of behavioral and emotional-related areas such as the limbic system, and especially the amygdala (which appears to play a very important role in loss aversion studies). Aside for the effects on trust and risk perception of oxytocin, we will mention serotonin in relation to intertemporal choice (the expected utility assigned by human subjects to events occurring at different times, as opposed to the assumed constancy-consistency of choice found in discounted utility), the presence of dopamine and increased activation of the dopamine reward pathway (especially the nucleus accumbens), as well as the BA8 area of the frontomedian cortex, the frontoparietal cortex and the mesial prefrontal cortex for difficult decision-making processes involving uncertainty. The latter is at the center of investigations regarding normal and abnormal (not necessarily in psychological-psychiatric terms) behavior, as with the generalized tendency to overweigh small probabilities and underweigh large ones in terms of showing risk-seeking behaviors, as evidenced by the studies by Tversky and Kahneman (1981). Moreover, when the balance between sheer uncertainty and risk appears to show 'heaviness' on the latter (as in gaming or gambling), we notice an increased activation of the insular cortex.

Finally, another important element in neuroeconomics is shared with psychology, especially in regard to the studies conducted by Bandura and Mischel on cognition, decision making and the connection (which is choice) between immediate and delayed reward. In such a process, the neural underpinnings appear to be the lateral prefrontal cortex, although with a ratio differential. More specifically, research studies suggest that the limbic system is (more) activated in the case of impulsive decisions, while the cortex is (more) activated in general aspects of the intertemporal decision process. In other words, the ratio of limbic to cortex activation decreased as a function of the amount of time (passed) until reward (obtained or perceived as such), and this would also explain the activation of other chemical components, especially hormones and neurotransmitters, as well as the production of cortisol and activation/deactivation of stress response in individual with drug addiction.

7.2.5 *Artificial Intelligence*

A thorough and critical examination of neuroscience, especially in regard to the mind–body problem, cannot possibly avoid the debate over artificial intelligence, a tremendously important topic in the contemporary scientific research. As we did in previous conversations, we will start with a definition of terminologies. First of all, debating artificial intelligence means being aware of a very particular comparison within the—for now assumed to be valid in theoretical terms—dichotomy/distinction between 'artificial' and 'natural'. The first term is pretty straight forward, as it originates in the Lat. *artificialis*, with the meaning of 'made with/by/out of art' (also in an Aristotelian 'techne' model, thereby scientific) and thus opposed/differentiated from nature. In this sense, 'artificial' also means "made, created, or produced by (human) beings rather than occurring naturally". Whether other (types of) animals beyond the human species could account for artificial in this case, is for now beyond the scope of this analysis. However, at least in English, the term was often used to define something first created as a copy or imitation of something natural. The second term, 'intelligence', is something more difficult to define in this context, although the etymology does not present particular difficulties in translation. As a direct derivative of the Lat. *intelligĕre*, the concept involves a specific way of 'reading between the lines', thus including 'comprehending' or 'perceiving'. Focusing for a moment on the first verb, if we follow the translation of 'understanding' in terms of "taking (each element, alone, and in combination and/or mutual interaction) together" we don't see too many complications in relating the term to a computational activity. This obviously implies a non-specific value/structure/meaning to human intelligence, in comparison to an artificial one, given that both deal with aspects such as collection, compilation, comparison, combination and calculation. Leaving aside for now philosophical considerations on utilitaristic/utilitarian and mechanistic aspects of such activity, artificial intelligence such as the one provided not only by the most advanced computer systems, but also by simple electronic calculators, appears in this area faster and more precise—in other terms, better—than the intelligence of human beings. With some elements possibly shared by both comprehension and perception, such as capacity for logic, learning (also, but not limited to a psychological-behavioral sense), planning, creativity and problem solving, the main problem in identifying perception has to do with the connections with 'the self' as in self-perception, self-image, self-awareness, all the way to multiple types (which is in itself an assumption, albeit with strong scientific evidence, for instance in the differentiation of cerebral areas and functions) of knowledge, wisdom and (deeper? broader?) understanding, also in the translational sense of 'foundation' or 'principle', thus connecting with 'inclination' (as a translation for the German *Neigung*) and motivation, orientation. Moreover, even from a strictly scientific point of view, a universal and universally accepted definition of intelligence is far from being 'a thing'. Certainly, throughout history there have been multiple attempts to provide an accurate definition in this sense. The Board of Scientific Affairs of the American Psychological Association gave in this sense this definition of the term:

> Individuals differ from one another in their ability to understand complex ideas, to adapt effectively to the environment, to learn from experience, to engage in various forms of reasoning, to overcome obstacles by taking thought. Although these individual differences can be substantial, they are never entirely consistent: a given person's intellectual performance will vary on different occasions, in different domains, as judged by different criteria. Concepts of "intelligence" are attempts to clarify and organize this complex set of phenomena. Although considerable clarity has been achieved in some areas, no such conceptualization has yet answered all the important questions, and none commands universal assent. Indeed, when two dozen prominent theorists were recently asked to define intelligence, they gave two dozen, somewhat different, definitions.[3]

[3] Cit. Neisser, U., Boodoo, G. Bouchard, T.J., Boykin, A. W., Brody, N., Ceci, S.J., Halpern, D.F, Loehlin, J.C., Perloff, R., Sternberg, R J., & Urbina, S. 1996. Intelligence: Knowns and unknowns. *American Psychologist.* 51: 77–101. doi:https://doi.org/10.1037/0003-066x.51.2.77

Within the realm of human intelligence, there are also certain parameters which are used to better define the term, including the volume, speed, acceleration (frequency, velocity) and/or structure of the working memory and/to the capacity for sequential activation and activity forecast, the hierarchy system of information-processing neural activities, including neurogenesis and axonal structuration, and the more complex (philosophically speaking, at least) activities of biographical memory and consciousness. In the United States, a popular definition of intelligence is the famous definition offered in 1994 by Mainstream Science on Intelligence:

> [Intelligence is] A very general mental capability that, among other things, involves the ability to reason, plan, solve problems, think abstractly, comprehend complex ideas, learn quickly and learn from experience. It is not merely book learning, a narrow academic skill, or test-taking smarts. Rather, it reflects a broader and deeper capability for comprehending our surroundings – "catching on", "making sense" of things, or "figuring out" what to do.[4]

Among the very interesting skills that human intelligence provides, and is therefore at the center of the developments of artificial intelligence technologies, we find pattern recognition (for instance, the ability to recognize facial expressions on one side and familiar faces on the other), and communication, including language understanding and speech production, as well as concept/idea formation. To be more precise, given the current scientific knowledge, at least until new technological advancements would reach the same levels and possibly surpass them, much of human intelligence is shared by other animals, especially mammals, as well. On a theoretical level, human intelligence can be understood within specific frameworks, for instance, the 'theory of multiple intelligences' by Gardner, which in turn refers to eight main abilities as bases for related 'intelligencies': linguistic, logical-mathematical, spatial, musical, bodily, kinesthetic, interpersonal and intrapersonal.[5] Other wide known theories include Bandura's theory of self-efficacy and cognition (with further developments by Walter Mischel), the Intelligence Compensation Theory or ICT by Wood and Englert, the Investment theory based on the Cattell–Horn–Carroll theory, the so-called LI effect or Latent inhibition by Lubow and

[4] Gottfredson, L.S. (1997). Foreword to "intelligence and social policy", in *Intelligence, (24)*1, pp. 1–12, doi:https://doi.org/10.1016/S0160-2896(97)90010-6

[5] Gardner, H. 1996. *Multiple Intelligences: The Theory in Practice, A Reader*. New York, NY: Basic Books.

Moore, the Parieto-frontal integration theory of intelligence, Piaget's theory and Neo-Piagetian theories, the PASS theory of intelligence by Luria, the Process, Personality, Intelligence & Knowledge theory or PPIK, and the Triarchic theory of intelligence by Sternberg. Aside from the specific content and differences between theories, for the perspective of our analysis we are interested in possibly defining, when and if possible, the main components that could possibly separate, in qualitative and/or quantitative terms, artificial intelligence from human (or more generally, animal) intelligence. This could also mean 'simply' characterizing those differences as stages or levels, albeit this decision would lay or ascribed, possibly artificial in themselves assumptions and values/goals, thus drawing a connection—once again—between intelligence, knowledge and awareness—consciousness, also in neurobiological terms (Fig. 7.3). For

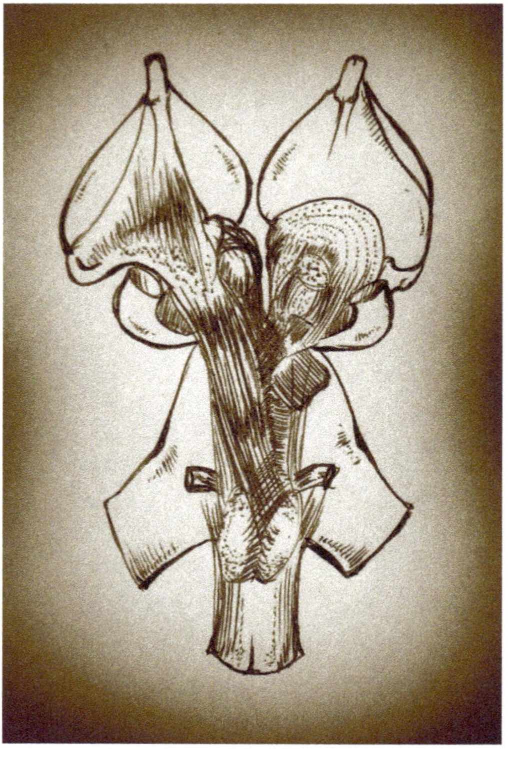

Fig. 7.3 View of the Reticular Activating System (RAS), with thalamus and corpus subthalamicus, substantia nigra, medial and lateral lemniscus (including the nucleus and the decussation), decussation of the superior peduncle, reticular formation, the internal arcuate fibers, and the olive. The RAs play a very special role in the processes involved in alertness, arousal, attention, consciousness and habituation

instance, we could quantify intelligence *a* (I*a*) as 'less complex' or 'slower' than intelligence *b* (I*b*). Then we would also decide (observe?) whether these intrinsic qualities are essential or existential, in the sense of result or (by) product of (external/internal) environmental factors, to determine whether the outcomes are 'produced' or 'defining' the 'being a' from (the)[6] 'being *b*'.

7.2.6 Sense, Purpose, Meaning

We finally reached the last part of our analysis, thereby investigating 'Sense, Purpose, Meaning' in relation to critical neuroscience and the mind–body problem, but also to determine its application to the sense, the purpose and the meaning of the research discussed so far. At the beginning of our discussion in Chap. 1, we made sure to stress how through this analysis we never claimed to have found a *final solution* (to the mind–body problem) because we do not think that there could be such thing as something 'final' given the very etymological sense of the term which we agreed upon, namely the combination of 'termination' and 'end' but also 'perimeter', 'boundary', 'function', 'goal' and truly 'Sense, Purpose, Meaning' which, by definition (again, a de-*fin*-ition) varies from individual to individual, subjectively and personally. In this sense, we argue that there can be as many "*my own* final solution to the mind-body problem" as there are individuals; even more, as there can be perspectives or viewpoints, even traits, even personalities, in each individual. To be sure, this view has been for centuries at the center of a harsh criticism of the very scope and practice of philosophy, it the sense that many have viewed philosophy as an empty effort, at least from the scientific (in this context fully intended as mechanistic, utilitarian, functional, empirical, material-observable problem-solving oriented, etc.), to provide any real progress. In our analysis we will not tackle the epistemological validity or integrity of philosophy given these premises, as our intention is not to provide evidence or defense for any related position (which we argue, could be certainly viewed as truly philosophical nevertheless, thus making the opposite argument a true fallacy), but to 'simply' provide a 'scientific re-examination of the mind-body problem' and related issues such as perception, cognition, awareness, (artificial vs. human) intelligence, hard problem of consciousness and con-

[6] Of course, the presence/absence of the substantivizing article would give raise to enormous implications, staring from Heideggerian perspectives.

science, *only* (thus limiting our *searchlight* to the related areas) in connection to our understanding of '*the* mind'. Now, as there are as many minds as there are individuals (leaving aside for a moment the problems of corpus callosotomy, dissociative identity disorder, /multiple personalities, demonic, *dia*bolic vs. *sym*bolic possessions, good vs. bad angels, etc.) we would like to embrace here a 'relatively relativistic' (thus limited, concept-constrained, not absolute, universal and general-izable) view according to which (certain/specific) philosophical positions, and more generally, worldviews might in turn have (certain/specific) neural underpinnings *and* vice versa. Thus, we would like to expand philosophy to philosophy itself and neuroscience to neuroscience itself. In other words, the relatively relativistic view here is that an individual might embrace a certain stance on the mind–body problem or the hard problem of consciousness exactly because of (which is, let us restate it again, not equivalent with 'caused by') a 'certain state of mind in the individual's brain'. This is of course a theoretical assumption, and it might remain such, given that it involves (an) observation from within, "(an) observation which is itself observing itself observing"—yes, adding a third level to double hermeneutics, possibly something like *'Triple Hermeneutics' necessary for the Triple-S Model (TH-TS): Self, Soul, Spirit*—and therefore not subject-able to the scientific method it attempts to use. However, this realization, just like the aforementioned epistemological considerations on philosophy itself, wouldn't have to be "indicative of epistemological failure or logical fallacy"; on the contrary, it might provide the exact key for solving the problem, that is, its 'insolvability due to particularity'. We could even link this argument to what we have previously said on the *sensus divinitatis*, the relation between sin and mental disorder, and even Pascal's wager. Thus, a 'believing brain' would be right in perceiving that god (or free will, or a mind-soul separated from the body) existed, just as much as a 'non-believing brain' would be right in perceiving that god (or free will, or a mind-soul separated from the body) did not exist. We could certainly theoretically extend this relativistic (many would argue very postmodernist) view to pretty much anything we can think of, understanding that the cognitive process underlying this view would be itself influenced by itself, that is, by the neural processes thereby *in process*. This application of nature vs. nurture-mutual influence, 'mind over matter' vs. 'matter over mind' would however, in our view, not necessarily lead to the dismissal of any absolute or universal truths we could (*hope* to) find, but to the realization that there are 'ulterior perceptual modalities' the content of which can only be understood (read:

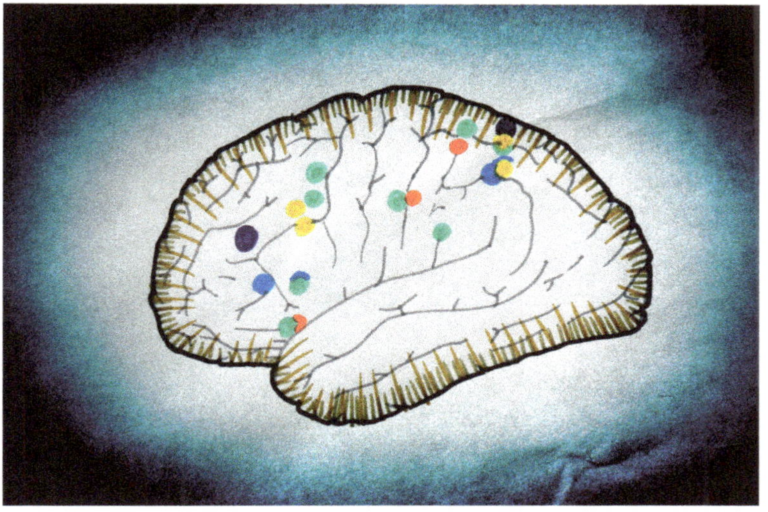

Fig. 7.4 Rendering of the comparative analysis of neural correlates of consciousness in humans according to the meta-analysis by Rees et al. (2002), showing localized areas (in red), Lumer et al. (1998, in aquamarine), Kleinschmidt et al. (1998, in light blue), Portas et al. (2000, in yellow) and Beck et al. (2001, in dark blue)

felt, perceived, interjected, intel-lected and interjected) with a different mindset, a different state of mind, ultimately a different state of consciousness (Fig. 7.4).

We are certainly aware of the difficulties of demonstrating in empirical terms the very existence of (a)—a somewhat easier task—these conscious states beyond their diagnostic labeling, and (b) this very *metaphysical* content we are expected to perceive. We attempted to present the most relevant research areas and fields investigating these varied layers of perception-consciousness based on neuroscience, and we certainly *believe*—that is, our *applied opinion* has been formed in connection, and in part as a result of the aforementioned processes—that there is enough (this time objective, albeit—thankfully,[7] we could say—limited to subjective evaluation and validation) *evidence* of the existence of these states, of multiple modalities of perceiving, understanding, and being, and of a non-

[7] As in gracefully, both full of grace and due to *gratia*.

physically-based-matter-based form of mind, soul, spirit. In this context we would also like to mention that the very concept of (having, possessing, be defined by an) opinion is connected to the Lat./Ital. *opinàre*, thus literally touching, reaching and creating, more specifically:

(a) Touching with the eye, reaching with the eye and creating with the eye (based on the Sansc. Root **op* as in the Gr. ὄψις, thus [ap]perception, grasping, seeing)
(b) Touching with the mind, reaching with the mind and creating with the mind (based on the Sansc. Root **ap* as found in *Opus/Opera*)

Given these premises, our perception *is* our opinion, and given the foundational impossibility of avoiding this subjective component of understanding reality, we should not try to eliminate (our) subjectivity in reaching a more objective understanding, the same way we should not try to eliminate the place effect in reaching a more universal (statistically intended) healing method. Let us restate to clarify, this is definitely not a post-modernist position involving a rejection of 'anything absolute and absolute*ly everything*' but the (possibly: exact) opposite, it is a truly Traditional (some would argue even reactionary or anti-modernist) view that understands the subjective experience as primary pathway (even in a purely neurological sense) to reach absolute, universal truth. It is finding infinite doors to an incredibly vast room beyond. To relate all these considerations to some practical examples, we can think of a theoretical understanding of the role such subjective perception plays in defining what we could qualify as an "impaired [ethical vs. moral] pathway" due to our human nature, which is truly *sinful* in the aforementioned sense, thus lacking the ability to perceive the whole picture, that is, 'the picture as perceived by the subjectivities of others', again in a Levinasian sense.

How can we then quantify these subjective experiences and judgments as universally valid, when faced with the difficult task of creating a series of guidelines to serve more noble (once this hierarchy of ethical nobility is in place, which in itself warrants further discussion) goals, to protect and serve the general population, or humanity as a whole? A good example to remember in this context, is the famous Stanley Milgram's experiment, also referred to as the 'obedience experiment', which began in July 1961 at Yale University. In this series of social-behavioral psychology experiments, Dr. Milgram was interested in answering questions on the role of authority and obedience to unethical, immoral and possibly dangerous and hostile behavior in the general population. More specifically, three types of individuals were involved in the study; the teacher, the learner and the experimenter. The subject of the study—always playing the role of the teacher, a role picked a priori without the subject's knowledge—was required to administer a shock to the learner (an actor) for each wrong answer, in 15-volt increments, to the maximum 45-volt shock three times in succession, after which the learner would cease to answer or reply in any way to the questions asked (Milgram 1974). The main controversy regarding the ethics of the experiment is related to the 'inflicted insight' suffered by the subjects-teachers, which is considered part of the more general methodology within 'deceptive debriefing' with potential future psychologically-emotionally harmful consequences (Levine 1988). More in detail, this criticism stems from the 'Right to full disclosure' according to which the researcher not only fully describes the specifics of the research to each participant/subject prior to the beginning of the experiment, but also grants the right to withdraw or refuse participation (Polit and Beck 2014, p. 84).

Notwithstanding these aspects of ethical research, the bigger questions might be focused on whether granting this right is of competence of the researcher himself/herself or whether it should be monitored, promoted and controlled 'from above'. This 'aboveness' could start through the legal path of guidelines, rules and regulations such as the ones found in professional Codes of Ethics (Polit and Beck 2014, p. 81). A good example in this sense, the Belmont Report, focuses on the three main ethical principles of "Respect for Persons", "Beneficence" and "Justice" (U.S. Department of Health & Human Services 1979) to address these issues. In our case, the question arising from the considerations above would be 'Beneficence *for whom*?' In other words, we could argue that even if the results of the experiments, as well as the experiment itself, could be beneficial for the general scientific research and therefore—which is not always a *segue*, from the perspective of correlation vs. causation—to the health of the general public, the experiment would fail to be beneficial to the subjects involved. Moreover, as we have seen, the experiment could also be potentially harmful to the subjects. Finally, aside from the details of the claimed unethical behavior of the researcher in planning and administering the study, another controversy stems from the very nature of this type of study. Some argue that the results were not completely correct, both in the sense of imprecise collection and in the sense of purposeful omission (Perry 2012); others argue that the results do not add much to the 'common sense knowledge' that anyone can commit wrongful act if:

(a) Under pressure or coercion
(b) Under perceived or real authority
(c) If they are made to believe that they are doing the *right* thing (*the combination of a & b*)

Certainly, in this debate lies the very difference between scientific psychology and 'pop' psychology, in the sense that every claim needs to be verified according to evidence-based methodology. Furthermore, many experiments and meta-analysis showed on multiple occasions that 'common sense' is not that 'common'. More specifically, although there are still many questions about the ethics of this study, Stanley Milgram's biggest contribution was the demolition of the assumed 'exceptionalism' of Nazi evil. He was able to add more evidence to the fact—I use this term with intent and purpose—that wrongful acts are not committed by certain cultures, nations, ethnicities or races (although we do not thereby negate the possible influence of particularisms in the context of collective unconscious in psychoanalytic perspectives). This is especially true in the context of WWII. First of all because many developments of medical ethics originate in the debate within the Nuremberg trial and the Adolf Eichmann trial, but also because it clearly challenges the assumption that the horrible war crimes, genocide and holocaust perpetrated by the Nazis won't vs. wouldn't fall within the ethically and morally wrong behavior of human beings in general.

In conclusion and paradoxically[8] the ethically wrong procedure of Stanley Milgram's Experiment contributed to the realization of the *possibility* of ethical wrongdoing within and across perceived opposite sides. Beside considerations proper to 'the banality of evil' (Arendt 1963) and 'the evil of banality' (Minnich 2016), we should also remember that both (*even*) Sartre and Camus warned us against the risks of 'bad philosophies' which promote the full abandonment of reason/rationality/rational methods, as in the case of (form the existentialist judgment) Heidegger,

[8]Although clearly this is not the case, from the perspective of Traditional –capital *T*- systems, both ethnocultural and religious, where the expectation of a universal capacity for evil in (human) beings is very often standard and basis for ritualistic disciplines, theologies, and technologies.

Husserl, Jaspers, Shestov and especially Kierkegaard, who didn't just abandon reason, but 'turned himself to God' and Husserl, who elevated reason, ultimately arriving at ubiquitous Platonic forms and an abstract god (Tomasi 2016). Under this analysis, embracing *the absurd* means acknowledging the contradiction between the desire of human reason and the unreasonable world. As it is well known, Sartre defended (his) affirmation of the reality of every truth in a naturalistic manner, as the confrontation of human beings with the possibility of choice. Therefore, since man is "responsible for what he is and at the same time he is responsible for other human beings, he always chooses the better for himself and therefore for humanity".[9]

The possibility of free will, free won't and free choice then is to be understood in a higher sense of personal opinion, as discussed above. The ethical-structural-functional 'goodness' of the (human) brain as a universally-similar neural element which is both sensor and deliverer of physical and metaphysical experiences helps navigating such notions of good vs. evil. We mentioned that the 'aboveness' of ethics in research could start through the legal path of guidelines, rules and regulations, but we now have to admit that there is an ulterior 'aboveness' which involves a higher, more profound sphere. An element of transcendental existence, which many have attributed to a divine, spiritual, religious entity. An entity[10] which communicates *with us through us*, more specifically to the complex apparatus of mind–body processes (in our analysis, especially neurological ones) to *create*—in this context, we agree with the poetic element therein—our own, true, both objective and subjective experience. A mind–body experience. A mind–body evidence. An experience which, to be completely understood, needs to be fully integrated. In conclusion, we hope that our re-examination of the mind–body problem provided experience and evidence, in ontological and practical (in the clinical sense) terms for the fact that we, human beings, can be 'out of tune' with our self, and this tuning activity, which is also a neurological mechanism, a psychological process, and an existential need, can be supported by "(an) appropriate application of (an) appropriate philosophy" to shed light on our true *nature and nurture.*

[9] Sartre, J-P. 1997. *Existentialism and Humanism*, translation and introduction by P. Mairet, Haskell House Publishers, Brooklyn, NY, p. 29.

[10] Given the aforementioned considerations, at this level, the possible *pluralis majestatis* of the term (i.e. entities), whether confirmatory or purely semantic, does not appear to represent an ontological problem, in the "bigger picture."

References and Further Readings[11]

Bibliography

Arendt. 1963. *Eichmann in Jerusalem: A report on the banality of evil*. New York: Viking Press.

Beck, A.T., R.A. Steer, R.Ball, W.F. Ranieri. 1996. Comparison of Beck Depression Inventories-IA and -II in Psychiatric Outpatients. *J Pers Assess*. 67: 588–597.

Christopher, J.C., D.C. Wendt, J. Marecek, and D.M. Goodman. 2014. Critical cultural awareness. Contributions to a globalizing psychology. *American Psychologist* 69 (7): 645–655. Washington, DC: APA, American Psychological Association.

Coleman, A.M.E. 2013. American journal of hospice & palliative medicine: Physician attitudes toward advanced directives: A literature review of variables impacting on physicians attitude toward advance directives. *Prime National Publishing Corporation*. https://doi.org/10.1177/1049909112464544.

Flood, M.M., and M. Dresher. 1951. *A preference experiment*. Santa Monica: RAND Research Paper.

Gardner, H. 2006. *Multiple intelligences: New horizons in theory and practice*. New York: Basic Books.

Greenhalgh, T., and B. Hurwitz. 1999. *Narrative based medicine: Why study narrative?* US National Library of Medicine, National Institutes of Health. *PubMed BMJ* 318 (7175): 48–50.

Hinderer, K.A., and M.C. Lee. 2014. Assessing a nurse-led advance directive and advance care planning seminar. *Applied Nursing Research* 27 (1): 84–86.

Johnson, Y. 2009. *Ethics and social welfare: The bioethical underpinnings of advance directives*. London: Routledge.

Kearney-Nunnery, R. 2012. *Advancing your career: Concepts of professional nursing*. 5th ed. Philadelphia: FA Davis.

Kleinschmidt, A., C. Büchel, S. Zeki, and R.S.J. Frackowiak. 1998. Human brain activity during spontaneously reversing perception of ambiguous pictures. *Proceedings of the Royal Society of London – Series B: Biological Sciences* 265: 2427–2433.

Kroning, M. 2014. Advance directives education: A critical need. *Journal of Christian Nursing: A Quarterly Publication of Nurses Christian Fellowship* 31 (4): 220.

Leder, D. 1990. Clinical interpretation: The hermeneutics of medicine. In: Theoretical medicine March 1990, 11, 1, Dordrecht: Kluwer Academic Publishers.

Levine, Robert. 1988. *Ethics and regulation of clinical research*. New haven: Yale University Press.

Lumer, E.D., K.J. Friston, and G. Rees. 1998. Neural correlates of perceptual rivalry in the human brain. *Science* 280 (5371): 1930–1934.

Marini, M.G. 2012. *Medicina Narrativa per una sanità sostenibile*. Milan, I: Edizione Lupetti.

Marini, M.G. 2012. *Narrative Medicine as a tool to detect the burden of illness*. Milan, I: ISTUD.

Ma, H.K. 2013. The moral development of the child: An integrated model. *Front Public Health* 11 (18): 1–57.

Milgram, S. 1974. *Obedience to authority: An experimental view*. New York: HarperCollins.

Minnich, E. 2016. *The evil of banality*. Lanham: Rowman & Littlefield.

Mitchell, Jennifer K, MSN, A.N.P.-B.C., G.N.P.-B.C. 2011. Nursing informatics 101. *ONS Connect* 26 (4): 8–12.

Moore. 1903. Meaning in memory and in attention. *Psychological Review* 10 (3): 267–283.

Mujkanović, L. 2016. *Nursing research studies*. Ethics, community, and healthcare. Hooksett: SNHU.

Perry, G. 2012. *Behind the shock machine: The untold story of the notorious Milgram psychology experiments*. Sidney: New Press.

Polit, D.F., and C.T. Beck. 2014. *Essentials of nursing research: Appraising evidence for nursing practice*. 8th ed. Philadelphia: Wolters Kluwer Health/ Lippincott Williams & Wilkins.

Portas, C.M., B.A. Strange, K.J. Friston, R.J. Dolan, and C.D. Frith. 2000. How does the brain sustain a visual percept? *Proceedings of the Royal Society of London – Series B: Biological Sciences* 267: 845–850.

Rampley, M. 2017. *The seductions of Darwin: Art, evolution, neuroscience*. University Park: The Pennsylvania State University Press.

Rees, G., G. Kreiman, and C. Koch. 2002. Neural correlates of consciousness in humans. *Nature Reviews Neuroscience* 3: 261–270.

Schicktanz, S., and M. Schweda. 2009. One man's trash is another man's treasure: Exploring economic and moral subtexts of the "organ shortage" problem in public views on organ donation. *Journal of Medical Ethics* 35 (8): 473–476.

Schleim, S. 2014. Critical neuroscience – Or critical science? A perspective on the perceived normative significance of neuroscience. *Frontiers in Human Neuroscience* 8: 336. https://doi.org/10.3389/fnhum.2014.00336.

Tomasi, D.L. 2016. *Medical philosophy. Philosophical analysis of patient self-perception in diagnostics and therapy*. New York: Ibidem Verlag/Columbia University Press.
Tversky, A., and D. Kahneman. 1981. The framing of decisions and the psychology of choice. *Science* 211 (4481): 453–458.
U.S. Department of Health & Human Services. 1979. *The Belmont report. Ethical principles and guidelines for the protection of human subjects of research*. Washington, DC: The National Commission for the Protection of Human Subjects of Biomedical and Behavioral Research, Department of Health, Education, and Welfare, Office of the Secretary. Retrieved from: http://www.hhs.gov/ohrp/regulations-and-policy/belmont-report/index.html
Verene, D.P. 1997. *Philosophy and the return to self-knowledge*. New Haven: Yale University Press.

References and Further Readings

Bibliography

Achinstein, P. 1983. *The nature of explanation*. Oxford: Oxford University Press.

American Bar Association (ABA). 2015. *Law for older Americans – Healthcare advance directives: What is the patient self-determination act*. Retrieved October 15, from: http://www.americanbar.org.ezproxy.snhu.edu/groups/public_education/resources/lawissues_for_consumers/patient_self_determination_act.html

American Psychological Association. 2016. Measuring socioeconomic status and subjective social status. Washington, DC: APA. Retrieved from: https://www.apa.org/pi/ses/resources/class/measuring-status.aspx

Andersen, H. 2001a. The history of reductionism versus holism approaches to scientific research. *Endeavor* 25: 153–156.

Antiseri, D., and V. Cagli. 2008. *Dialogo sulla diagnosi. Un filosofo e un medico a confronto*. Roma: Armando.

Antiseri, D., and M. Timio. 2000. *La medicina basata sulle evidenze*. Milano: Edizioni Memoria.

Antiseri, D., and G. Vattimo. 2008. *Ragione filosofica e fede religiosa nell'era post-moderna*. Soveria Mannelli: Rubbettino Editore.

Aristotle. 1966a. *Metaphysics*. Trans. H.G. Apostle. Bloomington: Indiana University Press.

Baglivi, G. 1723. *Practice of physick*. 2nd ed. London: Midwinter.

Barnett, J.E., and A.J. Shale. 2013. Alternative techniques. Today's psychologists are increasingly integrating complementary and alternative medicine techniques into their work with clients. Here's an overview of the most popular

D. L. Tomasi, *Critical Neuroscience and Philosophy*,
https://doi.org/10.1007/978-3-030-35354-4

treatments, the research on their efficacy and the ethical concerns they raise. *Monitor on Psychology* 44 (4): 48. Washington, DC: American Psychological Association.

Bartlett, E. 1844. *Essay on the philosophy of medical science.* Philadelphia: Lea & Blanchard.

Bassman, L. 1998a. *The whole mind: The definitive guide to complementary treatment for mind, mood, and emotion.* Novato: New World Library.

Bauschatz, P.C. 1982. *The well and the tree.* Amherst: The University of Massachusetts Press.

Beauchamp, T., and J.F. Childress. 2001. *Principles of biomedical ethics.* 5th ed. Oxford: Oxford University Press.

Beauregard, M., and D. O'Leary. 2008a. *The spiritual brain. A neuroscientist's case for the existence of the soul.* New York: HarperCollins.

Behnke, S.H. 2014a. Always a cultural perspective. The society of Indian psychologists comments on the APA Ethics code. *Ethically Speaking, Monitor on Psychology* 45 (10): 60–64.

Bellamoli, E., F. Alessandrini, and G. Zoccatelli. 2012. *L'applicazione del neuroimaging per valutare gli effetti neurobiologici degli interventi terapeutici.* http://issuu.com/dipartimentodipendenzeverona/docs/brochure_neuroscienze#

Bem, D.J. 2011a. Feeling the future: Experimental evidence for anomalous retroactive influences on cognition and affect. *Journal of Personality and Social Psychology* 100 (3): 407–425.

Bem, D.J., P.E. Tressoldi, T. Rabeyron, and M. Duggan. 2014a. *Feeling the future: A meta-analysis of 90 experiments on the anomalous anticipation of random future events* (SSRN scholarly paper no. ID 2423692). Rochester: Social Science Research Network.

Bendeck Sotillos, S. 2013a. *Psychology and the Perennial Philosophy*, Psychology and the perennial philosophy: Studies in comparative religion. Bloomington: World Wisdom.

Beneduce, R. 2007. *Etnopsichiatria. Sofferenza mentale e alterità fra Storia, dominio e cultura.* Roma: Carocci editore.

Bersoff, D.N. 1996. *Ethical conflicts in psychology.* Washington, DC: American Psychological Association.

Bettini, M., ed. 1991. *La maschera, il doppio e il ritratto. Strategie dell'identità* (the mask, the double and the portrait: Strategies of identity). Rome-Bari: Editori Laterza.

Bialosky, J.E., S.Z. George, M.E. Horn, et al. 2014. *Spinal manipulative therapy-specific changes in pain sensitivity in individuals with low back pain* (NCT01168999). *The Journal of Pain* 15 (2): 136–148.

Black, D.A.K. 1968. *The logic of medicine.* Edinburgh: Oliver & Boyd.

Block, N., O. Flanagan, and G. Guzeldere, eds. 1998a. *The nature of consciousness.* Cambridge, MA: MIT Press.

Blois, M.S. 1990a. Medicine and the nature of vertical reasoning. *New England Journal of Medicine* 318: 847–851.

Bobrow, D.G., ed. 1985. *Qualitative reasoning about physical systems.* Cambridge, MA: MIT Press.

Bock, G.R., and J.A. Goode, eds. 1998. *The limits of reductionism in biology.* London: Wiley.

Boncinelli, E. 2011. *La vita della nostra mente* (Our mind's life). Rome: Editori Laterza.

Bonelli, R., R.E. Dew, H.G. Koenig, D.H. Rosmarin, and S. Vasegh. 2012a. Religious and spiritual factors in depression: Review and integration of the research. *Depression Research and Treatment* 2012: 1–8.

Boorse, C. 1975. On the distinction between disease and illness. *Philosophy and Public Affairs* 5: 49–68.

———. 1987. Concepts of health. In *Health care ethics: An introduction*, ed. D. VanDeVeer and T. Regan, 359–393. Philadelphia: Temple University Press.

Boshuizen, H.P.A., and H.G. Schmidt. 1992a. On the role of biomedical knowledge in clinical reasoning by experts, intermediates, and novices. *Cognitive Science* 16 (2): 153–184.

Brattberg, G. 2008. Self-administered EFT (emotional freedom techniques) in individuals with fibromyalgia: A randomized trial. *Integrative Medicine: A Clinician's Journal* 7 (4): 30–35.

Braveman, P.A., C. Cubbin, S. Egerter, D.R. Williams, and E. Pamuk. 2010. Socioeconomic disparities in health in the United States: What the patterns tell us. *American Journal of Public Health* 100 (Suppl 1): S186–S196. https://doi.org/10.2105/AJPH.2009.166082.

Breggin, P.R. 1991a. *Toxic psychiatry: Why therapy, empathy, and love must replace the drugs, electroshock, and biochemical theories of the "new psychiatry"*. New York: St. Martin's Press.

Brianese, G., ed. 1998. *"Congetture e confutazioni" di Popper e il dibattito epistemologico post-popperiano* (Popper's conjectures and refutations and the post-Popper epistemological debate). I ed. – XIII. Turin: Paravia.

Briquel, D. 1998. The role of women in the traditions surrounding the Etruscan kings of Rome. *Comptes Rendus Des Seances De L Academie Des Inscriptions & Belles-Lettres* 2: 397–414.

Britton, W.B., N.E. Lepp, H.F. Niles, et al. 2014. A randomized controlled pilot trial of classroom-based mindfulness meditation compared to an active control condition in sixth-grade children. *Journal of School Psychology* 52 (3): 263–278.

Brody, H. 1992. *The healer's power.* New Haven: Yale University Press.

Brooks, L.R., G.R. Norman, and S.W. Allen. 1991. Role of specific similarity in a medical diagnostic task. *Journal of Experimental Psychology: General* 120 (3): 278–287.

Brothers, L. 2001a. *Mistaken identity: The mind-body problem reconsidered.* New York: SUNY Press.

Brown, L., and J. Jellison. 2011. *Auditory perception of emotion in sung and instrumental music in children with autism spectrum disorders*, AMTA published journals posters research catalog. Silver Spring: American Music Therapy Association.

Buchler, J. 1955a. *Nature and judgment*. New York: Columbia University Press.

Bunge, M. 1989. *Treatise on basic philosophy, Vol. 8: Ethics: The good and the right*. Dordrecht/Boston: Reidel.

Cacciari, M. 2000. *L'ubiquità del centro* (the ubiquity of the centre). *Tema Celeste*, no. 79–80, Milan: Gabrius, p. 94.

Cacioppo, J.T., G.G. Berntson, and J. Decety. 2010. Social neuroscience and its relation to social psychology. *Social Cognition* 28: 675–684.

Cameron, N. 1986. The methods of auxological anthropology. In *Human growth: A comprehensive treatise*, ed. F. Falkner and J.M. Tanner, vol. 3. New York: Plenum Press.

Caplan, A.L. 1986. Exemplary reasoning? A comment on theory structure in biomedicine. *Journal of Medicine and Philosophy* 11: 93–105.

———. 1992a. Does the philosophy of medicine exist? *Theoretical Medicine* 13: 67–77.

Cappuccio, A., S. Napolitano, F. Menzella, et al. 2019. Use of narrative medicine to identify key factors for effective doctor–patient relationships in severe asthma. *Multidisciplinary Respiratory Medicine* 14: 26. https://doi.org/10.1186/s40248-019-0190-7.

Carlat, D. 2010. *Unhinged: The trouble with psychiatry. A Doctor's revelations about a profession in crisis*. New York: Simon & Schuster.

Carlson, N.R. 1994. *Physiology of behavior*. 5th ed. Needham Heights: Allyn and Bacon/Paramount Publishing.

Carpenter, S. 2012. That gut feeling. With a sophisticated neural network transmitting messages from trillions of bacteria, the brain in your gut exerts a powerful influence over the one in your head, new research suggests. *Monitor on Psychology* 43 (8): 50–55. Washington, DC: American Psychological Association.

Carruthers, P. 2004a. *The nature of mind*. London: Routledge.

Carter, K.C. 2003a. *The rise of causal concepts of disease: Case histories*. Burlington: Ashgate.

Cassell, E.J. 2004. *The nature of suffering and the goals of medicine*. 2nd ed. New York: Oxford University Press.

Castel, A., M. Salvat, J. Sala, and M. Rull. 2009. Cognitive-behavioural group treatment with hypnosis: A randomized pilot trial in fibromyalgia. *Contemporary Hypnosis* 26: 48–59.

Cavalli-Sforza, L.L., and M. Feldman. 1981. *Cultural transmission and evolution: A quantitative approach*. Princeton: Princeton University Press.

Cefalù, W.T., Z.E. Floyd, J.M. Stephens, et al. 2014a. Botanicals and translational medicine: A paradigm shift in research approach. *Nutrition* 30 (7–8S): S1–S68. NIH PubMed.

Chabris, C., and D. Simons. 2010. *The invisible gorilla*. New York: Crown Publishing Group.

Chalmers, D. 1996a. *The conscious mind*. Oxford: Oxford University Press.

Chandrasekaran, B., J.W. Smith, and J. Sticklen. 1989a. Deep models and their relation to diagnosis. *Artificial Intelligence in Medicine* 1: 29–40.

Chapman, G.B., and A.S. Elstein. 2000. Cognitive processes and biases in medical decision making. In *Decision making in health care: Theory, psychology, and applications*, ed. G.B. Chapman and A. Frank, 183–210. Cambridge: Cambridge University Press.

Cherkin, D.C., K.J. Sherman, J. Kahn, et al. 2011. A comparison of the effects of 2 types of massage and usual care on chronic low-back pain: A randomized, controlled trial. *Annals of Internal Medicine* 155 (1): 1–9.

Churchland, P.S. 2002. *Brain-wise: Studies in Neurophilosophy*. Cambridge, MA: MIT Press.

Churchland, P.M. 2007a. *History of the concept of mind*. Farnham: Ashgate Publishers.

Churchland, P.S. 2013a. *Touching a nerve. The self as brain*. New York: W.W. Norton & Company.

Clay, R.A. 2014a. Random Sample. James E. Maddux, PhD. *Monitor on Psychology* 45 (11). Washington, DC: American Psychological Association.

Clouser, K.D., and B. Gert. 1990. A critique of principlism. *Journal of Medicine and Philosophy* 15: 219–236.

Cochrane, K. 2013. *All the rebel women: The rise of the fourth wave of feminism*. London: Guardian Books.

Coiera, E. 2000. When conversation is better than computation. *Journal of the American Medical Informatics Association* 7 (3): 277–286.

Collingwood, R.G. 1940. *An essay on metaphysics*. Oxford: Clarendon Press.

Collins, F.S. 2006a. *The language of God. A scientist presents evidence for belief*. New York: Free Press Simon & Schuster.

Conboy, L.A., E. Macklin, J. Kelley, et al. 2010. Which patients improve: Characteristics increasing sensitivity to a supportive patient-practitioner relationship. *Social Science & Medicine* 70 (3): 479–484.

Consolmagno, G. 2008. *God's mechanics: How scientists and engineers make sense of religion*. San Francisco: Wiley.

Cook, C. 2004a. Addiction and spirituality. *Addiction* 99: 539–551.

Cotton, S., Y.H. Roberts, J. Tsevat, et al. 2010. Mind-body complementary alternative medicine use and quality of life in adolescents with inflammatory bowel disease. *Inflammatory Bowel Disease* 16 (3): 501–506.

Coulter, A. 1999. Paternalism or partnership? *British Medical Journal* 319: 719–720.

Crespin, D.J., K.H. Griffin, J.R. Johnson, C. Miller, M.D. Finch, R.L. Rivard, S. Anseth, and J.A. Dusek. 2015. Acupuncture provides short-term pain relief for patients in a Total joint replacement program. *Pain Medicine* 16 (6): 1195–1203.

Crichton, P. 2004. When philosophy meets psychiatry. *The Lancet* 363 (9410): 741, 28 February, Amsterdam: Elsevier.
Cronenwett, L., G. Sherwood, and S.B. Gelmon. 2009. Improving quality and safety education: The QSEN learning collaborative. *Nursing Outlook* 57 (6): 304–312. https://doi.org/10.1016/j.outlook.2009.09.004.
Crossing the Quality Chasm. 2001. *A new health system for the 21st century.* Washington, DC: National Academies Press.
Crumbaugh, J.C., and L.T. Maholick. 1964a. An experimental study in existentialism: The psychometric approach to Frankl's concept of noogenic neurosis. *Journal of Clinical Psychology* 20: 200–207.
Culver, C.M., and B. Gert. 1982a. *Philosophy in medicine: Conceptual and ethical issues in medicine and psychiatry.* New York: Oxford University Press.
Cutter, P. 1979. *Problem solving in clinical medicine.* Baltimore: Williams & Wilkins.
Daschle, T. 2008. *Critical: What we can do about the health-care crisis.* New York: Thomas Dunne Books.
Davis, R.B. 1995. The principlism debate: A critical overview. *Journal of Medicine and Philosophy* 20: 85–105.
Davis-Floyd, R., and G. St. John. 1998a. *From doctor to healer: The transformative journey.* New Brunswick: Rutgers University Press.
Dawkins, R. 1989a. *The selfish gene.* 2nd ed. Oxford: Oxford University Press.
De Boer, H.M. 2010. Epilepsy stigma: Moving from a global problem to global solutions. *Seizure. The Journal of the British Epilepsy Association* 19 (10): 630–636. https://doi.org/10.1016/j.seizure.2010.10.017.
De Santillana, G. 1961. *The origins of scientific thought: From Anaximander to Proclus, 600 BC to 300 AD.* London: Weidenfeld & Nicolson.
De Santillana, G., and F. Enriques. 1932. Storia del pensiero scientifico. In *Il mondo antico.* Zanichelli: Bologna.
DeAngelis, T. 2013. Therapy gone wild. More psychologist are using the wilderness as a backdrop and therapeutic too in their work. *Monitor in Psychology* 44 (8): 48–52. Washington, DC: American Psychological Association.
Del Gaizo, M. 1909. *L'opera scientifica di Giovanni Alfonso Borelli e la scuola di Roma nel secolo XVII.* Rome: Memorie della Pontificia Accademia Romana dei Nuovi Licei.
Desbordes, G., L.T. Negi, T.W.W. Pace, et al. 2012a. Effects of mindful-attention and compassion meditation training on amygdala response to emotional stimuli in an ordinary, non-meditative state. *Frontiers in Human Neuroscience* 6: 292.
Dickinson, H., F. Campbell, F. Beyer, et al. 2008a. Relaxation therapies for the management of primary hypertension in adults: A Cochrane review. *Journal of Human Hypertension* 22 (12): 809–820.
Diener, E. 2000a. Subjective well-being: The science of happiness and a proposal for a national index. *American Psychologist* 55 (1): 34–43.

Diener, E., and S. Oishi. 2003a. Personality, culture and subjective wellbeing: Emotional and cognitive evaluations of life. *Annual Review of Psychology* 54: 403–425.

Diener, E., R.A. Emmons, R.J. Larsen, and S. Griffin. 1985a. The satisfaction with life scale. *Journal of Personality Assessment* 49: 71–75.

Dilthey, W. 1883. *Einleitung in die Geisteswissenschaften. Versuch einer Grundlegung für das Studium der Gesellschaft und der Geschichte*. Vol. 1. Leipzig: Duncker & Humblot.

Doody, R.S., J.C. Stevens, C. Beck, R.M. Dubinsky, J.A. Kaye, L. Gwyther, R.C. Mohs, L.J. Thal, P.J. Whitehouse, S.T. DeKosky, and J.L. Cummings. 2001. Practice parameter: Management of dementia (an evidence-based review). *Neurology* 56 (9): 1154–1166.

Dossey, L., D. Chopra, and R. Roy. 2010a. *The mythology of science-based medicine*, Huffpost healthy living. New York: The Huffington Post.

Doubilet, P., and B. McNeil. 1985. Clinical decision making. *Medical Care* 23: 648–662.

Dowie, J., and A.S. Elstein, eds. 1988. *Professional judgment: A reader in clinical decision making*. Cambridge: Cambridge University Press.

Dretske, F. 1971. *Conclusive reasons in Australasian journal of philosophy*. Vol. 49, 1–22. London: Routledge.

Drewerman, E. 1999a. … *und es geschah so. Die moderne Biologie und die Frage nach Gott*. Zürich: Patmos.

———. 2003a. *Religiös bedingte neurotische Erkrankungen*. Lengerich: Pabst Science Publishers.

———. 2010a. *Wir glauben, weil wir lieben. Woran ich glaube*. Verlag: Patmos.

Dummett, M.A.E. 1991a. *The logical basis of metaphysics*. Cambridge: Harvard University Press.

Duncan, L.E., A.R. Pollastri, and J.W. Smoller. 2014a. Mind the gap. Why many geneticists and psychological scientists have discrepant views about gene – Environment interaction (GxE) research. *American Psychologist* 69 (3): 218–324. Washington, DC: APA, American Psychological Association.

Dusek, J. 2015. *Effectiveness of integrative therapies on pain and anxiety in hospitalized oncology patients and cardiovascular, joint replacement patients*. A series in integrative medicine. Jointly sponsored by Vanderbilt University School of Medicine Department of Medicine, Division of General Internal Medicine and Public Health and the Consortium of Academic Health Centers for Integrative Medicine (CAHCIM).

Eagleman, D., and J. Downar. 2016. *Brain and behavior: A cognitive neuroscience perspective*. Oxford: Oxford University Press.

Earhart, G.M. 2009a. Dance as therapy for individuals with Parkinson disease. *European Journal of Physical and Rehabilitation Medicine* 45 (2): 231–238.

Eccles, J. 2015a. *How the self controls its brain*. Berlin: Springer Verlag.

Edelman, G.M., and G. Tononi. 2000. *A universe of consciousness: How matter becomes imagination*. New York: Basic Books.

Eliade, M. 1984. *Immagini e Simboli* (images and symbols), 1st ed. Milan: Jaca Book.

———. 1985. *Miti Sogni e Misteri* (myths, dreams and mysteries), 1st ed. Milan: Rusconi Editore.

Ellis, A. 1994. *Reason and emotion in psychotherapy: Comprehensive method of treating human disturbances*. New York: Citadel Press.

Ellison, C.G., and J.S. Levin. 1998a. The religion-health connection: Evidence, theory, and future directions. *Health Education & Behavior* 25 (6): 700–720.

Elsassar, W.M. 1998. *Reflections on a theory of organisms: Holism in biology*. Baltimore: Johns Hopkins University Press.

Elstein, A.S., L.S. Shulman, and S.A. Sprafka. 1978. *Medical problem solving: An analysis of clinical reasoning*. Cambridge: Harvard University Press.

Elstein, A.S., B. Kleinmuntz, M. Rabinowitz, R. McAuley, J. Murakami, P.S. Heckerling, et al. 1993. Diagnostic reasoning of high- and low-domain-knowledge clinicians: A reanalysis. *Medical Decision Making* 13 (1): 21–29.

Emanuel, E.J., and L.L. Emanuel. 1992a. Four models of the physician-patient relationship. *JAMA* 267 (16): 2221–2226.

Endres, H.G., M. Zenz, C. Schaub, A. Molsberger, M. Haake, K. Streitberger, G. Skipka, and C. Maier. 2005. German acupuncture trials (GERAC) address problems of methodology associated with acupuncture studies. *Schmerz* 19 (3): 201–4–206, 208–10.

Engel, G.L. 1977. The need for a new medical model: A challenge for biomedicine. *Science* 196: 129–136.

Engelhardt, H.T., Jr., ed. 2000. *Philosophy of medicine: Framing the field*. Dordrecht: Kluwer.

Engelhardt, H.T., Jr., and E.L. Erde. 1980. Philosophy of medicine. In *A guide to culture of science, technology, and medicine*, ed. P.T. Durbin, 364–461. New York: Free Press.

Engelhardt, H.T., Jr., and K.W. Wildes. 2004. Philosophy of medicine. In *Encyclopedia of bioethics*, ed. S.G. Post, 3rd ed., 1738–1742. New York: Macmillan.

Ernst, E. 2004a. Anthroposophical medicine: A systematic review of randomised clinical trials. *Wien, A: Wiener klinische Wochenschrift* 116 (4): 128–130.

Ernst, E., and S. Singh. 2008a. *Trick or treatment? Alternative medicine on trial*. London: Bantam Press.

Evans, A.S. 1993. *Causation and disease: A chronological journey*. New York: Plenum.

Evans, M., P. Louhiala, and P. Puustinen, eds. 2004a. *Philosophy for medicine: Applications in a clinical context*. Oxon: Radcliffe Medical Press.

Evidence-Based Medicine Working Group. 1992. Evidence-based medicine: A new approach to teaching the practice of medicine. *Journal of the American Medical Association* 268: 2420–2425.

Fahlböck, A. 2006. (Neuro-)Psychotherapie und therapeutisches Milieu. In *Klinische Neuropsychologie*, ed. Johann Lehrner, Gisela Pusswald, and Elisabeth Fertl. Heidelberg: Springer.

Feinstein, A.R. 1967. *Clinical judgment*. Huntington: Krieger.

Feltovich, P.J., P.E. Johnson, J.H. Moller, and D.B. Swanson. 1984. LCS: The role and development of medical knowledge in diagnostic expertise. In *Readings in medical artificial intelligence: The first decade*, ed. W.J. Clancey and E.H. Shortliffe, 275–319. Reading: Addison-Wesley.

Fenwick, P., and E. Fenwick. 2012a. *The truth in the light*. Guildford: White Crow Books.

Ferber, A., C. Jimenez, A. O'Reilly Herrera, and D. Samuels. 2009. *The matrix reader: Examining the dynamics of oppression and privilege*. New York: McGraw-Hill.

Ferré, F., J.J. Kockelmans, and J.E. Smith. 1982a. *The challenge of religion. Contemporary readings in philosophy of religion*. New York: The Seabury Press.

Finch, P.M. 2007. The evidence funnel: Highlighting the importance of research literacy in the delivery of evidence informed complementary health care. *Journal of Bodywork and Movement Therapies* 11: 78–81.

Fleming, M.H. 1991. Clinical reasoning in medicine compared with clinical reasoning in occupational therapy. *American Journal of Occupational Therapy* 45: 988–996.

Flyvbjerg, B. 1993a. Aristotle, foucault and progressive phronesis: Outline of an applied ethics for sustainable development. In *Applied ethics*, ed. Earl R. Winkler and R. Coombs Jerold. Cambridge: Blackwell Publishers.

Fogassi, et al. 2005. Parietal lobe: From action organization to intention understanding. In *Science*. Washington, DC: American Association for the Advancement of Science.

Foglia, L. 2005a. Rappresentazioni Mentali. Modelli teorici a confronto. *Arkete* I (2): 17–42.

Foglia, L., and R. Grush. 2010a. Neural reuse: A fundamental organizational principle of the brain. *Behavioral and Brain Sciences* 33: 245–313.

———. 2011a. The limitations of a purely enactive (non-representational) account of imagery. *Journal of Consciousness Studies* 18 (5–6): 35–43.

Foucault, M. 1966a. *The order of things: An archaeology of the human sciences* (Les Mots et les choses: Une archéologie des sciences humaines). New York: Pantheon Books.

Fox, C. 2006a. A brain-based philosophy of life. *The Lancet* 5 (3): 212. Amsterdam: Elsevier.

Frankl, V.E. 1984a. *Man's search for meaning*. 3rd ed. New York: Washington Square Press. (original work published 1946).

Freedman, E.B. 2009. Race and the politics of identity in U.S. feminism. In *The matrix reader: Examining the dynamics of oppression and privilege*, ed. A. Ferber, C.M. Jimenez, A.O. Herrera, and D.R. Samuels, 362–375. New York: McGraw-Hill.

Freeman, L.W. 2009a. *Mosby's complementary & alternative medicine: A research-based approach*. 3rd ed. St. Louis: Mosby Elsevier.

Fuchs, T., N. Birbaumer, W. Lutzenberger, J. Gruzelier, and J. Kaiser. 2003a. Neurofeedback treatment for attention-deficit/hyperactivity disorder in children: A comparison with methylphenidate. *Applied Psychophysiology and Biofeedback* 28: 1–12.

Fulford, K.W.M. 1989. *Moral theory and medical practice*. Cambridge: Cambridge University Press.

———. 1991a. The potential of medicine as a resource for philosophy. *Theoretical Medicine* 12: 81–85. Dordrecht: Kluwer Academic Publishers.

Gadamer, H.-G. 1960a. *Wahrheit und Methode. Grundzüge einer philosophischen Hermeneutik*. Tübingen: Mohr Siebeck [Unveränd. Nachdr. d. 3. erw. Aufl. Tübingen 1975].

Galimberti, U. 1983. *Il corpo* (the body). In *Universale economica Saggi 2003*, XII ed. Milan: Feltrinelli.

Gardiner, P. 2003. A virtue ethics approach to moral dilemmas in medicine. *Journal of Medical Ethics* 29: 297–302.

Gawande, A. 2002a. *Complications: A surgeon's notes on an imperfect science*. New York: Metropolitan Books.

———. 2007. *Better: A surgeon's notes on performance*. New York: Picador.

———. 2014a. *Being mortal: Medicine and what matters in the end*. London: Macmillan.

Gergen, K.J., R. Josselson, and M. Freeman. 2015a. The promises of qualitative inquiry. *American Psychologist* 70 (1): 1–60. Washington, DC: American Psychological Association, pp. 1–9.

Gert, B., C.M. Culver, and K.D. Clouser. 1997. *Bioethics: A return to fundamentals*. Oxford: Oxford University Press.

Gettier, E. 1963. Is justified true belief knowledge? In *Analytic philosophy*, ed. P. Martinich, and D. Sosa. 2001. Oxford: Blackwell Publishers.

Giardina, S., and A.G. Spagnolo. 2014. *I medici e il camice bianco. Note storico-culturali e implicazioni per la formazione dello studente di medicina*. Medicina e Morale 2014/2, Rome: Università Cattolica del Sacro Cuore.

Giddens, A. 1984. *The constitution of society. Outline of a theory of structuration*. Oakland: The University of California Press.

Gillies, D.A. 2005. Hempelian and Kuhnian approaches in the philosophy of medicine: The Semmelweis case. *Studies in History and Philosophy of Science Part C: Studies in History and Philosophy of Biological and Biomedical Science* 36: 159–181.

Gillon, R. 1986a. *Philosophical medical ethics*. New York: Wiley.

Giovetti, P., ed. 2000. *L'uomo e il mistero* (man and mystery), 1st ed., vol. 8. Rome: Edizioni Mediterranee.

Goldacre, B. 2007. Benefits and risks of homoeopathy. *The Lancet* 370 (9600): 1672–1673. Amsterdam: Elsevier.

Goldenberg, G. 2008a. Chapter 7, how the mind moves the body: Lessons from apraxia. In *Oxford handbook of human action. Social cognition and social neuroscience*, ed. E. Morsella, J.A. Bargh, and P.M. Gollwitzer, 136. Oxford: Oxford University Press.

Goldman, A. 1967. *A causal theory of knowing in the journal of philosophy*. Vol. 64. New York: Columbia University Press.

Goldman, G.M. 1990. The tacit dimension of clinical judgment. *Yale Journal of Biology and Medicine* 63: 47–61.

Goldman, L., and D. Ausiello. 2004. *Cecil textbook of medicine*. Philadelphia: Saunders.

Goller, H. 2015a. *Wohnt Gott im Gehirn?: Warum die Neurowissenschaften die Religion nicht erklären*. Kevelaer: Butzon & Bercker.

Golub, E.S. 1997. *The limits of medicine: How science shapes our hope for the cure*. Chicago: University of Chicago Press.

Gomora, X. 2003. *Ansichten eines Wilden über die zivilisierten Menschen* (perspectives of a wild man on a civilized society). Klagenfurt: NOI-Verlag.

Goodyear-Smith, F., and S. Buetow. 2001. Power issues in the doctor-patient relationship. *Health Care Analysis* 9: 449–462.

Gorowitz, S., and A. McIntyre. 1978. Toward a theory of medical fallibility. *Journal of Medicine and Philosophy* 1: 51–71.

Gorski, T.N. 2002. The Eisenberg data: Flawed and deceptive. Quackwatch, article posted on March 16. Available on http://www.quackwatch.org/11Ind/eisenberg.html

Graesser, A.C., M. Singer, and T. Trabasso. 1994. Constructing inferences during narrative comprehension. *Psychological Review* 101: 371–395.

Grawe, K., and F. Caspar. 2012. Allgemeine Psychotherapie. In *Praxis der Psychotherapie. Ein integratives Lehrbuch*, ed. W. Senf and M. Broda, 33–46. Stuttgart: Thieme.

Green, Z., and M.J. Stiers. 2002. *Multiculturalism and group therapy in the United States: A social constructionist perspective*. Antwerp: Springer.

Griffin, D.R., ed. 1998. *Reenchantment of science: Postmodern proposals*. New York: State University of New York Press.

Grof, S. 2000. *Psychology of the future: Lessons from modern consciousness research*. Albany: State University of New York Press.

———. 2006a. *The ultimate journey: Consciousness and the mystery of death*. Ben Lomond: Multidisciplinary Association for Psychedelic Studies (MAPS).

Grof, S., and C. Grof. 2010. *Holotropic breathwork: A new approach to self-exploration and therapy*. Albany: State University of New York Press.

Groopman, J. 2007a. *How doctors think*. New York: Houghton Mifflin.

Grün, A., and W. Jäger. 2013a. *Das Geheimnis jenseits aller Wege. Was uns eint, was uns trennt*. Münsterschwarzach: Vier Türme.

Guénon, R. 1962. *Simboli della Scienza sacra* (symbols of the sacred science), VII ed, 2005. Milan: Adelphi, Original: *Symboles fondamentaux de la Science sacrée*, 1962 Paris: Gallimard.

Gungov, A.L. 2012a. *Logic in medicine: Approaches to patient safety*. Sofia: Avangard.
———. 2013a. Diagnostics in a logical perspective. In *Normativity: Contemporary challenges*, ed. A.L. Gungov, V. Kaneva, and O. Kasabov, 2003. Sofia: St. Kliment Ohridski University Press.
Guyatt G., D. Rennie, M.O. Meade, and D.J. Cook. 2014. *Users' guides to the medical literature: A manual for evidence-based clinical practice*, 2nd ed. JAMA & Archives Journals, American Medical Association, New York: McGraw-Hill.
Haas, M., A. Spegman, D. Peterson, et al. 2010. Dose response and efficacy of spinal manipulation for chronic cervicogenic headache: A pilot randomized controlled trial. *Spine Journal* 10 (2): 117–128.
Habermas, J. 1999. *Wahrheit und Rechtfertigung*. Frankfurt am Main: Suhrkamp.
———. 2002. In *Religion and rationality: Essays on reason, god, and modernity*, ed. E. Medieta. Cambridge: MIT Press.
Halpern, J. 2001. *From detached concern to empathy: Humanizing medical practice*. New York: Oxford University Press.
Hameroff, S.R. 2006a. The entwined mysteries of anesthesia and consciousness. *Anesthesiology* 105 (2): 400–412. https://doi.org/10.1097/00000542-200608000-00024. PMID 16871075.
Hammrick, H., and M. Garfunkel. 1991. Editor's column-clinical decisions: How much analysis and how much judgment? *Journal of Pediatrics* 118: 67.
Hampton, J.R. 2002a. Evidence-based medicine, opinion-based medicine, and real-world medicine. *Perspectives in Biology and Medicine* 45: 549–568.
Hardin, E.E., C. Robitschek, L.Y. Flores, R.L. Navarro, and M.W. Ashton. 2014. The cultural lens approach to evaluating cultural validity of psychological theory. *American Psychologist* 69 (7): 656–668. Washington, DC: APA, American Psychological Association.
Harman, G.H. 1965. The inference to the best explanation. *Philosophical Review* 74: 88–95.
Haug, M.R., and B. Lavin. 1983. *Consumerism in medicine: Challenging physician authority*. Beverly Hills: Sage Publications.
Haynes, S. 2002. The Etruscan woman. The relationship between the sexes and the position of women in early Etruria (9th–5th centuries B.C.). *Journal of Roman Studies* 9 (2): 210–212.
Haynes, J.D., G. Roth, M. Stadler, and H.J. Heinze. 2003. Neuromagnetic correlates of perceived contrast in primary visual cortex. *Journal of Neurophysiology* 89 (5): 2655–2666. https://doi.org/10.1152/jn.00820.2002.
Häyry, H. 1991. *The limits of medical paternalism*. London: Routledge.
Hempel, C.G. 1965. *Aspects of scientific explanation and other essays in the philosophy of science*. New York: Free Press.
Hempel, C.G., and P. Oppenheim. 1948. Studies in the logical of explanation. *Philosophy of Science* 15: 135–175.
Herman, P.M., O. Szczurko, K. Cooley, et al. 2008. Cost-effectiveness of naturopathic care for chronic low back pain. *Alternative Therapies in Health and Medicine* 14 (2): 32–39.

Heyting, A. 1956. *Intuitionism. An introduction*. Amsterdam: North-Holland Publishing.

Hill, A.B. 1965a. The environment and disease: Association or causation? *Proceedings of the Royal Society of Medicine* 58: 295–300.

Hiroshi, H. 1998. On Vis medicatrix naturae and Hippocratic idea of Physis. *Memoirs of School of Health Sciences, Faculty of Medicine, Kanazawa University* 22: 45–54.

Hofmann, E. 2003. *Progressive Muskelentspannung, ein Trainingsprogramm*. 2nd ed. Göttingen: Hogrefe.

Hollifield, M., N. Sinclair-Lian, T.D. Warner, and R. Hammerschlag. 2007. Acupuncture for posttraumatic stress disorder: A randomized controlled pilot trial. *The Journal of Nervous and Mental Disease* 195 (6): 504–513.

Holyoak, K., and R.G. Morrison. 2012a. *The Oxford handbook of thinking and reasoning*, Oxford library of psychology. Oxford: Oxford University Press.

Hölzel, B.K., J. Carmody, M. Vangel, et al. 2011a. Mindfulness practice leads to increases in regional brain gray matter density. *Psychiatry Research: Neuroimaging* 191 (1): 36–43.

Howick, J.H. 2011. *The philosophy of evidence-based medicine*. Hoboken: Wiley-Blackwell.

Hudziak, J.J., M.D. Albaugh, et al. 2014a. Cortical thickness maturation and duration of music training: Health-promoting activities shape brain development. *JAACAP* 11: 1153–1161.

Illari, P.M., F. Russo, and J. Williamson, eds. 2011a. *Causality in the sciences*. New York: Oxford University Press.

Illich, I. 1976a. *Limits to medicine; medical Nemesis: The expropriation of health*. London: Marion Boyars Publishers.

Illingworth, P.M.L. 1988. The friendship model of physician/patient relationship and patient autonomy. *Bioethics* 2: 22–36.

Jackson, M. 2005a. *Existential anthropology – Events, exigencies, and effects (methodology and history in anthropology)*. California: Berghahn Books.

James, D.N. 1989. The friendship model: A reply to Illingworth. *Bioethics* 3: 142–146.

Jansen, K.L.R. 1998a. *The ketamine model of the near death experience: A central role for the NMDA receptor*, Journal Article. London: The Maudsley Hospital, Denmark Hill.

Jedel, S., A. Hoffman, P. Merriman, et al. 2014. A randomized controlled trial of mindfulness-based stress reduction to prevent flare-up in patients with inactive ulcerative colitis. *Digestion NIH PubMed* 89: 142–155.

Jensen, M.P., D.M. Ehde, K.J. Gertz, B.L. Stoelb, T.M. Dillworth, A.T. Hirsh, and G.H. Kraft. 2011. Effects of self-hypnosis training and cognitive restructuring on daily pain intensity and catastrophizing in individuals with multiple sclerosis and chronic pain. *International Journal of Clinical and Experimental Hypnosis* 59: 45–63.

Johansson, I., and N. Lynøe. 2008a. *Medicine and philosophy: A twenty-first century introduction*. Frankfurt: Ontos Verlag.
Jonas, W.B. 2001. The evidence house: How to build an inclusive base for complementary medicine. *Western Journal of Medicine* 175 (2): 79–80.
Jonsen, A.R. 2000. *A short history of medical ethics*. New York: Oxford University Press.
Jonsen, A., M. Siegler, and W. Winslade. 2002a. *Clinical ethics: A practical approach to ethical decisions in clinical medicine*. New York: McGraw-Hill.
Joseph, G.M., and V.L. Patel. 1990a. Domain knowledge and hypothesis generation in diagnostic reasoning. *Medical Decision Making* 10 (1): 31–46.
Jung, C.G. 1963a (1990 ed.) *Mysterium Coniunctionis. Untersuchungen über die Trennung und Zusammensetzung der seelischen Gegensätze in der Alchemie*. Olten und Freiburg im Breisgau: Walter-Verlag.
Kadane, J.B. 2005. Bayesian methods for health-related decision making. *Statistics in Medicine* 24: 563–567.
Kahneman, D., and A. Tversky, eds. 2000. *Choices, values and frames*. New York: Cambridge University Press.
Kahneman, D., P. Slovic, and A. Tversky. 1982. *Judgment under uncertainty: Heuristics and biases*. New York: Cambridge University Press.
Kaku, M. 2011. *Physics of the future: How science will shape human destiny and our daily lives by the year 2100*. New York: Doubleday.
———. 2014a. *The future of the mind: The scientific quest to understand, enhance, and empower the mind*. New York: Doubleday.
Kallio, S., and A. Revonsuo. 2003. Hypnotic phenomena and altered states of consciousness: A multilevel framework of description and explanation. *Contemporary Hypnosis* 20: 111–164. Peer commentaries in Contemporary Hypnosis (2005), 22(1), 1–55.
Kaliman, P., M.J. Álvarez-López, M. Cosín-Tomás, et al. 2014a. Rapid changes in histone deacetylases and inflammatory gene expression in expert meditators. *Psychoneuroendocrinology* 40: 96–107.
Kantarjian, H., and E. Rosenberg. 2014. *An unhealthy system. Compared to other nations, Americans overpay for their health care and get little in return*. U.S. News & World Report, May 30.
Kaptchuk, T.J. 2000a. *The web that has no weaver: Understanding Chinese medicine*. Chicago: Contemporary Books.
Kassirer, J.P. 1989. Diagnostic reasoning. *Annals of Internal Medicine* 110 (11): 893–900.
Kato, K., P.F. Sullivan, B. Evengård, and N.L. Pedersen. 2006. Importance of genetic influences on chronic widespread pain. *Arthritis Rheumatoides* 54 (5): 1682–1686. https://doi.org/10.1002/art.21798. PMID 16646040.
Katz, J. 2002. *The silent world of doctor and patient*. Baltimore: Johns Hopkins University Press.

Kay, J., and R.L. Kay. 2003a. Ethnocultural issues in psychotherapy. In *Psychiatry: Therapeutics, chapter 1: Individual psychoanalytic psychotherapy*, II ed. Chichester: Wiley.

Kessler, R.C., P. Berglund, O. Demler, R. Jin, K.R. Merikangas, and E.E. Walters. 2005. Lifetime prevalence and age-of-onset distributions of DSM-IV disorders in the National Comorbidity Survey Replication. *Archives of General Psychiatry* 62 (6): 593–602.

Keysers, C. 2010. Mirror Neurons. Current Biology 19 (21): R971–R973. https://doi.org/10.1016/j.cub.2010.03.013.

Kiecolt-Glaser, J.K., J.E. Graham, W.B. Malarkey, et al. 2008a. Olfactory influences on mood and autonomic, endocrine, and immune function. *Psychoneuroendocrinology* 33 (3): 328–339.

Kiecolt-Glaser, J.K., L. Christian, H. Preston, et al. 2010. Stress, inflammation, and yoga practice. *Psychosomatic Medicine* 72 (2): 113–121.

Kienle, S.G., H. Kiene, and H.-U. Albonico. 2006a. *Anthroposophische Medizin in der klinischen Forschung. Wirksamkeit, Nutzen, Wirtschaftlichkeit, Sicherheit.* Stuttgart: Schattauer.

Kihlstrom, J.F. 2002a. Mesmer, the Franklin Commission, and hypnosis: A counterfactual essay. *The International Journal of Clinical and Experimental Hypnosis* 50 (4): 407–419.

King, L.S. 1978. *The philosophy of medicine.* Cambridge: Harvard University Press.

Kleinman, A. 1988. *The illness narratives: Suffering, healing and the human condition.* New York: Basic Books.

Klinger, B. 2015a. *Non-pharmacological interventions to chronic pain. An evidence-based review of the use of integrative approaches including mind-body therapies and acupuncture to the management of chronic pain conditions.* Burlington: University of Vermont College of Medicine/Laura Mann Integrative Healthcare Lecture Series.

Knight, J.A. 1982a. The minister as healer, the healer as minister. *Journal of Religion and Health* 21: 100–114.

Knobe, J., and S. Nichols. 2008. *Experimental philosophy.* Oxford: Oxford University Press.

Knobe, J., and J. Prinz. 2008. Intuitions about consciousness: Experimental studies. In *Phenomenology and cognitive science.* Chapel Hill: University of North Carolina.

Kohn, L.T., J. Corrigan, and M.S. Donaldson. 2000. *To err is human: Building a safer health system.* Washington, DC: National Academies Press.

Kong, J., T.J. Kaptchuk, G. Polich, et al. 2009a. Expectancy and treatment interactions: A dissociation between acupuncture analgesia and expectancy evoked placebo analgesia. *NeuroImage* 45 (3): 940–949.

Konner, M. 1993. *Medicine at the crossroads: The crisis in health care.* New York: Pantheon Books.

Kovács, J. 1998. The concept of health and disease. *Medicine, Health Care and Philosophy* 1: 31–39.

Koyré, A. 1966. *La rivoluzione astronomica. Copernico, Keplero, Borelli*. Milan: Feltrinelli.

Kuipers, B., and J.P. Kassirer. 1984a. Causal reasoning in medicine: Analysis of a protocol. *Cognitive Science* 8 (4): 363–385.

Kulkarni, A.V. 2005. The challenges of evidence-based medicine: A philosophical perspective. *Medicine, Health Care and Philosophy* 8: 255–260.

Kunstler, R., F. Greenblatt, and N. Moreno. 2004. Aromatherapy and hand massage: Therapeutic recreation interventions for pain management. *Therapeutic Recreation Journal* 38: 133–147.

Lad, V.D. 2002. *Textbook of Ayurveda: Fundamental principles of Ayurveda*. Vol. 1. Albuquerque: Ayurvedic Press.

Lamme, V.A.F. 2000. Neural mechanisms of visual awareness: A linking proposition. *Brain and Mind* 1: 385–406.

———. 2003. Why visual attention and awareness are different. *Trends in Cognitive Sciences* 7: 12–1810.

Lampit, A., H. Hallock, C. Suo, S.L. Naismith, and M. Valenzuela. 2015. Cognitive training-induced short-term functional and long-term structural plastic change is related to gains in global cognition in healthy older adults: A pilot study. *Frontiers in Aging Neuroscience* 7: 14. https://doi.org/10.3389/fnagi.2015.00014.

Lang, E.V., K.S. Berbaum, S. Faintuch, et al. 2006. Adjunctive self-hypnotic relaxation for outpatient medical procedures: A prospective randomized trial with women undergoing large core breast biopsy. *Pain* 126 (1-3): 155–164. PubMed.

Larson, J.S. 1991. *The measurement of health: Concepts and indicators*. New York: Greenwood Press.

Laureys, S., and G. Tononi. 2009. *The neurology of consciousness: Cognitive neuroscience and neuropathology*. Cambridge: Academic Press.

Lee, D. 2013. Decision making: From neuroscience to psychiatry. *Neuron* 78 (2): 233–248. https://doi.org/10.1016/j.neuron.2013.04.008.

Le Fanu, J. 2002. *The rise and fall of modern medicine*. New York: Carroll & Graf.

Ledley, R.S., and L.B. Lusted. 1959. Reasoning foundations of medical diagnosis. In *Science*, vol. 130, 9–21.

Lee, M.S., M.H. Pittler, and E. Ernst. 2008. Effects of reiki in clinical practice: A systematic review of randomized clinical trials. *International Journal of Clinical Practice* 62 (6): 947–954.

Lee, M.S., B. Oh, and E. Ernst. 2011. Qigong for healthcare: An overview of systematic reviews. *JRSM Short Rep* 2 (2): 7.

Leontyev, A.N., A.R. Luria, and A. Smirnov, eds. 1966. *Psychological research in the USSR*. Moscow: Progress Publishers.

Leung, L. 2012a. Neurophysiological basis of acupuncture-induced analgesia – An updated review. *Journal of Acupuncture and Meridian Studies* 5 (6): 261–270.

Leurent, B., I. Nazareth, J. Bellón-Saameño, M.I. Geerlings, H. Maaroos, S. Saldivia, I. Švab, F. Torres-González, M. Xavier, and M. Kin. 2013a. Spiritual and religious beliefs as risk factors for the onset of major depression: An international cohort study. In *Psychological medicine*. London: Cambridge University Press.

Levi, B.H. 1996. Four approaches to doing ethics. *Journal of Medicine and Philosophy* 21: 7–39.

Levinas, E. 1961a. *Totality and infinity. An essay on exteriority*. 3rd ed. Pittsburg: Duquesne University Press. 1969.

Levine, J. 1999. Conceivability, identity, and the explanatory gap. In *Towards a science of consciousness III: The third Tucson discussions and debates*, ed. Stuart R. Hameroff, Alfred W. Kaszniak, and David Chalmers. Boston: The MIT Press.

Liberati, A., and P. Vineis. 2004. Introduction to the symposium: What evidence based medicine is and what it is not. *Journal of Medical Ethics* 30: 120–121.

Libet, B. 1985. Unconscious cerebral initiative and the role of conscious will in voluntary action. *The Behavioral and Brain Sciences* 8 (4): 529–566.

Libet, B. 1999a. Do we have a free will? *Journal of Consciousness Studies* 5: 49.

Licciardone, J.C., S. Buchanan, K.L. Hensel, et al. 2010. Osteopathic manipulative treatment of back pain and related symptoms during pregnancy: A randomized controlled trial. *American Journal of Obstetrics and Gynecology* 202 (1): 43.e1–43.e8.

Lipton, P. 2004a. *Inference to the best explanation*. 2nd ed. New York: Routledge.

Little, M. 1995. *Humane medicine*. Cambridge: Cambridge University Press.

Little, A.C., B.C. Jones, and L.M. DeBruine. 2014a. *Primacy in the effects of face exposure: Perception is influenced more by faces that are seen first*, Archives of Scientific Psychology, 43–47. Washington, DC: APA, American Psychological Association.

Loewy, E.H. 2002. Bioethics: Past, present, and an open future. *Cambridge Quarterly of Healthcare Ethics* 11: 388–397.

Loftus, S.F. 2006a. *Language in clinical reasoning: Using and learning the language of collective clinical decision making*. Sydney: Faculty of Health Sciences, School of Physiotherapy, University of Sydney.

Looijen, R.C. 2000. *Holism and reductionism in biology and ecology: The mutual dependence of higher and lower level research programmes*. Dordrecht: Kluwer.

Lowen, A. 1975a. *Bioenergetics*. New York: Coward, McCann & Geoghegan.

Lutz, A., J. Brefczynski-Lewis, T. Johnstone, et al. 2008a. Regulation of the neural circuitry of emotion by compassion meditation: Effects of meditative expertise. *PLoS One* 3 (3): e1897.

Lyons-Ruth, K., P. Pechtel, S.A. Yoon, C.M. Anderson, and M.H. Teicher. 2016. Disorganized attachment in infancy predicts greater amygdala volume in adulthood. *Behavioural Brain Research* 308: 83–93. https://doi.org/10.1016/j.bbr.2016.03.

Ma, Y. 2010. *Women journalists and feminism in China, 1898–1937*. Amherst: Cambria Press.

Maier, B., and W.A. Shibles. 2010a. *The philosophy and practice of medicine and bioethics: A naturalistic-humanistic approach*. New York: Springer.

Malcolm, N. 1996. *Bosnia: A short history*. New York: New York University Press.

Marchianò, G., ed. 1991. *La religione della terra* (The religion of the Earth). Como: Red Edizioni.

Marcum, J.A. 2005a. Metaphysical presuppositions and scientific practices: Reductionism and organicism in cancer research. *International Studies in the Philosophy of Science* 19: 31–45.

———. 2008a. *An introductory philosophy of medicine: Humanizing modern medicine*. New York: Springer.

Marcum, J.A., and G.M.N. Verschuuren. 1986. Hemostatic regulation and Whitehead's philosophy of organism. *Acta Biotheoretica* 35: 123–133.

Maslow, A.H. 1943a. *A theory of human motivation*. Originally published in *Psychological Review* 50 (4): 370–396. Washington, DC: APA.

Massa-Pairault, F. 2009. The Etruscan. Gender relations and status of women in early Etruria. *Latomus* 68 (4): 1082–1084. https://doi.org/10.2307/2856364.

Matthews, J.N.S. 2000. *An introduction to randomized controlled clinical trials*. London: Arnold.

May, W.F. 2000. *The physician's covenant: Images of the healer in medical ethics*. 2nd ed. Louisville: Westminster John Knox Press.

McDaniel, S.H., and F.V. deGruy III. 2014. An introduction to primary care and psychology. *American Psychologist* 69 (4): 325–331. Washington, DC: American Psychological Association.

McDonald, W. Ian, Alistair Compston, Gilles Edan, Donald Goodkin, Hans-Peter Hartung, Fred D. Lublin, Henry F. McFarland, Donald W. Paty, Chris H. Polman, Stephen C. Reingold, Magnhild Sandberg-Wollheim, William Sibley, Alan Thompson, Stanley Van Den Noort, Brian Y. Weinshenker, and Jerry S. Wolinsky. 2001. Recommended diagnostic criteria for multiple sclerosis: Guidelines from the international panel on the diagnosis of multiple sclerosis. *Annals of Neurology* 50 (1): 121–127.

McDougle, S.D., K.M. Bond, and J.A. Taylor. 2017. Implications of plan-based generalization in sensorimotor adaptation. *Journal of Neurophysiology* 118: 383–393.

McKenzie, E., et al. 1998a. *Healing Reiki*. Hamlyn: Hamlyn Health & Well Being.

Meehl, P.E. 1954. *Clinical versus statistical prediction: A theoretical analysis and a review of the literature*. Minneapolis: University of Minnesota Press.

Mikula, S., I. Trotts, J. Stone, and E.G. Jones. 2007a. Internet-enabled high-resolution brain mapping and virtual microscopy. *NeuroImage* 35 (1): 9–15.

Miller, A. 2014a. Friends wanted. New research by psychologists uncovers the health risks of loneliness and the benefits of strong social connections. *Monitor on Psychology* 45 (1): 54–58. Washington, DC: American Psychological Association.

Mitchell, M. 2012. An analysis of common arguments against advance directives. *Nursing Ethics* 19 (2): 245–251. https://doi.org/10.1177/0969733011416398.

Mlsna, Lucas J. 2010a. Stem cell based treatments and novel considerations for conscience clause legislation. *Indiana Health Law Review* 8 (2): 471–496. Bloomington: Indiana University Robert H. McKinney School of Law.

Mokdad, A.H., J.S. Marks, D.F. Stroup, and J.L. Gerberding. 2000. Actual causes of death in the United States. *Journal of the American Medical Association* 291 (10): 1238–1245. https://doi.org/10.1001/jama.291.10.1238.

Mommsen, Th. 1942a. Petrarch's conception of the 'dark ages'. *Speculum* 17 (2): 226–242. Cambridge, MA: Medieval Academy of America.

Montgomery, K. 2006a. *How doctors think: Clinical judgment and the practice of medicine*. New York: Oxford University Press.

Montgomery, G.H., D.H. Bovbjerg, J.B. Schnur, D. David, A. Goldfarb, et al. 2007. A randomized clinical trial of a brief hypnosis intervention to control side effects in breast surgery patients. *Journal of National Cancer Institute* 99: 1304–1312.

Moody, R. 1975a. *Life after life: The investigation of a phenomenon – Survival of bodily death*. San Francisco: Harper San Francisco.

Morabia, A. 1996. P.C.A. Louis and the birth of clinical epidemiology. *Journal of Clinical Epidemiology* 49: 1327–1333.

Morrison, J.H., and M.G. Baxter. 2012. The ageing cortical synapse: Hallmarks and implications for cognitive decline. *Nature Reviews Neuroscience* 13 (4): 240–250. https://doi.org/10.1038/nrn3200.

Morsella, E., and T.A. Poehlman. 2013. The inevitable contrast: Conscious vs. unconscious processes in action control. *Frontiers in Psychology* 4: 590. https://doi.org/10.3389/fpsyg.2013.00590.

Morton, R.A., and C.F. Valenzuela. 2016. Further characterization of the effect of ethanol on voltage-gated Ca(2+) channel function in developing CA3 hippocampal pyramidal neurons. *Brain Research* 1633: 19–26.

Mossbridge, J., P. Tressoldi, and J. Utts. 2012a. Predictive physiological anticipation preceding seemingly unpredictable stimuli: A meta-analysis. *Frontiers in Psychology* 3: 390.

Murphy, E.A. 1997. *The logic of medicine*. 2nd ed. Baltimore: The Johns Hopkins University Press.

Nagel, T. 1974a. What is it like to be a bat? *The Philosophical Review* 83: 434–450.

Napolitano, J.G., D.C. Lankin, T.N. Graf, et al. 2013a. HiFSA fingerprinting applied to isomers with near-identical NMR spectra: The silybin/isosilybin case. *Journal of Organic Chemistry* 78 (7): 2827–2839.

Naylor, T.H., W.H. Willimon, and M.R. Naylor. 1994a. *The search for meaning*. Nashville: Abingdon Press.

Nelson, J.M., and T.C. Nelson. 2014. Advance directives: Empowering patients at the end of life. *The Nurse Practitioner* 39 (11): 34–40.

Nesse, R.M. 2001. On the difficulty of defining disease: A Darwinian perspective. *Medicine, Health Care and Philosophy* 4: 37–46.

Nestoriuc, Y., A. Martin, W. Rief, and F. Andrasik. 2008a. Biofeedback treatment for headache disorders: A comprehensive efficacy review. *Applied Psychophysiology and Biofeedback* 33: 125–140.

Newberg, A.B. 2016a. *Principles of neurotheology*, Ashgate science and religion series. New York: Routledge.

Newberg, A.B., and M.R. Waldman. 2010a. *How God changes your brain. Breakthrough findings from a leading neuroscientist.* New York: Ballantine Books Trade Paperbacks.

Nguyen, H.T., J.G. Grzywacz, W. Lang, et al. 2010a. Effects of complementary therapy on health in a national U.S. sample of older adults. *Journal of Alternative and Complementary Medicine* 16 (7): 701–706.

Nicolelis, M.A.L. 2003a. Brain-machine interfaces to restore motor function and probe neural circuits. *Nature Reviews Neuroscience* 4 (5): 417–422.

———. 2011a. *Beyond boundaries: The new neuroscience of connecting brains with machines – And how it will change our lives.* New York: Times Books.

Nidich, S.I., M.V. Rainforth, D.A.F. Haaga, et al. 2009a. A randomized controlled trial on effects of the transcendental meditation program on blood pressure, psychological distress, and coping in young adults. *American Journal of Hypertension* 22 (12): 1326–1331.

Nordenfelt, L. 1995. *On the nature of health: An action-theory approach.* 2nd ed. Dordrecht: Kluwer.

Norman, G.R., and L.R. Brooks. 1997. The non-analytical basis of clinical reasoning. *Advances in Health Sciences Education* 2 (2): 173–184.

Norman, G.R., L.R. Brooks, C.L. Colle, and R.M. Hatala. 2010a. The benefit of diagnostic hypotheses in clinical reasoning: Experimental study of an instructional intervention for forward and backward reasoning. *Cognition and Instruction* 17 (4): 433–488. Hamilton: Lawrence Erlbaum Associates.

Novotney, A. 2013a. Music as medicine. Researchers are exploring how music therapy can improve health outcomes among a variety of patient populations, including premature infants and people with depression and Parkinson's disease. *Monitor on Psychology* 44 (10): 46–49. Washington, DC: American Psychological Association.

Nozick, R. 1981. *Philosophical explanations.* Boston: Harvard University Press.

O'Hear, A. 1998a. *The nature of consciousness.* Cambridge: Cambridge University Press.

OASH. 2010. *Disparities.* Washington, DC: Healthy People – U.S. Department of Health and Human Services. Retrieved from: https://www.healthypeople.gov/2020/about/foundation-health-measures/Disparities#1

Office of Disease Prevention and Health Promotion. 2016. *Healthy people 2020 topic and objectives list of potential health promotion topics.* Retrieved from: https://www.healthypeople.gov/2020/topics-objectives

Overby, P. 2005. The moral education of doctors. *New Atlantis* 10: 17–26.

Pagnini, A., ed. 2010a. *Filosofia della medicina. Epistemologia, ontologia, etica, diritto.* Roma: Carocci.

Pahal, J.S., and H.Z. Li. 2006a. The dynamics of resident-patient communication: Data from Canada. *Communication & Medicine* 3 (2): 161–170. New York: Mouton de Gruyter.

Papakostas, Y.G., and M.D. Daras. 2001a. Placebos, placebo effects, and the response to the healing situation: The evolution of a concept. *Epilepsia* 42: 1614–1625.

Parisi, L. 2017a. *Teologia e Neuroscienze.* Brescia: Editrice Queriniana.

Parker, M. 2002. Whither our art? Clinical wisdom and evidence-based medicine. *Medicine, Health Care and Philosophy* 5: 273–280.

Patel, V.L., and J.F. Arocha. 2001. The nature of constraints on collaborative decision making in health care settings. In *Linking expertise and naturalistic decision making*, ed. E. Salas and G. Klein, 383–405. Mahwah: Lawrence Erlbaum Associates.

Patel, V.L., and D.R. Kaufman. 1994a. Diagnostic reasoning and expertise. *Psychology of Learning and Motivation* 31: 137–252.

———. 1995a. Clinical reasoning and biomedical knowledge: Implications for teaching. In *Clinical reasoning in the health professions*, ed. J. Higgs and M. Jones, 117–128. Oxford: Butterworth Heinemann.

Patel, V.L., D.A. Evans, and D.R. Kaufman. 1989. Cognitive framework for doctor-patient communication. In *Cognitive science in medicine: Biomedical modeling*, ed. D.A. Evans and V.L. Patel, 257–312. Cambridge, MA: MIT Press.

Patel, V.L., J.F. Arocha, and J. Zhang. 2004a. Thinking and reasoning in medicine. In *Cambridge handbook of thinking and reasoning*, ed. K. Holyoak. Cambridge: Cambridge University Press.

———. 2010a. Medical reasoning and thinking. In *Oxford handbook of thinking & reasoning*, ed. K. Holyoak and Morrison. New York: Oxford University Press.

Pellegrino, E.D., and D.C. Thomasma. 1981. *A philosophical basis of medical practice: Toward a philosophy and ethic of the healing professions.* New York: Oxford University Press.

Phillips, C.V., and K.J. Goodman. 2004. The missed lessons of sir Austin Bradford Hill. *Epidemiologic Perspectives and Innovations* 1 (3): 3. https://doi.org/10.1186/1742-5573-1-3.

Phillips, R.L., L.A. Bartholomew, S.M. Dovey, G.E. Fryer, T.J. Miyoshi, and L.A. Green. 2004. Learning from malpractice claims about negligent, adverse events in primary care in the United States. *Quality and Safety in Health Care* 13 (2): 121–126. https://doi.org/10.1136/qshc.2003.008029.

Pickersgill, M. 2013. The social life of the brain: Neuroscience in society. *Current Sociology. La Sociologie Contemporaine* 61 (3): 322–340. https://doi.org/10.1177/0011392113476464.

Pinker, S. 2004. *The evolutionary psychology of religion.* Presented at the annual meeting of the Freedom from Religion Foundation, Madison, Wisconsin, October 29, on receipt of "The Emperor's New Clothes Award".

———. 2007a. *The mystery of consciousness.* New York: Time Magazine.

Plantinga, A.C., and D. Dennet. 2010a. *Science and religion.* Oxford: Oxford University Press.

Poh, E., T.J. Carroll, and J.A. Taylor. 2016. Effect of coordinate frame compatibility on the transfer of implicit and explicit learning across limbs. *The Journal of Neurophysiology* 116: 1239–1249.

Pole, S. 2006. *Ayurvedic medicine: The principles of traditional practice.* Philadelphia: Elsevier.

Polito, Jacquelyn M, R EEG T, RPSGT, R.S.T., M.H.A. 2012. Ethical considerations in internet use of electronic protected health information. *The Neurodiagnostic Journal* 52 (1): 34–41.

Pollen, D.A., S. Baker, D. Hinerfeld, J. Swearer, B.A. Evans, J.E. Evans, R. Caselli, E. Rogaeva, P. St George-Hyslop, and M. Moonis. 2010. Prevention of Alzheimer's disease in high risk groups: Statin therapy in subjects with PSEN1 mutations or heterozygosity for apolipoprotein E epsilon 4. *Alzheimer's Research & Therapy* 2: 31.

Post, S.G. 1994. Beyond adversity: Physician and patient as friends? *Journal of Medical Humanities* 15: 23–29.

Power, D'Arcy. 1897. *William Harvey: Masters of medicine.* London: T. Fisher Dunwin.

Prietula, M. 1981. Expertise and error in diagnostic reasoning. *Cognitive Science* 5: 235–283.

Procter, R. 2009. Definition of health informatics. National Institutes of Health: U.S. National Library of Medicine. Retrieved from: http://www.nlm.nih.gov/hsrinfo/informatics.html

Project of the ABIM Foundation, ACP-ASIM Foundation, and European Federation of Internal Medicine. 2002. Medical professionalism in the new millennium: A physician charter. *Annals of Internal Medicine* 136: 243–246.

Quante, M., and A. Vieth. 2002. Defending principlism well understood. *The Journal of Medicine and Philosophy* 27: 621–649.

Rahman, S.Z. 2003. *Were the "dark ages" really dark?* Grey matter (the co-curricular journal of Jawaharlal Nehru medical college). Aligarh: Aligarh Muslim University.

Raiffa, H. 1970. *Decision analysis.* Reading: Addison-Wesley.

Rajesh P. N. Rao. 2002. In *Probabilistic models of the brain: Perception and neural function (neural information processing)*, ed. B.A. Olshausen and M.S. Lewicki. Boston: MIT Press.

Ramachandran, V.S. 2005a. *A brief tour of human consciousness: From impostor poodles to purple numbers.* New York: Pearson.
Ramachandran, V.S., and S. Blakeslee. 1998a. *Phantoms in the brain: Probing the mysteries of the human mind.* New York: HarperCollins.
Ramoni, M.F., M. Stefanelli, L. Magnani, and G. Barosi. 1992. An epistemological framework for medical knowledge based system. *IEEE Transactions on Systems, Man, and Cybernetics* 22: 1361–1375.
Ramsey, K.M., B. Marcheva, A. Kohsaka, and J. Bass. 2007. The clockwork of metabolism. *Annual Review of Nutrition* 27: 219–240.
Rank, O. 1925. *Der Doppelgänger – Eine psychoanalytische studie* (The double: A psychoanalytic study), Wien: Turia+Kant, Wien 1993. Reprint der 1925 Wiener Ausgabe, Internat. Psychoanalytischer Verlag, The Estate of Otto Rank, Becket.
Rao, R.P.N., A. Stocco, M. Bryan, D. Sarma, T.M. Youngquist, et al. 2014. A direct brain-to-brain interface in humans. *PLoS One* 9 (11): e111332. https://doi.org/10.1371/journal.pone.0111332.
Reason, J.T. 1990. *Human error.* Cambridge: Cambridge University Press.
Reeder, L.G. 1972. The patient-client as a consumer: Some observations on the changing professional-client relationship. *Journal of Health and Social Behavior* 13: 406–412.
Reich, W. 1927. *Die Funktion des Orgasmus*, Revidierte Fassung 1982: Genitalität in der Theorie und Therapie der Neurose/Frühe Schriften II. Köln: Kiepenheuer & Witsch.
Reiser, S.J. 1978. *Medicine and the reign of technology.* Cambridge: Cambridge University Press.
Relman, A.S. 2007a. *A second opinion: Rescuing America's healthcare.* New York: Perseus Books.
Reznek, L. 1987. *The nature of disease.* London: Routledge & Kegan Paul.
Rho, K., S. Han, K. Kim, and M.S. Lee. 2006. Effects of aromatherapy massage on anxiety and self-esteem in Korean elderly women: A pilot study. *International Journal of Neuroscience* 116: 1447–1455.
Riger, S. 2007. Feminism and community psychology: Compelling convergences. In *APA handbook of community psychology: Theoretical foundations, core concepts, and emerging challenges*, ed. M.A. Bond, I. Serrano-García, C.B. Keys, M. Shinn, M.A. Bond, I. Serrano-García, et al., 129–147. Washington, DC: American Psychological Association.
Ritenbaugh, C., R. Hammerschlag, S.F. Dworkin, et al. 2012. Comparative effectiveness of traditional Chinese medicine and psychosocial care in the treatment of temporomandibular disorders-associated chronic facial pain. *Journal of Pain* 13 (11): 1075–1089.
Rizzi, D.A., and S.A. Pedersen. 1992. Causality in medicine: Towards a theory and terminology. *Theoretical Medicine* 13: 233–254.
Rizzolatti, G., and L. Craighero. 2004. The mirror-neuron system. *Annual Review of Neuroscience* 27: 169–192.

Rizzolatti, G., and C. Sinigaglia. 2006a. *So quel che fai. Il cervello che agisce e i neuroni specchio*. Milan: Raffaello Cortina Editore.

———. 2010a. The functional role of the parieto-frontal mirror circuit: Interpretations and misinterpretations. *Nature Reviews Neuroscience* 11 (4): 264–274.

Robiner, W.N., K.E. Dixon, J.L. Miner, and B.A. Hong. 2014. Psychologists in Medical Schools and Academic Medical Centers. Over 100 years of growth, influence, and partnership. *American Psychologist* 69 (3): 217–315. Washington, DC: APA, American Psychological Association.

Robinson, P., and N.C. Ellis, eds. 2008. *Handbook of cognitive linguistics and second language acquisition*. London: Routledge.

Rosenblueth, A. 1970. *Mind and brain: A philosophy of science*. Cambridge, MA: MIT Press.

Rosenblueth, A., and N. Wiener. 1945. The role of models in science. *Philosophy in Science* 12: 316–321.

Roter, D. 2000. The enduring and evolving nature of the patient-physician relationship. *Patient Education and Counseling* 39: 5–15.

Rothman, K.J. 1976. Causes. *Journal of Epidemiology* 104: 587–592.

Rouder, J.N., and R.D. Morey. 2011a. A Bayes factor meta-analysis of Bem's ESP claim. *Psychonomic Bulletin & Review* 18 (4): 682–689.

Russell, B. 1912. *The problems of philosophy*, Barnes & Noble edition. New York: Barnes & Noble.

Rutherford, A., and M. Pettit. 2015a. Feminism and/in/as psychology: The public sciences of sex and gender. *History of Psychology* 18 (3): 223–237. https://doi.org/10.1037/a0039533.

Ryff, C.D., and B. Singer. 1998. Human health: New directions for the next millennium. *Psychological Inquiry* 9: 69–85.

Sackett, D.L., W.S. Richardson, W. Rosenberg, and R.B. Haynes. 1998. *Evidence-based medicine: How to practice and teach EBM*. London: Churchill Livingstone.

Safranski, R. 1999a. *Martin Heidegger – Between good and evil*. Cambridge, MA: Harvard University Press.

Salmon, W. 1984. *Scientific explanation and the causal structure of the world*. Princeton: Princeton University Press.

Saper, R.B., R.S. Phillips, A. Sehgal, et al. 2008a. Lead, mercury, and arsenic in U.S. and Indian-manufactured Ayurvedic medicines sold via the internet. *Journal of the American Medical Association* 300 (8): 915–923.

Saunders, J. 2000a. The practice of clinical medicine as an art and as a science. *Medical Humanities* 26: 18–22.

Schicktanz, S., T. Amelung, and J.W. Rieger. 2015. Qualitative assessment of patients' attitudes and expectations toward BCIs and implications for future technology development. *Frontiers in Systems Neuroscience* 9: 64. https://doi.org/10.3389/fnsys.2015.00064.

Schaffner, K.F., and H.T. Engelhardt Jr. 1998. Medicine, philosophy of. In *Routledge encyclopedia of philosophy*, ed. E. Craig, 264–269. London: Routledge.

Schiepek, G. 2004. *Neurobiologie der Psychotherapie*, Studienausgabe. 1st ed. Stuttgart: Schattauer.

Schönemann, D. 1999. Der Spiritismus. In: *Spiritismus. Sind paranormale Phänomene unumstößliche Beweise für ein Fortleben nach dem Tode?*. NOI International, Jahrgang 34 (128). Klagenfurt: NOI-Verlag, pp. 9–28.

———. 2002. Warum *verstehen wir uns gegenseitig so oft nicht?*, NOI International 139/140, I. Aufl. Klagenfurt: NOI-Verlag.

Schultze-Kraft, M., D. Birman, M. Rusconi, C. Allefeld, K. Görgen, S. Dähne, B. Blankertz, and J.D. Haynes. 2016. The point of no return in vetoing self-initiated movements. *Proceedings of the National Academy of Sciences* 113: 1080–1085. https://doi.org/10.1073/pnas.1513569112.

Schwartz, W.B., G.A. Gorry, J.P. Kassirer, and A. Essig. 1973. Decision analysis and clinical judgment. *American Journal of Medicine* 55: 459–472.

Scotton, B.W., A.B. Chinen, and J.R. Battista. 1996a. *Textbook of transpersonal psychiatry and psychology*. New York: Basic Books.

Searle, J. 2015a. *Seeing things as they are: A theory of perception*. Oxford: Oxford University Press.

Seifert, J. 2004a. *The philosophical diseases of medicine and their cures: Philosophy and ethics of medicine, vol. 1: Foundations*. New York: Springer.

Senn, S. 2007a. *Statistical issues in drug development*. 2nd ed. Hoboken: Wiley.

Shavers, V.L. 2007. Measurement of socioeconomic status in health disparities research. *Journal of the National Medical Association* 99 (9): 1013–1023.

Sheehan, D.P., ed. 2006a. *Frontiers of time: Retrocausation – Experiment and theory*. Melville: American Institute of Physics.

Silver, J., M.E. Schwab, and P.G. Popovich. 2014. Central nervous system regenerative failure: Role of oligodendrocytes, astrocytes, and microglia. *Cold Spring Harbor Perspectives in Biology* 7 (3): a020602. https://doi.org/10.1101/cshperspect.a020602.

Simon, J.R. 2010. Advertisement for the ontology of medicine. *Theoretical Medicine and Bioethics* 31: 333–346.

Sirven, J.I., and P.O. Shafer. 2014. *What is epilepsy?* Landover: Epilepsy Foundation. Retrieved from: http://www.epilepsy.com/learn/epilepsy-101/what-epilepsy

Slagter, H.A., A. Lutz, L.L. Greischar, A.D. Francis, D. Nieuwenhuis, J.M. Davis, and R.J. Davidson. 2007a. Mental training affects distribution of limited brain resources. *PLOS Biology* 5 (6): e138.

Slayton, S.C., J. D'Archer, and F. Kaplan. 2010a. Outcome studies on the efficacy of art therapy: A review of findings. *Art Therapy: Journal of the American Art Therapy Association* 27 (3): 108–111.

Smart, J.J.C. 1963. *Philosophy and scientific realism*. London: Routledge & Kegan Paul.

Smith, B.G. 2000. *Global feminisms since 1945*. London: Psychology Press.

Smuts, J. 1926. *Holism and evolution*. New York: Macmillan.

Soch, J., J.D. Haynes, and C. Allefeld. 2016. How to avoid mismodelling in GLM-based fMRI data analysis: Cross-validated Bayesian model selection. *NeuroImage* 141: 469–489. https://doi.org/10.1016/j.neuroimage.2016.07.047.

Solomon, M.J., and R.S. McLeod. 1998. Surgery and the randomized controlled trial: Past, present and future. *Medical Journal of Australia* 169: 380–383.

Spagnolo, A.G. 1997. *Bioetica nella ricerca e nella prassi medica*. Torino: Edizioni Camilliane.

———. 2008a. Quality of life and ethical decisions in medical practice. *Journal of Medicine and The Person* 6 (3): 118–122.

———. 2014. La valutazione della ricerca in bioetica. *Medicina e Morale* 3: 355–365. Rome: Università Cattolica del Sacro Cuore.

Spodick, D.H. 1982. The controlled clinical trial: Medicine's most powerful tool. *The Humanist* 42 (12–21): 48.

Stefanelli, M., and M.F. Ramoni. 1992a. Epistemological constraints on medical knowledge-based systems. In *Advanced models of cognition for medical training and practice*, ed. D.A. Evans and V.L. Patel, vol. 97, 3–20. Heidelberg: Springer-Verlag.

Stegenga, J. 2011. Is meta-analysis the platinum standard of evidence? *Studies in History and Philosophy of Biological and Biomedical Sciences* 42 (4): 497–507. https://doi.org/10.1016/j.shpsc.2011.07.003. PMID 22035723.

Steger, M.F., P. Frazier, S. Oishi, and M. Kaler. 2006a. The meaning in life questionnaire: Assessing the presence of and search for meaning in life. *Journal of Counseling Psychology* 53: 80–93.

Steiner, R. 1912. *Von Jesus zu Christus* (From Jesus to Christ). I. Aufl. Berlin. 1980 Dornach: Rudolf Steiner Verlag.

———., ed. 2001. *Das esoterische Christentum und die geistige Führung der Menscheit* (Exoteric Christianity and the spiritual guidance of humanity). IV Taschenbuchausgabe aus der 4. Gesamtausgabe 1995. Dornach: Rudolf Steiner Nachlaßverwaltung.

Steinhart, E.C. 2014a. *Your digital afterlives: Computational theories of life after death (Palgrave Frontiers in philosophy of religion)*. New York: Palgrave Macmillan.

Steinpach, R. 2006. (Aufl.; Schönemann, D., Red.). *Wieso wir nach dem Tode leben und welchen Sinn das Leben hat*. In: *Der Tod, ein Übergang; das Jenseits, eine andere Realität*. NOI International, Jahrgang 41 (154). Klagenfurt: NOI-Verlag, pp. 8–43.

Stempsey, W.E. 2004. The philosophy of medicine: Development of a discipline. *Medicine, Health Care and Philosophy* 7: 243–251.

Stern, J.E., and J.A. Filosa. 2013. Bidirectional neuro-glial signaling modalities in the hypothalamus: Role in neurohumoral regulation. *Autonomic Neuroscience: Basic & Clinical* 175 (1–2): 51–60. https://doi.org/10.1016/j.autneu.2012.12.009.

Stevenson, C.L. 1944. *Ethics and language*. New Haven: Yale University Press.

Stevenson, I. 1974a. *Twenty cases suggestive of reincarnation*. 2nd ed. Charlottesville: University Press of Virginia.

Stocco, A., C. Lebiere, and J.R. Anderson. 2010a. Conditional routing of information to the cortex: A model of the basal ganglia's role in cognitive coordination. *Psychological Review* 117 (2): 541–574.

Storm, L., P.E. Tressoldi, and L. Di Risio. 2010a. Meta-analysis of free-response studies, 1992–2008: Assessing the noise reduction model in parapsychology. *Psychological Bulletin* 136 (4): 471–485.

Strassman, R. 2001a. *DMT: The spirit molecule: A doctor's revolutionary research into the biology of near-death and mystical experiences*. Rochester: Park Street Press.

Straus, S.E., and F.A. McAlister. 2000a. Evidence-based medicine: A commentary on common criticisms. *Canadian Medical Association Journal* 163: 837–840.

Summer, W.G. 1963. *Social Darwinism: Selected essays*. Ed. Stow Persons. Englewood Cliff: Prentice-Hall.

Svenaeus, F. 2000. *The hermeneutics of medicine and the phenomenology of health: Steps towards a philosophy of medical practice*. Dordrecht: Kluwer.

Szalavitz, M. 2010. *Mayo Clinic vs. WebMD: Another perspective*. New York: TIME Magazine. February 7, 2011. Retrieved from: http://healthland.time.com/2011/02/07/mayo-clinic-vs-webmd-another-perspective/

Tallis, R.C. 2006. Doctors in society: Medical professionalism in a changing world. *Clinical Medicine* 6: 7–12.

Tasman, A., J. Kay, and J.A. Lieberman. 2003a. *Psychiatry. Second edition. Therapeutics*. Chichester: Wiley.

Tate, M., G. Herbet, S. Moritz-Gasser, J. Tate, and H. Duffau. 2014. Probabilistic map of critical functional regions of the human cerebral cortex: Broca's area revisited. *Brain* 137 (10): 2773–2782.

Tauber, A.I. 2005a. *Patient autonomy and the ethics of responsibility*. Cambridge, MA: MIT Press.

Teyber, E. 2000a. *Interpersonal process in psychotherapy. A relational approach*. 4th ed. Belmont: Wadsworth/Thomson Learning.

Thagard, P. 1999. *How scientists explain disease*. Princeton: Princeton University Press.

The Heritage Foundation. 2014. *The impact of Obamacare. From the Frontlines of our health care crisis*. Washington, DC: The Heritage Foundation. Retrieved from: http://www.heritage.org/research/projects/impact-of-obamacare

The Multi-Society Task Force on PVS. 1994a. Medical aspects of the persistent vegetative state – Second of two parts. *The New England Journal of Medicine (NEJM)* 330: 1499–1508.

The Week Staff. 2012. *ObamaCare survives the Supreme Court: 5 takeaways.* New York: The Week. Retrieved from: http://theweek.com/articles/474219/obamacare-survives-supreme-court-5-takeaways

Tillich, P.J. 1983a. *Gesammelte Werke*, 14 Bände + 6 Ergänzungs- und Nachlassbände, 1958–1983, Hg. Renate Albrecht, Stuttgart: Evangelisches Verlagswerk, (new edition by Verlag Walter de Gruyter, Berlin).

Tjeltveit, A.C., and M.C. Gottlieb. 2012a. Avoiding ethical missteps. By drawing on the science of prevention, psychologists can develop skills, relationships and personal qualities to bolster ethical resilience and minimize risks related to unethical behavior. *Monitor on Psychology* 43 (4). Washington, DC: American Psychological Association 44 (4): 68–74.

Todorov, I., N. Kornell, M. Larsson Sundqvist, and F.U. Jönsson. 2014a. Phrasing questions in terms of current (not future) knowledge increases preferences for cue-only judgments of learning. *Archives of Scientific Psychology* 1: 7–13. Washington, DC: APA, American Psychological Association.

Tonelli, M.R. 1998. The philosophical limits of evidence-based medicine. *Academic Medicine* 73: 1234–1240.

Tononi, G. 2012. *PHI: A voyage from the brain to the soul.* New York: Pantheon Books.

Toombs, S.K. 1993. *The meaning of illness: A phenomenological account of the different perspectives of physician and patient.* Dordrecht: Kluwer.

Tracey, J.G., B.E. Wampold, J.W. Lichtenberg, and R.K. Goodyear. 2014a. Expertise in psychotherapy. An elusive goal? *American Psychologist* 69 (3): 217–315. Washington, DC: APA, American Psychological Association, pp. 225–226.

Trosseau, A. 1869. *Lectures on clinical medicine.* The new Sydenham society, 1869 submitted by A.L. Wyman: Filler. Medicine: Art or science. 2 (320): 1322.

Tucker, N.H. 1999. *Presidential message art vs science. Jacksonville Medicine* 50 (12): 234–240.

Tye, M. 1995a. *Ten problems of consciousness.* Cambridge, MA: MIT Press.

United States District Court for the District of New Hampshire. 2015. *Janice Farley vs. United States of America.* Landya McCafferty United States District Judge. Case 1:13-cv-00261-LM Document 58 Filed 4/3/15. http://3mc77e18jo7n1uk8m71my8ml.wpengine.netdna-cdn.com/wp-content/uploads/2015/04/STROKE-Decision-Memorandum-and-Order1.pdf

Unschuld, P.U. 2010a. *Medicine in China: A history of ideas.* 2nd ed. Berkeley: University of California Press.

Van Dalen, D. 2012a. *L.E.J. Brouwer – Topologist, intuitionist, philosopher: How mathematics is rooted in life.* London: Springer.

Van der Steen, W.J., and P.J. Thung. 1988a. *Faces of medicine: A philosophical study.* Dordrecht: Kluwer.

Van Gijn, J. 2005. From randomized trials to rational practice. *Cardiovascular Diseases* 19: 69–76.

Van Inwagen, P. 2006a. *The problem of evil.* Oxford: Oxford University Press.

———. 2017a. *Thinking about free will.* Cambridge: Cambridge University Press.

Van Lommel, P. 2011a. *Consciousness beyond life: The science of the near-death experience.* New York: HarperCollins.

Vassallo, N. 2002. Epistemologia. In *Storia della filosofia analitica*, ed. F. D'Agostini and N. Vassallo. Torino: Einaudi.

Vattimo, G., and S. Zabala. 2011. *Hermeneutic communism: From Heidegger to Marx.* New York: Columbia University Press.

Veatch, R.M. 1991. *The patient-physician relations: The patient as partner, part 2.* Bloomington: Indiana University Press.

Velanovich, V. 1994. Does philosophy of medicine exist? A commentary on Caplan. *Theoretical Medicine* 15: 88–91.

Vickers, A.J., and K. Linde. 2014. Acupuncture for chronic pain. *JAMA: The Journal of the American Medical Association* 311 (9): 955–956.

Vickers, A.J., A.M. Cronin, A.C. Maschino, G. Lewith, H. MacPherson, N.E. Foster, K.J. Sherman, C.M. Witt, K. Linde, and Acupuncture Trialists' Collaboration. 2012. Acupuncture for chronic pain: Individual patient data meta-analysis. *Archives of Internal Medicine* 172 (19): 1444–1453.

Villar, F. 1996. *Gli Indoeuropei e le origini dell'Europa* (the Indoeuropeans and the origins of Europe), Bologna: Società editrice Il Mulino, 1997 – Original edition: *Los Indoeuropeos y los origenes de Europa.* Lenguaje e historia, Madrid: Gredos, Madrid.

Vis, J., and H. Boynton. 2008a. Spirituality and transcendent meaning making: Possibilities for enhancing posttraumatic growth. *Journal of Religion & Spirituality in Social Work* 27: 69–86.

Volk, S. 2011a. *Fringe-ology: How I tried to explain away the unexplainable – And couldn't.* New York: HarperCollins.

Vygotsky, L.S. 1978a. *Mind in society.* Cambridge, MA: Harvard.

Wasserstein, R.L., and N.A. Lazar. 2016a. The ASA's statement on p-values: Context, process, and purpose. *The American Statistician* 70 (2): 129–133.

Weatherall, D. 1996a. *Science and the quiet art: The role of medical research in health care.* New York: Norton.

Weindling, P.J. 2004a. *Nazi medicine and the Nuremberg trials. From medical war crimes to informed consent.* Houndmills: Palgrave Macmillan.

Westen, D., and J. Weinberger. 2005a. In praise of clinical judgment: Meehl's forgotten legacy. *Journal of Clinical Psychology* 61: 1257–1276.

Whitaker, R. 2001a. *Mad in America: Bad science, bad medicine, and the enduring mistreatment of the mentally ill.* New York: Perseus Publishing.

———. 2010a. *Anatomy of an epidemic: Magic bullets, psychiatric drugs, and the astonishing rise of mental illness in America.* New York: Crown.

Whitbeck, C. 1981. A theory of health. In *Concepts of health and disease: Interdisciplinary perspectives*, ed. A.L. Caplan, H.T. Engelhardt Jr., and J.J. McCartney, 611–626. London: Addison-Wesley.
Wider, K. 2013. Sartre and Spinoza on the nature of mind. *Continental Philosophy Review* 46: 4. New York: Springer.
Wildes, K.W. 2001. The crisis of medicine: Philosophy and the social construction of medicine. *Kennedy Institute of Ethics Journal* 11: 71–86.
Williams, D.R., S.A. Mohammed, J. Leavell, and C. Collins. 2010. Race, socioeconomic status, and health: Complexities, ongoing challenges, and research opportunities. *Annals of the New York Academy of Sciences* 1186: 69–101.
Wilson, C. 1971. *Das Okkulte*, (Original: The Occult) Lizenzausgabe für Parkland Verlag, Köln/Deutschland, 2004 Neu Isenburg: Melzer Verlag.
Wilson, R.A., and L. Foglia. 2011. *Embodied cognition*, Stanford encyclopedia of philosophy. Palo Alto: Stanford University Press.
Witt, C.M., N.E.J. Berling, N.T. Rinpoche, M. Cuomo, and S.N. Willich. 2009. Evaluation of medicinal plants as part of Tibetan medicine prospective observational study in Sikkim and Nepal. *Journal of Alternative & Complementary Medicine* 15 (1): 59–65.
Wittgenstein, L. 1958a. *Preliminary studies for the "philosophical investigations", generally known as the blue and Brown books*. Hoboken: Blackwell Publishers.
Woodward, J. 2003. *Making things happen: A theory of causal explanation*. Oxford: Oxford University Press.
World Health Organization. 2013. *WHO Health Report. Health system attainment and performance in all member states*. Retrieved from: http://www.who.int/whr/2013/report/en/
Wortman, J., M.B. Donnellan, and R.E. Lucas. 2014a. *Can physical warmth (or coldness) predict trait loneliness? A replication of Bargh and Shalev (2012)*, Archives of scientific psychology, 13–19. Washington, DC: APA, American Psychological Association.
Wrathall, M. 2006a. *How to read Heidegger*, 100. New York: W. W. Norton.
Wulff, H.R., S.A. Pedesen, and R. Rosenberg. 1990a. *Philosophy of medicine: An introduction*. 2nd ed. Oxford: Blackwell.
Yeh, G.Y., E.P. McCarthy, P.M. Wayne, et al. 2011. Tai chi exercise in patients with chronic heart failure: A randomized controlled trial. *Archives of Internal Medicine* 171 (8): 750–757.
Zaner, R.M. 1981a. *The context of self: A phenomenological inquiry using medicine as a clue*. Athens: Ohio University Press.
Zauzmer, J. 2017a. *Why a Yale neuroscientist decided to change careers – And is now becoming a priest*. Washington, DC: The Washington Post.
Žižek, S., and B. Gunjević. 2012a. *God in pain: Inversions of apocalypse*. New York: Seven Stories Press.

Zürrer, R. 1989a. *Reinkarnation. Die umfassende Wissenschaft der Seelenwanderung* (Reincarnation: The omni-comprehensive science of the migration of the body), IV. Aufl. 2000. Neuhausen: Govinda-Verlag.

References

Andersen, H. 2001b. The history of reductionism versus holism approaches to scientific research. *Endeavor* 25: 153–156.

Aristotle. 1966b. *Metaphysics.* Trans. H.G. Apostle. Bloomington: Indiana University Press.

Bassman, L. 1998b. *The whole mind: The definitive guide to complementary treatment for mind, mood, and emotion.* Novato: New World Library.

Bauschatz, P.C. 1984. *The well and the tree.* Amherst: The University of Massachusetts Press.

Beauregard, M., and D. O'leary. 2008b. *The spiritual brain. A neuroscientist's case for the existence of the soul.* New York: HarperCollins.

Behnke, S.H. 2014b. Always a cultural perspective. The society of Indian psychologists comments on the APA Ethics Code. *Ethically Speaking, Monitor on Psychology* 45 (10).

Bem, D.J. 2011b. Feeling the future: Experimental evidence for anomalous retroactive influences on cognition and affect. *Journal of Personality and Social Psychology* 100 (3): 407–425.

Bem, D.J., P.E. Tressoldi., T. Rabeyron, and M. Duggan. 2014b. *Feeling the future: A meta-analysis of 90 experiments on the Anomalous anticipation of random future events* (SSRN Scholarly Paper No. ID 2423692). Rochester: Social Science Research Network.

Bendeck Sotillos, S. 2013b. *Psychology and the Perennial Philosophy*, Psychology and the perennial philosophy: Studies in comparative religion. Bloomington: World Wisdom.

Block, N., O. Flanagan, and G. Guzeldere, eds. 1998b. *The nature of consciousness.* Cambridge, MA: MIT Press.

Blois, M.S. 1990b. Medicine and the nature of vertical reasoning. *New England Journal of Medicine* 318: 847–851.

Bonelli, R., R.E. Dew, H.G. Koenig, D.H. Rosmarin, and S. Vasegh. 2012b. Religious and spiritual factors in depression: Review and integration of the research. *Depression Research and Treatment* 2012: 1–8.

Boshuizen, H.P.A., and H.G. Schmidt. 1992b. On the role of biomedical knowledge in clinical reasoning by experts, intermediates, and novices. *Cognitive Science* 16 (2): 153–184.

Breggin, P.R. 1991b. *Toxic psychiatry: Why therapy, empathy, and love must replace the drugs, electroshock, and biochemical theories of the "new psychiatry".* New York: St. Martin's Press.

Brothers, L. 2001b. *Mistaken identity: The mind-body problem reconsidered.* New York: SUNY Press.

Buchler, J. 1955b. *Nature and judgment.* New York: Columbia University Press.

Caplan, A.L. 1992b. Does the philosophy of medicine exist? *Theoretical Medicine* 13: 67–77.

Carruthers, P. 2004b. *The nature of mind.* London: Routledge.

Carter, K.C. 2003b. *The rise of causal concepts of disease: Case histories.* Burlington: Ashgate.

Cefalù, W.T., Z.E. Floyd, J.M. Stephens, et al. 2014b. Botanicals and translational medicine: A paradigm shift in research approach. *Nutrition* 30 (7–8S): S1–S68. NIH PubMed.

Chabris, C., and D. Simons. 2010. *The invisible gorilla.* New York: Crown Publishing Group.

Chalmers, D. 1996b. *The conscious mind.* Oxford: Oxford University Press.

Chandrasekaran, B., J.W. Smith, and J. Sticklen. 1989b. Deep models and their relation to diagnosis. *Artificial Intelligence in Medicine* 1: 29–40.

Choudhury, S., and J. Slaby. 2012. *Critical neuroscience. A handbook of the social and cultural contexts of neuroscience.* Chichester: Wiley Blackwell.

Churchland, P.M. 2007b. *History of the concept of mind.* Farnham: Ashgate Publishers.

Churchland, P.S. 2013b. *Touching a nerve. The self as brain*, 11. New York: W.W. Norton & Company.

Clay, R.A. 2014b. Random sample. James E. Maddux, PhD. *Monitor on Psychology* 45 (11): 546–551. Washington, DC: American Psychological Association.

Collins, F.S. 2006b. *The language of god. A scientist presents evidence for belief.* New York: Free Press Simon & Schuster.

Cook, C. 2004b. Addiction and spirituality. *Addiction* 99: 539–551.

Crumbaugh, J.C., and L.T. Maholick. 1964b. An experimental study in existentialism: The psychometric approach to Frankl's concept of noogenic neurosis. *Journal of Clinical Psychology* 20: 200–207.

Culver, C.M., and B. Gert. 1982b. *Philosophy in medicine: Conceptual and ethical issues in medicine and psychiatry.* New York: Oxford University Press.

Davis-Floyd, R., and G. St. John. 1998b. *From doctor to healer: The transformative journey.* New Brunswick: Rutgers University Press.

Dawkins, R. 1989b. *The selfish gene.* 2nd ed. Oxford: Oxford University Press.

De Santillana, G. 1961. *The origins of scientific thought: From Anaximander to Proclus, 600 BC to 300 AD.* London: Weidenfeld & Nicolson.

Desbordes, G., L.T. Negi, T.W.W. Pace, et al. 2012b. Effects of mindful-attention and compassion meditation training on amygdala response to emotional stimuli in an ordinary, non-meditative state. *Frontiers in Human Neuroscience* 6: 292.

Dickinson, H., F. Campbell, F. Beyer, et al. 2008b. Relaxation therapies for the management of primary hypertension in adults: A Cochrane review. *Journal of Human Hypertension* 22 (12): 809–820.

Diener, E. 2000b. Subjective well-being: The science of happiness and a proposal for a national index. *American Psychology* 55 (1): 34–43.

Diener, E., and S. Oishi. 2003b. Personality, culture and subjective wellbeing: Emotional and cognitive evaluations of life. *Annual Review of Psychology* 54: 403–425.

Diener, E., R.A. Emmons, R.J. Larsen, and S. Griffin. 1985b. The satisfaction with life scale. *Journal of Personality Assessment* 49: 71–75.

Dossey, L., D. Chopra, and R. Roy. 2010b. *The mythology of science-based medicine*, Huffpost healthy living. New York: The Huffington Post.

Drewerman, E. 1999b. … *und es geschah so*. Patmos: Die moderne Biologie und die Frage nach Gott.

———. 2003b. *Religiös bedingte neurotische Erkrankungen*. Pabst: Science Publishers.

———. 2010b. *Wir glauben, weil wir lieben*. Patmos: Woran ich glaube.

Dummett, M.A.E. 1991b. *The logical basis of metaphysics*. Cambridge: Harvard University Press.

Duncan, L.E., A.R. Pollastri, and J.W. Smoller. 2014b. Mind the gap. Why many geneticists and psychological scientists have discrepant views about gene – Environment interaction (GxE) research. *American Psychologist* 69 (3): 218–324. Washington, DC: APA, American Psychological Association.

Earhart, G.M. 2009b. Dance as therapy for individuals with Parkinson disease. *European Journal of Physical and Rehabilitation Medicine* 45 (2): 231–238.

Eccles, J. 2015b. *How the self controls its brain*. Berlin: Springer Verlag.

Ellison, C.G., and J.S. Levin. 1998b. The religion-health connection: Evidence, theory, and future directions. *Health Education & Behavior* 25 (6): 700–720.

Emanuel, E.J., and L.L. Emanuel. 1992b. Four models of the physician-patient relationship. *Journal of American Medical Association* 267: 2221–2226.

Ernst, E. 2004b. Anthroposophical medicine: A systematic review of randomised clinical trials. *Wien, A: Wiener klinische Wochenschrift* 116 (4): 128–130.

Ernst, E., and S. Singh. 2008b. *Trick or treatment? Alternative medicine on trial*. London: Bantam Press.

Evans, M., P. Louhiala, and P. Puustinen, eds. 2004b. *Philosophy for medicine: Applications in aclinical context*. Oxon: Radcliffe Medical Press.

Fenwick, P., and E. Fenwick. 2012b. *The truth in the light*. Guildford: White Crow Books.

Ferré, F., J.J. Kockelmans, and J.E. Smith. 1982b. *The challenge of religion. Contemporary readings in philosophy of religion*. New York: The Seabury Press.

Flyvbjerg, B. 1993b. Aristotle, Foucault and progressive phronesis: Outline of an applied ethics for sustainable development. In *Applied ethics*, ed. Earl R. Winkler and R. Coombs Jerold. Cambridge: Blackwell Publishers.

Foglia, L. 2005b. Rappresentazioni Mentali. Modelli teorici a confronto. *Arkete* I (2): 17–42.

Foglia, L., and R. Grush. 2010b. Neural reuse: A fundamental organizational principle of the brain. *Behavioral and Brain Sciences* 33: 245–313.

———. 2011b. The limitations of a purely enactive (non-representational) account of imagery. *Journal of Consciousness Studies* 18 (5–6): 35–43.

Foucault, M. 1966b. *The order of things: An archaeology of the human sciences* (Les Mots et les choses: Une archéologie des sciences humaines). New York: Pantheon Books.

Fox, C. 2006b. A brain-based philosophy of life. *The Lancet* 5 (3): 212. Amsterdam: Elsevier.

Frankl, V.E. 1984b. *Man's search for meaning*. 3rd ed. New York: Washington Square Press. (Original work published 1946).

Freeman, L.W. 2009b. *Mosby's complementary & alternative medicine: A research-based approach*. 3rd ed. St. Louis: Mosby Elsevier.

Fuchs, T., N. Birbaumer, W. Lutzenberger, J. Gruzelier, and J. Kaiser. 2003b. Neurofeedback treatment for attention-deficit/hyperactivity disorder in children: A comparison with methylphenidate. *Applied Psychophysiology and Biofeedback* 28: 1–12.

Fulford, K.W.M. 1991b. The potential of medicine as a resource for philosophy. *Theoretical Medicine* 12: 81–85. Dordrecht: Kluwer Academic Publishers.

Gadamer, H.-G. 1960b. *Wahrheit und Methode. Grundzüge einer philosophischen Hermeneutik*. Tübingen: Mohr Siebeck [Unveränd. Nachdr. d. 3. erw. Aufl. Tübingen 1975].

Gawande, A. 2002b. *Complications: A Surgeon's notes on an imperfect science*. New York: Metropolitan Books.

———. 2014b. *Being mortal: Medicine and what matters in the end*. London: Macmillan.

Gergen, K.J., R. Josselson, and M. Freeman. 2015b. The promises of qualitative inquiry. *American Psychologist* 70 (1): 1–60. Washington, DC: American Psychological Association, pp. 1–9.

Gillon, R. 1986b. *Philosophical medical ethics*. New York: Wiley.

Goldenberg, G. 2008b. Chapter 7, how the mind moves the body: Lessons from Apraxia. In *Oxford handbook of human action. Social cognition and social neuroscience*, ed. E. Morsella, J.A. Bargh, and P.M. Gollwitzer, 136. Oxford: Oxford University Press.

Goller, H. 2015b. *Wohnt Gott im Gehirn?: Warum die Neurowissenschaften die Religion nicht erklären*. Kevelaer: Butzon & Bercker.

Grof, S. 2006b. *The ultimate journey: Consciousness and the mystery of death*. Ben Lomond: Multidisciplinary Association for Psychedelic Studies (MAPS).

Groopman, J. 2007b. *How doctors think*. New York: Houghton Mifflin.

Grün, A., and W. Jäger. 2013b. *Das Geheimnis jenseits aller Wege. Was uns eint, was uns trennt*. Münsterschwarzach: Vier Türme.

Gungov, A.L. 2012b. *Logic in medicine: Approaches to patient safety*. Sofia: Avangard.

———. 2013b. Diagnostics in a logical perspective. In *Normativity: Contemporary challenges*, ed. A.L. Gungov, V. Kaneva, and O. Kasabov, 2003. Sofia: St. Kliment Ohridski University Press.

Hameroff, S.R. 2006b. The entwined mysteries of anesthesia and consciousness. *Anesthesiology* 105 (2): 400–412. https://doi.org/10.1097/00000542-200608000-00024. PMID 16871075.

Hampton, J.R. 2002b. Evidence-based medicine, opinion-based medicine, and real-world medicine. *Perspectives in Biology and Medicine* 45: 549–568.

Hill, A.B. 1965b. The environment and disease: Association or causation? *Proceedings of the Royal Society of Medicine* 58: 295–300.

Holyoak, K., and R.G. Morrison. 2012b. *The Oxford handbook of thinking and reasoning*, Oxford library of psychology. Oxford: Oxford University Press.

Hölzel, B.K., J. Carmody, M. Vangel, et al. 2011b. Mindfulness practice leads to increases in regional brain gray matter density. *Psychiatry Research: Neuroimaging* 191 (1): 36–43.

Hudziak, J.J., M.D. Albaugh, et al. 2014b. Cortical thickness maturation and duration of music training: Health-promoting activities shape brain development. *JAACAP* 11: 1153–1161.

Illari, P.M., F. Russo, and J. Williamson, eds. 2011b. *Causality in the sciences.* New York: Oxford University Press.

Illich, I. 1976b. *Limits to medicine; medical Nemesis: The expropriation of health.* London: Marion Boyars Publishers.

Jackson, M. 2005b. *Existential anthropology – Events, exigencies, and effects (methodology and history in anthropology).* California: Berghahn Books.

Jansen, K.L.R. 1998b. *The ketamine model of the near death experience: A central role for the NMDA receptor*, Journal Article. London: The Maudsley Hospital, Denmark Hill.

Johansson, I., and N. Lynøe. 2008b. *Medicine and philosophy: A twenty-first century introduction.* Frankfurt: Ontos Verlag.

Jonsen, A., M. Siegler, and W. Winslade. 2002b. *Clinical ethics: A practical approach to ethical decisions in clinical medicine.* New York: McGraw-Hill.

Joseph, G.M., and V.L. Patel. 1990b. Domain knowledge and hypothesis generation in diagnostic reasoning. *Medical Decision Making* 10 (1): 31–46.

Jung, C.G. 1963b. (1990 ed). *Mysterium Coniunctionis. Untersuchungen über die Trennung und Zusammensetzung der seelischen Gegensätze in der Alchemie*, Olten und Freiburg im Breisgau: Walter-Verlag.

Kaku, M. 2014b. *The future of the mind: The scientific quest to understand, enhance, and empower the mind.* New York: Doubleday.

Kaliman, P., M.J. Álvarez-López, M. Cosín-Tomás, et al. 2014b. Rapid changes in histone deacetylases and inflammatory gene expression in expert meditators. *Psychoneuroendocrinology* 40: 96–107.

Kaptchuk, T.J. 2000b. *The web that has no weaver: Understanding Chinese medicine.* Chicago: Contemporary Books.

Kay, J., and R.L. Kay. 2003b. Ethnocultural issues in psychotherapy. In *Psychiatry: Therapeutics, chapter 1: Individual psychoanalytic psychotherapy*, II ed. Chichester: Wiley.

Kiecolt-Glaser, J.K., J.E. Graham, W.B. Malarkey, et al. 2008b. Olfactory influences on mood and autonomic, endocrine, and immune function. *Psychoneuroendocrinology* 33 (3): 328–339.

Kienle, S.G., H. Kiene, and H.-U. Albonico. 2006b. *Anthroposophische Medizin in der klinischen Forschung. Wirksamkeit, Nutzen, Wirtschaftlichkeit, Sicherheit.* Stuttgart: Schattauer.

Kihlstrom, J.F. 2002b. Mesmer, the Franklin Commission, and hypnosis: A counterfactual essay. *The International Journal of Clinical and Experimental Hypnosis* 50 (4): 407–419.

Klinger, B. 2015b. *Non-pharmacological interventions to chronic pain. An evidence-based review of the use of integrative approaches including mind-body therapies and acupuncture to the management of chronic pain conditions.* Burlington: University of Vermont College of Medicine/Laura Mann Integrative Healthcare Lecture Series.

Knight, J.A. 1982b. The minister as healer, the healer as minister. *Journal of Religion and Health* 21: 100–114.

Kong, J., T.J. Kaptchuk, G. Polich, et al. 2009b. Expectancy and treatment interactions: A dissociation between acupuncture analgesia and expectancy evoked placebo analgesia. *NeuroImage* 45 (3): 940–949.

Kuipers, B., and J.P. Kassirer. 1984b. Causal reasoning in medicine: Analysis of a protocol. *Cognitive Science* 8 (4): 363–385.

Leung, L. 2012b. Neurophysiological basis of acupuncture-induced analgesia – An updated review. *Journal of Acupuncture and Meridian Studies* 5 (6): 261–270.

Leurent, B., I. Nazareth, j. Bellón-Saameño, M.I. Geerlings, H. Maaroos, S. Saldivia, I. Švab, F. Torres-González, M. Xavier, and M. Kin. 2013b. Spiritual and religious beliefs as risk factors for the onset of major depression: An international cohort study. In *Psychological medicine*. London: Cambridge University Press.

Levinas, E. 1961b. *Totality and infinity. An essay on exteriority.* 3rd ed. Pittsburg: Duquesne University Press. 1969.

Libet, B. 1999b. Do we have a free will? *Journal of Consciousness Studies* 5: 49.

Lipton, P. 2004b. *Inference to the best explanation.* 2nd ed. New York: Routledge.

Little, A.C., B.C. Jones, and L.M. DeBruine. 2014b. *Primacy in the effects of face exposure: Perception is influenced more by faces that are seen first*, Archives of scientific psychology, 43–47. Washington, DC: APA, American Psychological Association.

Loftus, S.F. 2006b. *Language in clinical reasoning: Using and learning the language of collective clinical decision making.* Sydney: Faculty of Health Sciences, School of Physiotherapy, University of Sydney.

Lowen, A. 1975b. *Bioenergetics.* New York: Coward, McCann & Geoghegan.

Lutz, A., J. Brefczynski-Lewis, T. Johnstone, et al. 2008b. Regulation of the neural circuitry of emotion by compassion meditation: Effects of meditative expertise. *PLoS ONE* 3 (3): e1897.

Maier, B., and W.A. Shibles. 2010b. *The philosophy and practice of medicine and bioethics: A naturalistic-humanistic approach*. New York: Springer.

Marcum, J.A. 2005b. Metaphysical presuppositions and scientific practices: Reductionism and organicism in cancer research. *International Studies in the Philosophy of Science* 19: 31–45.

———. 2008b. *An introductory philosophy of medicine: Humanizing modern medicine*. New York: Springer.

Maslow, A.H. 1943b. *A theory of human motivation*. Originally published in *Psychological Review* 50 (4): 370–396. Washington, DC: APA.

McKenzie, E., et al. 1998b. *Healing Reiki*. Hamlyn: Hamlyn Health & Well Being.

Mikula, S., I. Trotts, J. Stone, and E.G. Jones. 2007b. Internet-enabled high-resolution brain mapping and virtual microscopy. *NeuroImage* 35 (1): 9–15.

Miller, A. 2014b. Friends wanted. New research by psychologists uncovers the health risks of loneliness and the benefits of strong social connections. *Monitor on Psychology* 45 (1): 54–58. Washington, DC: American Psychological Association.

Mlsna, Lucas J. 2010b. Stem cell based treatments and novel considerations for conscience clause legislation. *Indiana Health Law Review* 8 (2): 471–496. Bloomington: Indiana University Robert H. McKinney School of Law.

Mommsen, Th. 1942b. Petrarch's conception of the 'dark ages'. *Speculum* 17 (2): 226–242. Cambridge, MA: Medieval Academy of America.

Montgomery, K. 2006b. *How doctors think: Clinical judgment and the practice of medicine*. New York: Oxford University Press.

Moody, R. 1975b. *Life after life: The investigation of a phenomenon – Survival of bodily death*. San Francisco: Harper San Francisco.

Mossbridge, J., P. Tressoldi, and J. Utts. 2012b. Predictive physiological anticipation preceding seemingly unpredictable stimuli: A meta-analysis. *Frontiers in Psychology* 3: 390.

Nagel, T. 1974b. What is it like to be a bat? *The Philosophical Review* 83: 434–450.

Napolitano, J.G., D.C. Lankin, T.N. Graf, et al. 2013b. HiFSA fingerprinting applied to isomers with near-identical NMR spectra: The silybin/isosilybin case. *Journal of Organic Chemistry* 78 (7): 2827–2839.

Naylor, T.H., W.H. Willimon, and M.R. Naylor. 1994b. *The search for meaning*. Nashville: Abingdon Press.

Nestoriuc, Y., A. Martin, W. Rief, and F. Andrasik. 2008b. Biofeedback treatment for headache disorders: A comprehensive efficacy review. *Applied Psychophysiology and Biofeedback* 33: 125–140.

Newberg, A.B. 2016b. *Principles of neurotheology*, Ashgate science and religion series. New York: Routledge.

Newberg, A.B., and M.R. Waldman. 2010b. *How God changes your brain. Breakthrough findings from a leading neuroscientist.* New York: Ballantine Books Trade Paperbacks.

Nguyen, H.T., J.G. Grzywacz, W. Lang, et al. 2010b. Effects of complementary therapy on health in a national U.S. sample of older adults. *Journal of Alternative and Complementary Medicine* 16 (7): 701–706.

Nicolelis, M.A.L. 2003b. Brain-machine interfaces to restore motor function and probe neural circuits. *Nature Reviews. Neuroscience* 4 (5): 417–422.

———. 2011b. *Beyond boundaries: The new neuroscience of connecting brains with machines – And how it will change our lives.* New York: Times Books.

Nidich, S.I., M.V. Rainforth, D.A.F. Haaga, et al. 2009b. A randomized controlled trial on effects of the transcendental meditation program on blood pressure, psychological distress, and coping in young adults. *American Journal of Hypertension* 22 (12): 1326–1331.

Norman, G.R., L.R. Brooks, C.L. Colle, and R.M. Hatala. 2010b. The benefit of diagnostic hypotheses in clinical reasoning: Experimental study of an instructional intervention for forward and backward reasoning. *Cognition and Instruction* 17 (4): 433–488. Hamilton: Lawrence Erlbaum Associates.

Novotney, A. 2013b. Music as medicine. Researchers are exploring how music therapy can improve health outcomes among a variety of patient populations, including premature infants and people with depression and Parkinson's disease. *Monitor on Psychology* 44 (10): 46–49. Washington, DC: American Psychological Association.

O'Hear, A. 1998b. *The nature of consciousness.* Cambridge: Cambridge University Press.

Pagnini, A., ed. 2010b. *Filosofia della medicina. Epistemologia, ontologia, etica, diritto.* Roma: Carocci.

Pahal, J.S., and H.Z. Li. 2006b. The dynamics of resident-patient communication: Data from Canada. *Communication & Medicine* 3 (2): 161–170. New York: Mouton de Gruyter.

Papakostas, Y.G., and M.D. Daras. 2001b. Placebos, placebo effects, and the response to the healing situation: The evolution of a concept. *Epilepsia* 42: 1614–1625.

Parisi, L. 2017b. *Teologia e Neuroscienze.* Brescia: Editrice Queriniana.

Patel, V.L., and D.R. Kaufman. 1994b. Diagnostic reasoning and expertise. *Psychology of Learning and Motivation* 31: 137–252.

———. 1995b. Clinical reasoning and biomedical knowledge: Implications for teaching. In *Clinical reasoning in the health professions*, ed. J. Higgs and M. Jones, 117–128. Oxford: Butterworth Heinemann.

Patel, V.L., J.F. Arocha, and J. Zhang. 2004b. Thinking and reasoning in medicine. In *Cambridge handbook of thinking and reasoning*, ed. K. Holyoak. Cambridge: Cambridge University Press.

———. 2010b. Medical reasoning and thinking. In *Oxford handbook of thinking & reasoning*, ed. K. Holyoak and Morrison. New York: Oxford University Press.

Pinker, S. 2007b. *The mystery of consciousness.* New York: Time Magazine.

Plantinga, A.C., and D. Dennet. 2010b. *Science and religion.* Oxford: Oxford University Press.

Ramachandran, V.S. 2005b. *A brief tour of human consciousness: From impostor poodles to purple numbers.* New York: Pearson.

Ramachandran, V.S., and S. Blakeslee. 1998b. *Phantoms in the brain: Probing the mysteries of the human mind.* New York: HarperCollins.

Rao, Rajesh P.N., B.A. Olshausen, and M.S. Lewicki, eds. 2002. *Probabilistic models of the brain: Perception and neural function (neural information processing).* Boston: MIT Press.

Relman, A.S. 2007b. *A second opinion: Rescuing America's healthcare.* New York: Perseus Books.

Rizzolatti, G., and C. Sinigaglia. 2006b. *So quel che fai. Il cervello che agisce e i neuroni specchio.* Milan: Raffaello Cortina Editore.

———. 2010b. The functional role of the parieto-frontal mirror circuit: Interpretations and misinterpretations. *Nature Reviews Neuroscience* 11 (4): 264–274.

Robiner, W.N., K.E. Dixon, J.L. Miner, and B.A. Hong. 2014. Psychologists in Medical Schools and Academic Medical Centers. Over 100 years of growth, influence, and partnership. *American Psychologist* 69 (3): 217–315. Washington, DC: APA, American Psychological Association.

Rouder, J.N., and R.D. Morey. 2011b. A Bayes factor meta-analysis of Bem's ESP claim. *Psychonomic Bulletin & Review* 18 (4): 682–689.

Rutherford, A., and M. Pettit. 2015b. Feminism and/in/as psychology: The public sciences of sex and gender. *History of Psychology* 18 (3): 223–237. https://doi.org/10.1037/a0039533.

Safranski, R. 1999b. *Martin Heidegger – Between good and evil.* Cambridge, MA: Harvard University Press.

Saper, R.B., R.S. Phillips, A. Sehgal, et al. 2008b. Lead, mercury, and arsenic in U.S. and Indian-manufactured Ayurvedic medicines sold via the internet. *Journal of the American Medical Association* 300 (8): 915–923.

Saunders, J. 2000b. The practice of clinical medicine as an art and as a science. *Medical Humanities.* 26: 18–22.

Scotton, B.W., A.B. Chinen, and J.R. Battista. 1996b. *Textbook of transpersonal psychiatry and psychology.* New York: Basic Books.

Searle, J. 2015b. *Seeing things as they are: A theory of perception.* Oxford: Oxford University Press.

Seifert, J. 2004b. *The philosophical diseases of medicine and their cures: Philosophy and ethics of medicine, vol. 1: Foundations.* New York: Springer.

Senn, S. 2007b. *Statistical issues in drug development*. 2nd ed. Hoboken: Wiley.
Sheehan, D.P., ed. 2006b. *Frontiers of time: Retrocausation – Experiment and theory*. Melville: American Institute of Physics.
Slagter, H.A., A. Lutz, L.L. Greischar, A.D. Francis, D. Nieuwenhuis, J.M. Davis, and R.J. Davidson. 2007b. Mental training affects distribution of limited brain resources. *PLOS Biology* 5 (6): e138.
Slayton, S.C., J. D'Archer, and F. Kaplan. 2010b. Outcome studies on the efficacy of art therapy: A review of findings. *Art Therapy: Journal of the American Art Therapy Association* 27 (3): 108–111.
Spagnolo, A.G. 2008b. Quality of life and ethical decisions in medical practice. *Journal of Medicine and The Person* 6 (3): 118–122.
Stefanelli, M., and M.F. Ramoni. 1992b. Epistemological constraints on medical knowledge-based systems. In *Advanced models of cognition for medical training and practice*, ed. D.A. Evans and V.L. Patel, vol. 97, 3–20. Heidelberg: Springer-Verlag.
Steger, M.F., P. Frazier, S. Oishi, and M. Kaler. 2006b. The meaning in life questionnaire: Assessing the presence of and search for meaning in life. *Journal of Counseling Psychology* 53: 80–93.
Steinhart, E.C. 2014b. *Your digital afterlives: Computational theories of life after death (Palgrave Frontiers in philosophy of religion)*. New York: Palgrave Macmillan.
Stevenson, I. 1974b. *Twenty cases suggestive of reincarnation*. 2nd ed. Charlottesville: University Press of Virginia.
Stocco, A., C. Lebiere, and J.R. Anderson. 2010b. Conditional routing of information to the cortex: A model of the basal ganglia's role in cognitive coordination. *Psychological Review* 117 (2): 541–574.
Storm, L., P.E. Tressoldi, and L. Di Risio. 2010b. Meta-analysis of free-response studies, 1992–2008: Assessing the noise reduction model in parapsychology. *Psychological Bulletin* 136 (4): 471–485.
Strassman, R. 2001b. *DMT: The Spirit molecule: A Doctor's revolutionary research into the biology of near-death and mystical experiences*. Rochester: Park Street Press.
Straus, S.E., and F.A. McAlister. 2000b. Evidence-based medicine: A commentary on common criticisms. In *Canadian Medical Association journal*, vol. 163, 837–840.
Tasman, A., J. Kay, and J.A. Lieberman. 2003b. *Psychiatry. Second edition. Therapeutics*. Chichester: Wiley.
Tauber, A.I. 2005b. *Patient autonomy and the ethics of responsibility*. Cambridge, MA: MIT Press.
Teyber, E. 2000b. *Interpersonal process in psychotherapy. A relational approach*. 4th ed. Belmont: Wadsworth/Thomson Learning.
The Multi-Society Task Force on PVS. 1994b. Medical aspects of the persistent vegetative state – Second of two parts. *The New England Journal of Medicine (NEJM)* 330: 1499–1508.

Tillich, P.J. 1983b. *Gesammelte Werke*, 14 Bände + 6 Ergänzungs- und Nachlassbände, 1958–1983, Hg. Renate Albrecht, Stuttgart: Evangelisches Verlagswerk, (new edition by Verlag Walter de Gruyter, Berlin).

Tjeltveit, A.C., and M.C. Gottlieb. 2012b. Avoiding ethical missteps. By drawing on the science of prevention, psychologists can develop skills, relationships and personal qualities to bolster ethical resilience and minimize risks related to unethical behavior. *Monitor on Psychology* 43 (4). Washington, DC: American Psychological Association 44 (4): 68–74.

Todorov, I., N. Kornell, M. Larsson Sundqvist, and F.U. Jönsson. 2014b. *Phrasing questions in terms of current (not future) knowledge increases preferences for cue-only judgments of learning*, Archives of scientific psychology 1: 7–13. Washington, DC: APA, American Psychological Association.

Tracey, J.G., B.E. Wampold, J.W. Lichtenberg, and R.K. Goodyear. 2014b. Expertise in psychotherapy. An elusive goal? *American Psychologist* 69 (3): 217–315. Washington, DC: APA, American Psychological Association, pp. 225–226.

Tye, M. 1995b. *Ten problems of consciousness*. Cambridge, MA: MIT Press.

Unschuld, P.U. 2010b. *Medicine in China: A history of ideas*. 2nd ed. Berkeley: University of California Press.

Van Dalen, D. 2012b. *L.E.J. Brouwer – Topologist, intuitionist, philosopher: How mathematics is rooted in life*. London: Springer.

Van der Steen, W.J., and P.J. Thung. 1988b. *Faces of medicine: A philosophical study*. Dordrecht: Kluwer.

Van Inwagen, P. 2006b. *The problem of evil*. Oxford: Oxford University Press.

———. 2017b. *Thinking about free will*. Cambridge: Cambridge University Press.

Van Lommel, P. 2011b. *Consciousness beyond life: The science of the near-death experience*. New York: HarperCollins.

Vis, J., and H. Boynton. 2008b. Spirituality and transcendent meaning making: Possibilities for enhancing posttraumatic growth. *Journal of Religion & Spirituality in Social Work* 27: 69–86.

Volk, S. 2011b. *Fringe-ology: How I tried to explain away the unexplainable – And couldn't*. New York: HarperCollins.

Vygotsky, L.S. 1978b. *Mind in society*. Cambridge, MA: Harvard.

Wasserstein, R.L., and N.A. Lazar. 2016b. The ASA's statement on p-values: Context, process, and purpose. *The American Statistician* 70 (2): 129–133.

Weatherall, D. 1996b. *Science and the quiet art: The role of medical research in health care*. New York: Norton.

Weindling, P.J. 2004b. *Nazi medicine and the Nuremberg trials. From medical war crimes to informed consent*. Houndmills: Palgrave Macmillan.

Westen, D., and J. Weinberger. 2005b. In praise of clinical judgment: Meehl's forgotten legacy. *Journal of Clinical Psychology* 61: 1257–1276.

Whitaker, R. 2001b. *Mad in America: Bad science, bad medicine, and the enduring mistreatment of the mentally ill*. New York: Perseus Publishing.

———. 2010b. *Anatomy of an epidemic: Magic bullets, psychiatric drugs, and the astonishing rise of mental illness in America*. New York: Crown.

Wittgenstein, L. 1958b. *Preliminary studies for the "philosophical investigations", generally known as the blue and Brown books*. Hoboken: Blackwell Publishers.

Wortman, J., M.B. Donnellan, and R.E. Lucas. 2014b. *Can physical warmth (or coldness) predict trait loneliness? A replication of Bargh and Shalev (2012)*, Archives of scientific psychology, 13–19. Washington, DC: APA, American Psychological Association.

Wrathall, M. 2006b. *How to read Heidegger*, 100. New York: W. W. Norton.

Wulff, H.R., S.A. Pedesen, and R. Rosenberg. 1990b. *Philosophy of medicine: An introduction*. 2nd ed. Oxford: Blackwell.

Zaner, R.M. 1981b. *The context of self: A phenomenological inquiry using medicine as a clue*. Athens: Ohio University Press.

Zauzmer, J. 2017b. *Why a Yale neuroscientist decided to change careers – And is now becoming a priest*. Washington, DC: The Washington Post.

Žižek, S., and B. Gunjević. 2012b. *God in pain: Inversions of apocalypse*. New York: Seven.

Zürrer, R. 1989b. *Reinkarnation. Die umfassende Wissenschaft der Seelenwanderung* (Reincarnation: The omni-comprehensive science of the migration of the body), IV. Aufl. 2000. Neuhausen: Govinda-Verlag.

Further Reading

Atul Gawande, Being Mortal, on Frontline. Available at: http://www.pbs.org/wgbh/pages/frontline/being-mortal/

Azarian, B. 2016. *A neuroscientist explains why artificially intelligent robots will never have consciousness like humans*. Available at: http://www.rawstory.com/2016/03/a-neuroscientist-explains-why-artificially-intelligent-robots-will-never-have-consciousness-like-humans/

Barker, J. (Smith, M.W., review). 2013. *Express yourself: Your mouth, your life. The power of positive talking*. New York: WebMD. Available at: http://www.webmd.com/balance/express-yourself-13/positive-self-talk

Brüder Grün. Glaube und Wissen im Zeitalter der Quantenphysik (Sternstunde Religion, 22.11.15). Available at: https://www.youtube.com/watch?v=XlaQusxtnzA

BioLogos Foundation. Available at: https://biologos.org/

Borri, A., ed. *Intervista ad Alessandro Pagnini*. In: Humanamente/Alessandro Pagnini. http://www.humanamente.eu/PDF/Intervista_Alessandro%20Pagnini_issue%209.pdf

Cain Travis, M. 2015. *The existence of the soul: Philosophy, not neuroscience*. Apologetics, Worldview, and a Pebble in the Shoe. Available at: https://hcchristian.wordpress.com/2015/04/25/the-existence-of-the-soul-philosophy-not-neuroscience/

Chopra, D. 2016. *Artificial intelligence will never rival the deep complexity of the human mind*. Available at: http://www.huffingtonpost.com/deepak-chopra/artificial-intelligence-human_b_10240122.html

Consortium of Academic Health Centers for Integrative Medicine. Available at: http://www.imconsortium.org/

Edzard Ernst Official Website. Available at: http://edzardernst.com/

Fasol, U. 2011. *Dio? No grazie, siamo scienziati!* Verona: Sentinelle del mattino. Available at: https://vimeo.com/23885035

Gehirn und Geist. Available at: http://www.gehirn-und-geist.de/

Gemeinsamer Bibliotheksverbund/Illich. Available at: http://www.gbv.de/dms/faz-rez/771011_FAZ_0054_L22_0002.pdf

Hall, H. 2008. *Death by medicine. Science-based medicine. Exploring issues & controversies in science & medicine*. The complete discussion on the writing by Carolyn Dean, Gary Null, and others is available at http://www.sciencebased-medicine.org/death-by-medicine/

INCF. 2005. *About us: INCF History*. The international neuroinformatics coordinating facility. Stockholm: Karolinksa Institutet. Available at: https://www.incf.org/about-us/history

Integral World Website. Available at: http://www.integralworld.net/

International Association for Near Death Studies. Available at: http://iands.org/home.html

Interview by Alex Tsakiris with Patricia Churchland. Available at: https://www.youtube.com/watch?v=7a6ZaivvCn

Journal of Translational Medicine. Available at: http://www.translational-medicine.com

Kuhn, R.L. 2014. *Closer to truth*. Available at: http://www.closertotruth.com/

Künstliche Intelligenz: Overhyped oder unterschätzt? *CeBIT future talk*. With Wolfgang Wahlster. 14.3.2016. Available at: https://www.youtube.com/watch?v=77QhkWNOqS8

Loftus, S.F. 2006c. *Language in clinical reasoning: Using and learning the language of collective clinical decision making*. See in particular Chapter 2 -Clinical Reasoning in Medicine: Practice and Education. Available at: http://ses.library.usyd.edu.au/bitstream/2123/1165/3/03chapter2.pdf

Meaning and Happiness. Available at: http://www.meaningandhappiness.com/

National Center for Complementary and Integrative Health at the U.S. National Institutes of Health (NIH). Available at: https://nccih.nih.gov/

Near-Death Experience: Consciousness and Neuroscience (Pascal Michael, Aberdeen University). Available at: https://www.youtube.com/watch?v=TAjYpNlwkBQ

Neue Medizin. Available at: http://neue-medizin.com/

Neuroethics, Center for Bioethics at the College of Physicians and Surgeons of Columbia University, in collaboration with the Columbia Center for New Media Teaching & Learning (CCNMTL). Available at: http://ccnmtl.columbia.edu/projects/neuroethics/index.html

Open Sciences. Available at: http://www.opensciences.org/

Pim van Lommel Official Website. Available at: http://www.pimvanlommel.nl/?home_eng, and analysis, available at: http://www.skepticalinvestigations.org/Mediaskeptics/vanLommel.html

Quackwatch. Available at: www.quackwatch.org

Reasons to Believe Official Website. Available at: http://www.reasons.org/

Rettew, D.C. 2014. *A prescription for music: Study finds musical training linked to enhanced brain maturation in children*. University of Vermont Medical Center Blog, December 17.

Rupert Sheldrake Website. Available at: http://www.sheldrake.org/

Sartori, G., and C.A. Defanti. 2012. *Le neuroscienze ed il Mistero del Libero Arbitrio*. Bergamo: Ordine dei Medici ed Odontoiatri della Provincia di Bergamo. Available at: https://www.youtube.com/watch?v=KWRkpU7bmTg

Science-Based Medicine. Available at: http://www.sciencebasedmedicine.org/

Talis, R. 2009. *Conscious computers are a delusion*. The Guardian. Available at: https://www.theguardian.com/commentisfree/belief/2009/sep/03/computers-artificial-intelligence

The integrated and fragmented nature of consciousness, with Giulio Tononi. Available at: https://www.youtube.com/watch?v=Sg4apVaKPT8

The Lancet. Available at: http://www.thelancet.com

Torley, V. 2013. *Do split-brain cases disprove the existence of an immaterial soul?* Available at: https://uncommondescent.com/intelligent-design/do-split-brain-cases-disprove-the-existence-of-an-immaterial-soul-part-two/

University of Florida Center for Spirituality and Health, Louis A. Ritz. Available at: http://www.ufspiritualityandhealth.org/directorsadvisors/ritz.asp

University of Tartu's Blog. Available at: http://blog.ut.ee/religious-people-more-prone-to-depression/

University of Vermont Medical Center Blog. Available at: https://medcenterblog.uvmhealth.org/children-health/a-prescription-for-music-study-finds-musicaltraining-linked-to-enhanced-brain-maturation-in-children/

World Science Festival. 2009. *Science faith and religion 2009*. Bill Blakemore hosts scientists Lawrence Krauss, Ken Miller and Guy Consolmagno, and philosopher Colin McGinn. Available at: https://www.youtube.com/watch?v=7o8xxIhMbnA

———. 2015. *Consciousness, explored and explained*. The integrated and fragmented nature of consciousness, with Giulio Tononi. Available at: https://www.youtube.com/watch?v=Sg4apVaKPT8

Index[1]

[1] Note: Page numbers followed by 'n' refer to notes.

D. L. Tomasi, *Critical Neuroscience and Philosophy*,
https://doi.org/10.1007/978-3-030-35354-4

K

L

M

N

O

P

Printed in the USA
CPSIA information can be obtained
at www.ICGtesting.com
CBHW051327100724
11407CB00005B/81

9 783030 35353